THE SECOND NREL CONFERENCE ON THERMOPHOTOVOLTAIC GENERATION OF ELECTRICITY

THE SECOND NREL CONFERENCE ON THERMOPHOTOVOLTAIC GENERATION OF ELECTRICITY

Colorado Springs, CO July 1995

EDITORS
John P. Benner
Timothy J. Coutts
David S. Ginley
*National Renewable
Energy Laboratory*

AIP CONFERENCE
PROCEEDINGS 358

American Institute of Physics   Woodbury, New York

Authorization to photocopy items for internal or personal use, beyond the free copying permitted under the 1978 U.S. Copyright Law (see statement below), is granted by the American Institute of Physics for users registered with the Copyright Clearance Center (CCC) Transactional Reporting Service, provided that the base fee of $6.00 per copy is paid directly to CCC, 222 Rosewood Drive, Danvers, MA 01923. For those organizations that have been granted a photocopy license by CCC, a separate system of payment has been arranged. The fee code for users of the Transactional Reporting Service is: 1-56396-509-7/ 96 /$6.00.

© 1996 American Institute of Physics

Individual readers of this volume and nonprofit libraries, acting for them, are permitted to make fair use of the material in it, such as copying an article for use in teaching or research. Permission is granted to quote from this volume in scientific work with the customary acknowledgment of the source. To reprint a figure, table, or other excerpt requires the consent of one of the original authors and notification to AIP. Republication or systematic or multiple reproduction of any material in this volume is permitted only under license from AIP. Address inquiries to Office of Rights and Permissions, 500 Sunnyside Boulevard, Woodbury, NY 11797-2999; phone 516-576-2268; fax: 516-576-2499; e-mail: rights@aip.org.

L.C. Catalog Card No. 95-83335
ISBN 1-56396-509-7
DOE CONF- 950795

Printed in the United States of America

CONTENTS

Preface ix

PAPERS

Opening Session

A Simple Parametric Study of TPV System Efficiency and Output Power Density Including a Comparison of Several TPV Materials 3
J. L. Gray and A. El-Husseini

Comparison of Selective Emitter and Filter Thermophotovoltaic Systems 16
B. S. Good, D. L. Chubb, and R. A. Lowe

IR Filters for TPV Converter Modules 35
W. E. Horne, M. D. Morgan, and V. S. Sundaram

Session I: Thermophotovoltaic Systems Design and Performance

Effect of Expanded Integration Limits and of Measured Infrared Filter Improvements on Performance of RTPV System 55
A. Schock, C. Or, and M. Mukunda

Small Radioisotope Thermophotovoltaic (RTPV) Generators 81
A. Schock, C. Or, and V. Kumar

Design of a TPV Generator with a Durable Selective Emitter and Spectrally Matched PV Cells 98
D. B. Sarraf and T. S. Mayer

Thermophotovoltaic Energy Converters Based on Thin Film Selective Emitters and InGaAs Photovoltaic Cells 109
N. S. Fatemi, R. H. Hoffman, D. M. Wilt, R. A. Lowe, L. M. Garverick, and D. Scheiman

Development of a Small Air-Cooled "Midnight Sun" Thermophotovoltaic Electric Generator 128
L. M. Fraas, H. H. Xiang, S. Hui, L. Ferguson, J. Samaras, R. Ballantyne, M. Seal, and E. West

Demonstration of a Candle Powered Radio Using GaSb Thermophotovoltaic Cells 134
D. J. Williams and L. M. Fraas

Laboratory Development TPV Generator 138
G. A. Holmquist, E. M. Wong, and C. H. Waldman

Extended Use of Photovoltaic Solar Panels 162
G. E. Guazzoni and M. F. Rose

Electricity from Wood Powder Report on a TPV Generator in Progress 177
L. Broman, K. Jarefors, J. Marks, and M. Wanlass

Solar Thermophotovoltaic (STPV) System with Thermal Energy Storage 181
    D. L. Chubb, B. S. Good, and R. A. Lowe
Testing and Modeling of a Solar Thermophotovoltaic Power System 199
    K. W. Stone, D. L. Chubb, D. M. Wilt, and M. W. Wanlass

Session II: Markets and Applications

Competing Technologies for Thermophotovoltaics 213
    M. F. Rose
Grid-Independent Residential Power Systems 221
    R. E. Nelson
Utility Market and Requirements for a Solar Thermophotovoltaic
System 238
    K. Stone and S. McLellan
Thermophotovoltaic Energy Conversion: Technology and Market
Potential 251
    L. J. Ostrowski, U. C. Pernisz, and L. M. Fraas

Session III: Optical System Development

A Small Particle Selective Emitter for Thermophotovoltaic Energy
Conversion 263
    D. L. Chubb and R. A. Lowe
Multiband Spectral Emitters Matched to MBE Grown Photovoltaic Cells 278
    E. M. Wong, P. N. Uppal, J. P. Hickey, C. H. Waldman, and G. A. Holmquist
Characteristics of Indium Oxide Plasma Filters Deposited by Atmospheric
Pressure CVD 290
    S. Dakshina Murthy, E. Langlois, I. Bhat, R. Gutmann, E. Brown,
    R. Dzeindziel, M. Freeman, and N. Choudhury
Characteristics of Degenerately Doped Silicon for Spectral Control
in Thermophotovoltaic Systems 312
    H. Ehsani, I. Bhat, J. Borrego, R. Gutmann, E. Brown, R. Dzeindziel,
    M. Freeman, and N. Choudhury
TPV Plasma Filters Based on Cadmium Stannate 329
    X. Wu, W. P. Mulligan, J. D. Webb, and T. J. Coutts
Thermophotovoltaic Devices Utilizing a Back Surface Reflector
for Spectral Control 339
    G. W. Charache, D. M. DePoy, P. F. Baldasaro, and B. C. Campbell
Measurement of Conversion Efficiency of Thermophotovoltaic Devices 351
    G. W. Charache, D. M. DePoy, M. Zierak, J. M. Borrego, P. F. Baldasaro,
    J. R. Parrington, M. J. Freeman, E. J. Brown, M. A. Postlethwait,
    and G. J. Nichols
TPV Cells with High BSR 361
    P. A. Iles and C. L. Chu

Session IV: TPV Cells I—InGaAs Cells

Lattice-Matched and Strained InGaAs Solar Cells
for Thermophotovoltaic Use 375
    R. K. Jain, D. M. Wilt, R. Jain, G. A. Landis, and D. J. Flood

$In_xGa_{1-x}As$ TPV Experiment-based Performance Models 387
    S. Wojtczuk

Molecular Beam Epitaxy of $In_{0.74}Ga_{0.26}As$ on InP for Low
Temperature TPV Generator Applications 394
    T. S. Mayer, W. Hwang, R. Kochhar, M. Micovic, D. L. Miller,
    and S. M. Lord

Session V: TPV Cells II

Polycrystalline-Thin-Film Thermophotovoltaic Cells 409
    N. G. Dhere

Characteristics of GaSb and GaInSb Layers Grown by Metalorganic
Vapor Phase Epitaxy 423
    H. Ehsani, I. Bhat, C. Hitchcock, J. Borrego, and R. Gutmann

Recombination Lifetime in Ordered and Disordered InGaAs 434
    R. K. Ahrenkiel, S. P. Ahrenkiel, and D. J. Arent

The Influence of Bandgap on TPV Converter Efficiency 446
    P. A. Iles, C. Chu, and E. Linder

Development of p-on-n GaInAs TPV Devices 458
    P. R. Sharps and M. L. Timmons

Session VI: Emitter Design and Testing

A Fluidized Bed Selective Emitter System Driven by a Non-
premixed Burner 469
    U. Ortabasi, K. O. Lund, and K. Seshadri

SiC IR Emitter Design for Thermophotovoltaic Generators 488
    L. M. Fraas, L. Ferguson, L. G. McCoy, and U. C. Pernisz

A New High Temperature Air-Stable TPV Emitter 495
    J. B. Milstein and R. G. Roy

Development of Thermophotovoltaic Array Testing Capabilities 502
    J. J. Lin, D. R. Burger, and R. L. Mueller

Author Index 525

PREFACE

Thermophotovoltaic (TPV) Energy Converters generate electricity by photovoltaic conversion of photons emitted from a radiant heat source. A wide range of fuels can drive the heat source including various fossil sources, nuclear, renewable, and even direct solar energy. The technology supports a diverse range of potential applications including:

- Remote electrical supplies
- Hybrid vehicles
- Co-generation
- Electric-grid independent appliances
- Aerospace and military power systems

Depending upon the source temperature, the conversion process has the potential to deliver fuel-to-electric power conversion efficiencies 20% higher than achieved with gasoline powered rotating machinery, with equipment that does not require moving parts. These attributes, along with recent advances in TPV component systems, contribute to a rekindled and rapidly growing interest in accelerating development of the technology.

One of the great successes of The First NREL Conference on TPV Generation of Electricity was that the attendees were drawn from all the sub-disciplines of the technology. This ensured a balanced conference with a wide diversity of views. The enthusiastic contributions of more than 40 technical papers from the community met the objective of compiling a comprehensive review covering all aspects of TPV in the proceedings volume published by the American Institute of Physics. This collection bolsters the case for TPV as an efficient, silent, low-maintenance, modular power source that could use any type of fuel by demonstrating that the technology is backed by a credible community working together in overcoming barriers.

The First Conference helped to show that TPV is a good potential choice for energy conversion in a wide range of applications. However, the Conference was largely directed toward assessing the technology and collecting an update on recent progress. It did not address some of the key aspects needed for a development plan, specifically, applications, markets, strategic outlook, benefits of the business, and national needs. The challenges for the second conference included not only reviewing the technology progress, but also discussing business and societal considerations and identifying enabling technology such that TPV would become the best choice for some specific applications.

The goals of The Second NREL Conference on Thermophotovoltaic Generation of Electricity focus on these issues, specifically:

- To build on the success of the first conference in this series by bringing together the TPV community, including researchers, systems engineers, potential users, federal and state agencies, and universities to develop common goals and coordinate action.
- To provide a forum that covers all aspects of TPV generation of electricity including combustion processes, emission of radiation, spectral control, conversion of radiation into electricity, and systems engineering.
- To discuss and assess potential use and associated TPV systems characteristics with technical experts in competing technologies and program managers of end use applications.
- To begin to develop a quantification of the benefits of TPV applied to various technology areas and focus on the opportunities.
- To include workshops, panel discussions, and working groups to define the status and future direction for TPV, and reach consensus on a national plan to coordinate development of the TPV area.

- To publish conference papers and workshop summaries as a volume of the American Institute of Physics Conference Proceedings Series. This serves to both collect and widely disseminate the most current information on TPV technology.

Sessions at the close of the first and second day focused discussion on establishing application specific technology requirements and evaluation of financial and market barriers. Breakout workshop sessions were used to collect perspective on development of a coordinated plan for development of TPV.

The key advantage of TPV, diversity of both fuel source and end-use application, also presents the major challenge in focusing resources to accelerate development. Namely, no single existing agency or federal program has a compelling need to assume sole responsibility for the technology. Thus, the TPV community must work together closely to coordinate the support they draw from diverse federal, state, and private interests. The Second Conference made it clear that the community can function cohesively.

The existence of The Second NREL Conference on Thermophotovoltaic Generation of Electricity implies that the First Conference was worth doing, that it was done rather well, and that there is more to do. The First Conference brought together interested parties who introduced themselves to each other, providing answers to the question of who is involved in this new technology and what are they doing. During the Second, these parties started to identify what contributions they could make and discussed projects with prospective partners to gauge what aspects they would hope to gain in alliances. This situation would seem to demand that a Third Conference on Thermophotovoltaic Generation of Electricity be held to assess the progress forthcoming and to continue to foster these partnerships.

This conference is sponsored by the National Renewable Energy Laboratory (NREL) under contract to the U.S. Department of Energy (DOE). Funding supporting the conference is provided by a grant from the NREL Director's Development Fund. Additional support is provided by the National Photovoltaics Program of DOE's Office of Energy Efficiency and Renewable Energy.

J. P. Benner
T. J. Coutts
D. S. Ginley

OPENING SESSION

A Simple Parametric Study of TPV System Efficiency and Output Power Density Including a Comparison of Several TPV Materials

Jeffery L. Gray and Ali El-Husseini

School of Electrical and Computer Engineering
Purdue University
West Lafayette, In 47907-1285

Abstract. This paper presents a parametric study of thermophotovoltaic (TPV) system efficiency and output power density based upon a simple model of the TPV system. The efficiencies presented here are based on thermodynamic limits. Some issues relating to the choice of TPV materials are considered. It is shown that the optimum TPV cell band gap depends not only on the emitter spectrum, but on the type and effectiveness of the spectral selection. Trade-offs between efficiency and output power are also illustrated. In addition, issues associated with creating a more detailed TPV system model are discussed.

INTRODUCTION

There has recently been renewed interest in thermophotovoltaic (TPV) generation of electricity. This interest has been sparked by recent technological advances in emitter and cell fabrication, as well as the prospect of potential military and commercial applications.

A bibliography of many of the published articles on TPV from 1950 to 1994 is available in [1]. The proceedings of The First NREL Conference on Thermophotovoltaic Generation of Electricity [2] and the First World Conference on Photovoltaic Energy Conversion [3] are excellent sources for information on many of the recent developments in TPV.

There have been many theoretical studies of the potential efficiency of TPV systems [4-9]. The objective of this paper is too investigate the fundamental limitations of TPV system performance in order to provide a framework for the development of a more detailed system model. The thermodynamic limits on system performance are investigated and some issues relating to the choice of semiconductor material for the TPV cell are discussed. Also discussed is the trade-off between system efficiency and output power density. The efficiencies predicted here are much higher than those of any current TPV systems. These limits do, however, provide a goal and can challenge the TPV community to close the gap. Detailed sytem models that include losses due to conduction, convection, and cooling the TPV array, as well as incorporating detailed numerical device models of the TPV cell, are needed in order to make more realistic system performance predictions. For the authors, this is a subject for future reseach.

A SIMPLE TPV SYSTEM

A schematic of a simple TPV system is shown in Figure 1. A TPV system can be viewed as consisting of three basic components: a spectral emitter, a spectral selector, and the TPV cell. It can be seen that a photovoltaic system, in which sunlight is converted into electricity, is really a special case of a TPV system. In photovoltaics, the sun is the emitter and the spectral selector cannot return energy back to its source. In a TPV system, there are many more degrees of freedom in the design of the system. The emitter, spectral selector, and cell material can all be independently chosen. This freedom, while an advantage from the designers point of view, makes the analysis and optimization of the system much more complex as compared to the design and optimization of photovoltaic systems.

In a TPV system, the input power to the emitter can come from a variety of sources: combustion (fossil fuel, wood, propane, methane, etc.), radioisotopes (primarily for space applications), and even the sun. A variety of materials can be used for the emitter. The radiation spectrum of the emitter can often be modeled by a black-body at temperature T_{BB}. Recall that the sun can be very closely modeled by a 5762 K black-body. TPV systems are typically proposed to have emitter temperatures much lower this, usually between 1000 K and 3000 K. The emitted

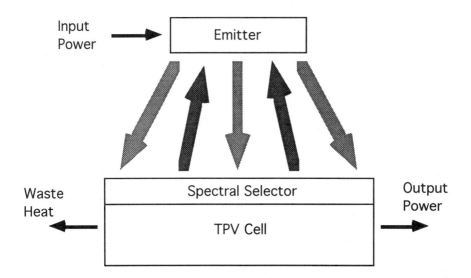

FIGURE 1. Schematic of a simple TPV system.

power of a black-body is proportional to T_{BB}^4, so higher power densities require higher temperatures.

The power spectrum of a 2000 K black-body is shown in Figure 2. Also shown is the maximum power that can be extracted from the spectrum using a TPV cell with a band gap of 0.5 eV. Notice that none of the energy less than the band gap can be converted into electricity. It is here that spectral selection is useful. There are three basic methods to achieve spectral selection. The first is through the use of a selective emitter that will only radiate energy in a band useful to the TPV cell. Another method is to insert a reflective spectral filter between the emitter and the cell. The filter is designed to reflect poorly utilized radiation back to the emitter where in can be reabsorbed, thus reducing the input power required to maintain the emitter temperature. The third method is to place a reflector at the back side of the TPV cell so that unabsorbed photons are returned to the emitter, again reducing the required input power. Of course, it may be beneficial to use these methods in combination. The final basic component is the thermophotovoltaic device which converts the radiated photons into electric current. For an efficient system, the band gap of the TPV cell must be matched to the spectrum of the emitter/spectral selector combination. Just as in solar photovoltaics, consideration of graded band gap cells and stacked cell systems can lead to a higher system efficiency.

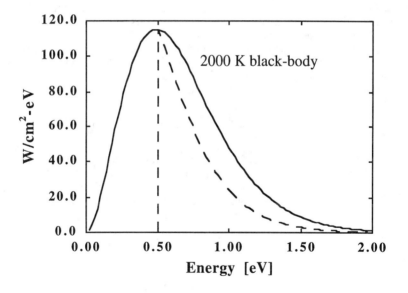

Figure 2. Power spectrum of a 2000 K black-body. The dashed line shows the useful power for a TPV cell with a band gap of 0.5 eV.

TPV System Efficiency Analysis

As can be seen in Figure 1, the source of the input power is not considered here. It is beyond the scope of this work to analyze this aspect of the system. For the purposes of this paper, the TPV system efficiency[1] will be defined as the ratio of the output power density, P_{OUT}, to the input power density, P_{IN}.

$$\eta_{SYS} = \frac{P_{OUT}}{P_{IN}} \tag{1}$$

In is important here to define a reference area since the area of the emitter, A_E, can be smaller than the total area of the TPV cells, A_E. Here, the system efficiency will be defined with respect to the area of the emitter, A_E. This leads to the definition of a dilution factor (or view factor), F.

$$F = \frac{A_E}{A_C} \tag{2}$$

In an ideal flat-plate system, $F = 1$. For a cylindrical geometry, $F = r_E/(r_E+r_{EC})$, where r_E is the radius of the emitter and r_{EC} is the distance between the emitter and the TPV array. The concept of the dilution factor is closely related to that of the concentration factor, C, in solar photovoltaics. While C must be less than about 46,300, the dilution factor must always be less than or equal to one, since the laws of physics do not allow light to be concentrated more strongly than at its source. This is an important consideration because TPV cells (like solar cells) tend to operate more efficiently at higher light concentrations. On the other hand, cooling of the TPV cell will be easier when F is small.

For simplicity, the net effect of all methods of spectral selection will be lumped into a single set of parameters. The emitter will be modeled as an ideal black-body at temperature $T_E = T_{BB}$ with an emitted total power density, P_E. The total power density returned to the emitter is P_R and can be expressed as follows,

$$P_R = \int_0^{E_L} p_E(E)dE + R\int_{E_L}^{E_H} p_E(E)dE + \int_{E_H}^{\infty} p_E(E)dE \tag{3}$$

where E_L is the low energy cutoff of the spectral selector, E_H is the high energy cutoff, and R is the cell reflectance for photon energies between E_L and E_H. R can account for reflection from the cell's front metal grid, as well the back surface reflector. Using these definitions, the input power density can be expressed as follows.

$$P_{IN} = P_E - \beta P_R \tag{4}$$

---

[1] This definition of system efficiency ignores the source of the input power. In a practical commercial system, the cost per watt (including amortized capital costs) is most important. If the cost of providing the input power, $G(P_{IN}, A_E)$, can be specified, the cost per watt of output power can be given as $\eta_{SYS} G(P_{IN}, A_E)$.

where β accounts for some inefficiency in returning power to the emitter. The TPV array is assumed to be at temperature T_C. The cost in power, if any, to keep the array at this temperature is neglected here.

While many simplifications have been made in defining this model, the resulting parameters defining the system are few enough to permit a useful study of how TPV system efficiency and output power density depend on these parameters to be conducted.

Carnot Efficiency Limit

Before analyzing the system efficiency using the parameterized model above, it is enlightening to examine the most fundamental limit on efficiency. A TPV system can be viewed as a special case of an irreversible thermodynamic engine operating between two temperatures, T_E and T_C. From Carnot's Theorem[2], the maximum possible efficiency of a TPV system is given by

$$\eta_{max} = 1 - \frac{T_C}{T_E} \quad (5)$$

For solar photovoltaics (T_E=5762 K and T_C=300 K), η_{max}=94.79%. However, approaching this limit requires the use of many stacked cells of different band gaps. For a TPV system with T_E=2000 K and T_C=325 K, η_{max}=83.75%. Again, this can be achieved using stacked cells. However, the greater flexibility of a TPV system will, in theory, allow this efficiency to be approached using a single band gap cell if spectral selection is used.

Radiative Efficiency Limit

A somewhat more realistic, although still optimistic, limit on TPV system efficiency can be calculated assuming radiative recombination is the limiting recombination mechanism in the TPV cell. In this case, the TPV system can be modeled as an endoreversible thermodynamic engine [10]. For the simple model presented here, the TPV cell maximum output power density is given by

$$P_{OUT} = \frac{2\pi V_{mp}}{c^2 h^3} \left[(1-R) \int_{E_L}^{E_H} \frac{E^2 dE}{\exp(E/kT_{BB}) - 1} - \frac{1}{F} \int_{E_G}^{\infty} \frac{E^2 dE}{\exp((E - V_{mp})/kT_C) - 1} \right] \quad (6)$$

Note that in equation (3), we have neglected the radiation emitted by the TPV cells and in equation (6), the incident radiation from other cells in the array (for F<1) has also been neglected. For reasonable temperatures, the neglected terms are small. In addition, their inclusion would only increase the calculated system

---

[2] *The efficiency of all reversible engines operating between the same two temperatures is the same, and no irreversible engine operating between the same two temperatures can have a greater efficiency than this.* Sade Carnot, 1824.

efficiency. The cell voltage at the maximum power point, V_{mp}, can be found by maximizing P_{OUT}. Using this formula, a parametric study of TPV system efficiency and output power density can be conducted.

No Spectral Selection

The simplest case, analogous to a solar photovoltaic system, is the case in which there is no spectral selection. Plots of system efficiency and output power density for $F=1$, $R=0$, $E_L=0$, $E_H=\infty$, and $T_C=300$ K as a function of TPV cell band gap for various black-body emitter temperatures (1500 K to 3000 K) are shown in Figures 3 and 4. There is a general correlation between system efficiency and output power density, favoring higher emitter temperatures. In addition, small band gap TPV cells have the highest potential efficiencies and output power densities. Increasing the cell temperature and/or using a dilution factor less than one gives the same general results, only with somewhat smaller efficiencies and output power densities.

Spectral Selection, High Pass

If all the photons with energy less than the band gap are returned to the emitter and reabsorbed, the predicted maximum efficiency rises significantly. More importantly, the dependence of system efficiency on TPV cell band gap is changed. In fact, wider band gap TPV cells are now favored, although at the cost of lower output power density. This is shown in Figures 5 and 6. The model parameters for these figures are $F=1$, $R=0$, $E_L=E_G$, $E_H=\infty$, $T_C=300$ K, and $\beta=1$. Note that the predicted efficiencies are now somewhat less sensitive to the choice of black-body emitter temperature. The crossovers observed in the efficiency plot are due to the change in position of the band gap energy with respect to the peak power energy of the black-body spectrum as the band gap varies.

Spectral Selection, Band Pass

The potential efficiency can be further increased through the use of a band pass filter. This was modeled using the parameters above with the exception being $E_H=E_G+0.1\ eV$ and is shown in Figures 7 and 8. The narrowness of the band pass filter increases the efficiency of the system, but greatly reduces the output power density. This is to be expected since there is much less power incident on the TPV cells. The potential efficiency for wider band gap cells is improved even over that for the high pass filter.

Non-Unity Return Efficiency

The increase in potential system efficiency with TPV cell band gap is due almost entirely to the assumption that all the reflected photons are returned to the black-

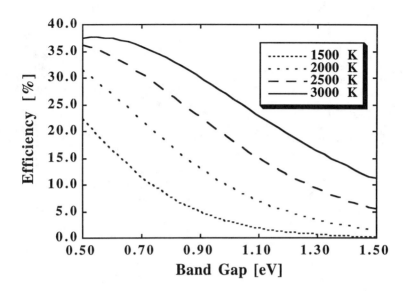

Figure 3. Predicted maximum TPV system efficiency with no spectral selection for various temperatures of the black-body emitter. The temperature of the TPV cell is 300 K.

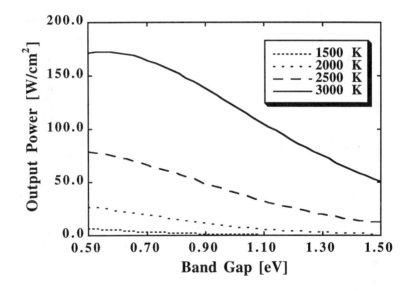

Figure 4. TPV system output power with no spectral selection for various temperatures of the black-body emitter. The temperature of the TPV cell is 300 K.

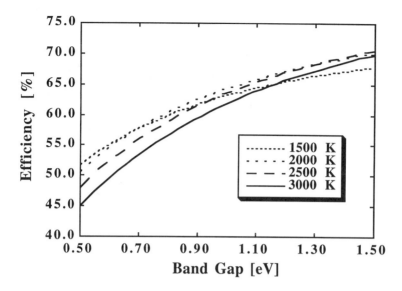

Figure 5. Predicted maximum TPV system efficiency with high pass ($E>E_G$) spectral selection for various temperatures of the black-body emitter. The TPV cell is at 300 K.

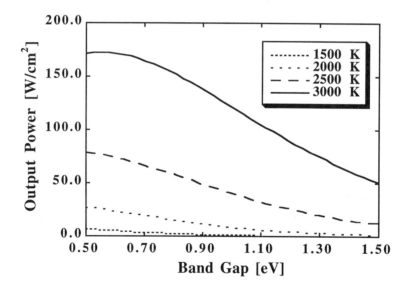

Figure 6. TPV system output power with high pass ($E>E_G$) spectral selection for various temperatures of the black-body emitter. The TPV cell is at 300 K.

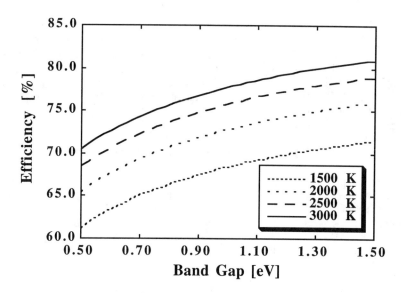

Figure 7. Predicted maximum TPV system efficiency with band pass ($E_G < E < E_G + 0.1$ eV) spectral selection for various temperatures of the black-body emitter. $T_C = 300$ K.

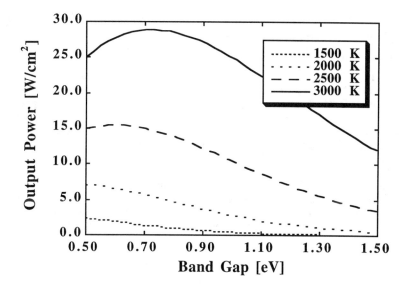

Figure 8. TPV system output power with band pass ($E_G < E < E_G + 0.1$ eV) spectral selection for various temperatures of the black-body emitter. $T_C = 300$ K.

body emitter with 100% efficiency. For realistic systems, this will not be the case. In Figures 9 and 10, the predicted system efficiencies and output power densities are shown for $F=1$, $R=0$, $E_L=E_G$, $E_H=\infty$, $T_C=300$ K, and $\beta=0.75$. Note that the potential efficiency is now less and that there is now an optimum band gap for maximum efficiency at a particular black-body emitter temperature (about 0.82 eV for $T_{BB}=3000$ K). Note, also, that this peak does not correspond to the peak in the output power density.

TPV CELL MATERIAL ISSUES

In the preceding analysis, only the band gap of the TPV material was considered. For the prediction of maximum potential efficiencies, this is fine. However, realistic system models must include detailed models for the TPV materials. In this section, some of these issues will be examined.

A good TPV cell will have a high collection efficiency regardless of the material used, thus the short-circuit current is mostly determined by the effective optical thickness of the cell. Good light trapping and/or high absorption coefficients will contribute to a high short-circuit current. The open-circuit voltage is much more sensitive to the choice of TPV cell material. The open-circuit voltage can be written as

$$V_{OC} = \frac{nkT}{q} \ln \frac{J_{SC}}{J_0} \qquad (7)$$

when $J_{SC} \gg J_0$. The short-circuit current is relatively insensitive to temperature, while the saturation current, J_0, is very sensitive through its dependence on the intrinsic carrier concentration, n_i. Since $J_0 \propto n_i^v$, where $v = 2$ for low injection and $v = 1$ for high injection, choosing a material with a small intrinsic carrier concentration will lead to a higher open-circuit voltage. The intrinsic carrier concentrations for a few TPV materials is shown in Table 1. While the general trend is for n_i to increase with decreasing band gap, notice that germanium deviates significantly from this trend. In fact, its intrinsic carrier concentration is about 5 times that of InGaAs with the same band gap. This is due to its larger effective densities-of-state. Thus, all else being equal, one might except germanium to be a poorer choice for the TPV cell material.

Also important for the choice of TPV cell material is the temperature dependence of the cell efficiency. This is greatly influenced by the temperature dependence of the open-circuit voltage. It is easy to show that the theoretical temperature dependence of the open-circuit voltage of a TPV cell can be approximated as

$$\frac{dV_{OC}}{dT} = -\frac{1}{T}(E_G - V_{OC}) \qquad (8)$$

It can be seen that an additional benefit of a high open-circuit voltage is a lesser sensitivity to cell operating temperature. This is an important consideration since the cost of cooling the TPV cells can be significant.

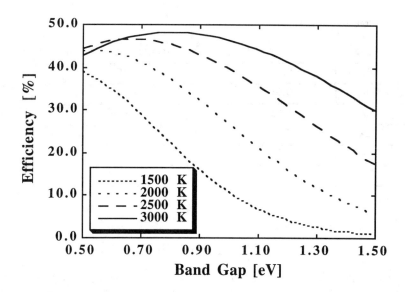

Figure 9. Predicted maximum TPV system efficiency with high pass spectral selection for various temperatures of the black-body emitter. The return efficiency, β, is 0.75.

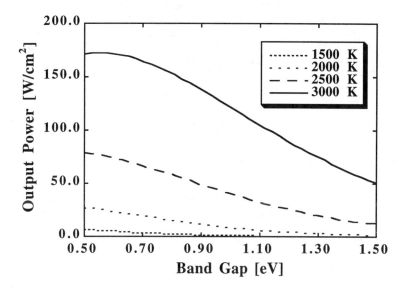

Figure 10. TPV system output power with high pass spectral selection for various temperatures of the black-body emitter. The return efficiency, β, is 0.75.

Table 1. TPV Material Parameters

Material	Band Gap (*eV*)	Intrinsic Carrier Density (cm^{-3})
InAs	0.36	7×10^{14}
InGaAs	0.66	5×10^{12}
Ge	0.66	2.4×10^{13}
GaSb	0.73	3×10^{11}
InGaAs	0.77	4×10^{11}
Si	1.12	1×10^{10}

CONCLUSIONS

A simple model for a TPV system with spectral selection was described and a parametric study was carried out. It was demonstrated that spectral selection can enhance system efficiency, but only at the expense of output power density. The efficiency with which returned photons are reabsorbed by the emitter plays an important role in choosing the optimum band gap of the TPV cell. In addition, the intrinsic carrier concentration of the TPV cell material was shown to be important in determining the cell efficiency and its temperature dependence.

Future work will involve creating a more detailed model of the TPV system. Losses associated with conduction and convection must be considered in a realistic model. Detailed models of real emitters and spectral selectors must also be incorporated. Finally, interfacing a detailed numerical device model [11,12] for the TPV cell with the detailed system model will yield a useful tool for the design and analysis of TPV systems.

REFERENCES

1. Broman, L., *Thermophotovoltaics Bibliography*, NREL/TP-412-6845.
2. Coutts, T. J. and Benner, J. P., Editors, *The First NREL Conference on Thermophotovoltaic Generation of Electricity*, AIP Conference Proceedings 321, AIP Press, 1994.
3. *1994 IEEE First World Conference on Photovoltaic Energy Conversion*, Conference Record of the Twenty-Fourth IEEE Photovoltaic Specialists Conference, Waikoloa, Hawaii, December 5-9, 1994.
4. Yeargan, J. R., Cook, R. G., and Sexton, F. W., "Thermophotovoltaic systems for electrical energy conversion," in *Conference Record of the Twelfth Photovoltaic Specialists Conference*, 1976, pp. 807-812.
5. Wurfel, P., and Ruppel, W., *IEEE Transactions on Electron Devices*, **ED-27**, 745-750, 1980.
6. Caruso, A. and Piro, G., *Solar Cells*, **19**, 123-130, 1986.
7. Swanson, R. M., "Recent developments in thermophotovoltaic conversion," in *Proceedings of the International Electron Devices Meeting*, 1980, pp. 186-189.

8. Woolf, L. D., *Solar Cells*, **19**, 19-38, 1986.
9. Woolf, L. D., "Solar thermophotovoltaic energy conversion," *in Conference Record of the Nineteenth IEEE Photovoltaic Specialists Conference*, 1987, pp. 427-432.
10. De Vos, A., *Solar Energy Materials and Solar Cells*, **31**, 75-93, 1993.
11. Gray, J. L., "ADEPT: A general purpose numerical device simulator for modeling solar cells in one-, two-, or three-dimensions," in *Conference Record of the Twenty-Second IEEE Photovoltaic Specialists Conference*, 1991, pp. 436-439.
12. Schwartz, R. J. and Gray, J. L., "The design and modeling of photovoltaic cells for TPV systems," in *The First NREL Conference on Thermophotovoltaic Generation of Electricity*, 1994, pp. 379-389.

Comparison of Selective Emitter and Filter Thermophotovoltaic Systems

Brian S. Good and Donald L. Chubb
NASA Lewis Research Center
Cleveland, Ohio

and

Roland A. Lowe
Kent State University
Kent, Ohio

Abstract

At the NASA Lewis Research Center we have developed a systems model for a general thermophotovoltaic (TPV) system. The components included in the model are a solar concentrator, a receiver, a thermal storage module, an emitter, a protective window, a filter, and a photovoltaic (PV) array. The system model requires the wavelength dependence of the optical properties of the components, together with the PV cell spectral response and the cell current-voltage characteristics. With these inputs, the system efficiency, the emitter or filter efficiencies, the PV cell efficiency, the emitter operating temperature, and the cell output power density are calculated.

In this paper we compare the performance of a variety selective emitter and filter TPV systems. The overall system model is based on the solar TPV system being developed jointly by McDonnell-Douglas and NASA. In the current study, the concentrator, receiver, and storage parameters are fixed; only the characteristics of the emitter/filter and the PV cell are varied.

We present computed results for the emitter/filter efficiency, the PV cell efficiency, and the cell output power, at a number of emitter operating temperatures, as functions of the PV cell bandgap energy.

Introduction

Thermophotovoltaic (TPV) systems potentially offer a number of features that make them attractive as power sources for both space and terrestrial applications. In order to characterize the performance potential of such systems, we have developed a computer model that incorporates detailed models of the individual system components. Using these component models, along with considerations of energy conservation, our system model predicts component efficiencies, the operating temperatures of some of the components, the overall system efficiency, and the system power output.

The models for the individual system components are simplified compared with, for example, our model for the performance of the selective emitter (1). However, we have made an effort to maintain approximately the same level of complexity and sophistication for all the component models.

Model Overview

The general form of the model system studied in this work is depicted in Fig. 1. The system consists of a solar concentrator, a receiver, a thermal storage subsystem, either a selective emitter or a filter, a protective window, and a photovoltaic cell or array. Radiation falls on the concentrator and is reflected through the aperture of the receiver and into the receiver cavity. The thermal storage system, if present, is located at the end of the cavity opposite the aperture, and the emitter, if present, is attached to the end of the storage unit opposite the receiver (or to the receiver directly if there is no storage). The emitter and filter spectrally modify the radiation so as to make it more suitable for conversion by the PV cell. A protective window lies between the receiver-storage-emitter combination and the filter, if one is present, and the last component in the chain is the PV cell.

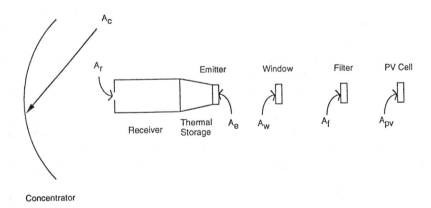

FIGURE 1. Schematic diagram of model system components.

The optical properties (i.e. emittance ϵ, reflectance ρ, and transmittance τ) of the components are assumed to vary with wavelength. The system model in its present form assumes that the wavelength dependence of the optical properties of the emitter, filter and window can be characterized using four bands; each property is assumed to be constant within a band, and the band limits may be different for each of the three components.

Fundamentally, the model incorporates a set of coupled equations for the steady-state radiation fluxes incident on the three components that are "down-stream" with respect to the receiver, that is, the radiation fluxes incident on the emitter/filter, the window, and the PV cell. These equations are written in terms of the flux incident on the concentrator, the receiver and emitter temperatures, and the component properties. The equations are solved iteratively for the emitter flux and the two temperatures, and the component efficiencies and the system output power density are computed.

Concentrator

The solar concentrator model uses parameters derived from the McDonnell-Douglas Stirling Dish (2). The concentrator is characterized by its area and efficiency. The concentrator is segmented, and its area can be varied by covering and uncovering individual segments. Varying the area varies the net flux into the concentrator, and consequently is one means by which the temperatures of the rest of the system components may be controlled.

Receiver

The receiver consists of an insulated cavity with an aperture through which radiation, reflected from the concentrator, enters. The receiver is assumed to be a grey or black body, characterized by its emissivity. The black body condition will result in the largest possible radiative energy loss through the aperture and is therefore the most conservative assumption. The receiver is assumed to be at a uniform temperature, in thermal equilibrium with the rest of the system, and operating under steady-state conditions.

When the system contains no thermal storage but does contain a selective emitter, the emitter is attached to the end of the receiver opposite the aperture. If the system contains thermal storage, it sits between the receiver and emitter, and is in thermal contact with both. In addition to its physical dimensions, the receiver is characterized by a thermal efficiency that reflects nonradiative thermal losses experienced by the receiver. At the present time, this thermal efficiency parameter is used to characterize the whole receiver-storage-emitter subsystem.

Thermal Storage

The thermal storage module is characterized by the thermal conductivity of the storage material, the length, and the areas of the storage subsystem as measured at the receiver and emitter ends. Allowing the storage system to be tapered results in additional concentration of the thermal flux as it passes through the storage module. This taper can be varied to control the relative temperatures of the receiver and the emitter. At the present time, the storage mass is assumed to exist in a single phase throughout; additional development on the effects of a liquid-solid phase change is currently under way.

Emitter

Studies have been run for two different configurations of emitter: (1) a singly-doped emitter with properties derived experiment, and (2) a hypothetical doubly-doped emitter. The singly-doped emitter model uses parameters derived from experimental Er-YAG emitters, as described elsewhere (3). The emitter is characterized using the four-band scheme described above. The optical properties that describe the emitter are the emittance ϵ, reflectance ρ, and transmittance τ. In this work the transmittances are assumed to be zero at all wavelengths, the emittances are inputs, and the reflectances are computed assuming that the emittance and reflectance sum to unity at all wavelengths. The emission band is the region defined by $\lambda_{e1} \leq \lambda \leq \lambda_{e2}$. The emission bandwidth is tied to λ_g, the wavelength corresponding to the bandgap energy E_g of the PV cell; the band limits are given by

$$\lambda_{e2} = \frac{hc_0}{eE_g} \qquad [1]$$

and

$$\lambda_{e1} = \lambda_{e2} \frac{1 - \frac{\Delta E_b}{E_b}}{1 + \frac{\Delta E_b}{E_b}} \qquad [2]$$

where h is Planck's constant, c_0 is the speed of light in vacuum, e is the fundamental electron charge and $\frac{\Delta E_b}{E_b}$ is the dimensionless emission bandwidth.

The doubly-doped emitter is characterized in a manner similar to the singly-doped emitter. At the present time, experimental emittances are not available for the doubly-doped emitter. Therefore, the emission bandwidth is taken to be twice that of the singly-doped emitter, with the same emittance values within this wide emission band as are used in the narrower, singly-doped-emitter band.

Window

A protective window is included in the system to protect the PV cell from damage, and to allow a vacuum interface at the emitter surface. The incorporation of the vacuum interface precludes heat transfer between the emitter and cell via conduction or convection, and results in a minimal temperature gradient across the thickness of the emitter, the presence of which has been shown to be detrimental to performance (1).

Filter

Two distinct model filters were considered. Both filters are described using the four-band scheme described above. The resonant-array filter (4) consists of a reflective metal film containing an array of small holes.

The plasma/interference filter consists of a number of alternating layers of high- and low-refractive-index materials, with a coating on the back that functions as a plasma cutoff filter (5).

In both cases, filter optical characteristics are taken from the literature. All filters are assumed to be lossless.

PV Cell

The PV cell is characterized by a bandgap energy and corresponding wavelength, the short-circuit current, the saturation current, the series resistance, and the junction ideality factor. The cell is modeled using the diode equation (6), as described in the next section.

Current Model Limitations

At the present time, the model does not contain a description of a power conditioning subsystem. It is expected that such a system can be easily characterized by an overall efficiency.

The model in its current state is strictly steady-state. No provisions have yet been made for including a time-dependent solar flux, nor for including transient effects due to the heating and cooling of the thermal storage mass. The thermal storage module is assumed to contain only a single phase of material, that is, no provisions have yet been made for incorporating the thermal effects associated with a phase change in the storage material. Work in these areas is under way in our laboratory.

Theoretical Development

Energy balance considerations yield coupled equations for the fluxes (energy per unit area per unit wavelength) incident on the window, filter and PV cell, as shown in Fig. 2. These fluxes depend on the power radiated by the receiver (and by the emitter if one is present), and consequently depend on the receiver and emitter temperatures. We therefore solve the coupled equations iteratively, by establishing initial guesses for the two temperatures, and solving the flux equations using these guesses. Subsequently, new temperatures and fluxes are computed in turn until convergence, at which time the component efficiencies and the output power density of the PV cell are computed.

The total system efficiency η_t is given by

$$\eta_T = \frac{P_{EL}}{P_{in}} = \frac{P_{EL}}{A_c Q_s} \qquad [3]$$

in which P_{EL} is the electrical output power produced by the PV cell, A_c is the concentrator area, and Q_s the solar flux. (Note that in the above, and in the following development, the wavelength-integrated fluxes (energies per unit area) are denoted using upper-case Qs, while the lower-case fluxes (energy per unit area per unit wavelength) are inidicated as lower-case quantities.)

The system efficiency may also be written as the product of the individual component efficiencies:

$$\eta_T = \eta_c \eta_r \eta_{th} \eta_{ef} \eta_{pv} \qquad [4]$$

where $\eta_c, \eta_r, \eta_{th}, \eta_{ef}$ and η_{pv} are the concentrator efficiency, the receiver efficiency, the thermal efficiency (that is, the sum of conduction and convection losses in the receiver, emitter, and storage system), the emitter-filter efficiency, and the PV cell efficiency. Note that there are no losses due to power conditioning in this model.

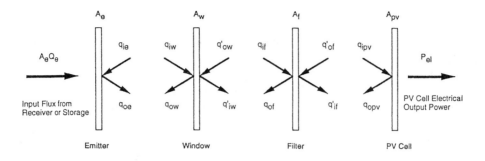

FIGURE 2. Fluxes appearing in model solution.

These efficiencies are defined in terms of energy balance:

$$\eta_c = \frac{P_r}{P_{in}} = \frac{\text{power entering receiver aperture}}{\text{solar power incident on concentrator}} \quad [5]$$

$$\eta_r = \frac{P_{th}}{P_r} = \frac{\text{power absorbed by receiver cavity}}{\text{power entering receiver aperture}} \quad [6]$$

$$\eta_{th} = \frac{P_{ef}}{P_{th}} = \frac{\text{power input to emitter or filter from receiver}}{\text{power absorbed by receiver cavity}} \quad [7]$$

$$\eta_{ef} = \frac{P_{pv}}{P_{ef}} = \frac{\text{input radiation on PV cells of } \lambda_s \leq \lambda_g}{\text{power input to emitter or filter from receiver}} \quad [8]$$

$$\eta_{pv} = \frac{P_{el}}{P_{pv}} = \frac{\text{eletrical power output from PV cells}}{\text{input radiation on PV cells of } \lambda_s \leq \lambda_g} \quad [9]$$

where λ_g is the wavelength corresponding to the PV cell bandgap energy, as defined in Eq. [1].

It should be noted that the cell efficiency is defined in terms of P_{pv}, the power associated with only the incident radiation of energy greater than the bandgap energy. This definition differs from the usual definition of PV cell efficiency, which is based on the total photon flux having $0 \leq \lambda \leq \infty$. In a TPV system with a perfect emitter or filter, all the incident light would be shifted in wavelength such that $\lambda < \lambda_g$, so that all the incident photons are capable of creating electron-hole pairs, and our definition of efficiency is simply a measure of how effectively the PV cell produces power from the wavelengths that are actually incident on it.

At the present time, the concentrator efficiency η_c and the thermal efficiency η_{th} are input quantities, assumed to be known experimentally. The other efficiencies are computed based on the fluxes described earlier.

The total flux into the system is determined by the area of the concentrator, A_c and the solar flux constant Q_s. The flux entering the receiver cavity is this incident flux, multiplied by the efficiency of the concentrator. The flux entering the receiver determines the receiver temperature T_r. This temperature, in turn, determines the flux propagating "downstream" via the following equation for the power radiated by the emitter:

$$P_e = \eta_{th}[P_r - \sigma_{sb}T_r^4 A_r] = \eta_{th}[\eta_c Q_s A_c - \sigma_{sb}T_r^4 A_r] \qquad [\,10\,]$$

in which σ_{sb} is the Stefan-Boltzmann constant, and the emitter power P_e and the receiver power P_r are given by

$$P_e = A_e Q_e = A_e \int_0^\infty \epsilon_e [e_b - q_{ie}] \qquad [\,11\,]$$

$$P_r = \eta_c A_c Q_s \qquad [\,12\,]$$

The radiation flux q_{ie} is as shown in Fig. 2, ϵ_e is the spectral emittance of the emitter (with $\rho_e = 1 - \epsilon_e$), and e_b is the black-body emissive power:

$$e_b = 2\pi k c_0^2 \lambda^5 [exp(hc_0/kT\lambda) - 1]^{-1} \qquad [\,13\,]$$

The fluxes are assumed to be uniform across the corresponding component face.

The receiver cavity as described by Eq. [10] is assumed to be either a grey or black body at a uniform temperature T_r, so that the radiation loss out the receiver aperture is proportional to T_r^4. Under black body conditions, this radiative loss is maximized, so that the assumption of a black body receiver is conservative.

The emitter is physically attached to the receiver backplate (or to the end of the thermal storage unit if one is present), with conductive heat transfer taking place through the backplate and storage system. The temperature of the emitter therefore depends on the receiver temperature through Eq. [14a] when there is no thermal storage present, and through Eq. [14b] when storage is present.

$$T_r = T_e + \frac{d_r}{k_r} Q_e \qquad [14a]$$

$$T_r = T_e + \left(\frac{d_l \log(r_t)}{k_l (r_t - 1)} + \frac{d_r}{k_r r_t} + \frac{d_e}{k_e} \right) Q_e \eta_{th} \qquad [14b]$$

Using Eqs. [10], [11] and [14a] or [14b], the emitter flux and the emitter and receiver temperatures are obtained iteratively. The receiver efficiency is given by

$$\eta_r = 1 - \frac{\sigma_{sb} T_r^4}{\eta_c Q_s} \left(\frac{A_r}{A_c} \right) \qquad [15]$$

and the emitter-filter efficiency by

$$\eta_{ef} = A_{pv} \int_0^{\lambda_g} q_{ie} d\lambda / A_e \int_0^\infty \epsilon_e [e_b - q_{ie}] d\lambda \qquad [16]$$

A simplified model for the PV cell model is used, based on the diode equation (6),

$$J = J_{sc} - J_s [exp(eV/kT_{PV} a_0) - 1] \qquad [17]$$

where J_{sc} is the short-circuit current, J_s the saturation current, V the cell output voltage and a_0 the junction ideality factor. The following approximation for the saturation current is used:

$$J_s = J_0 [exp(eE_g/kT_{PV}] \qquad [18]$$

where the constant J_0 is taken as $4 \times 10^4 A/cm^2$, a value typical of III-V materials (7). In any event, the results do not depend strongly on the choice of J_0; for TPV applications $J_{sc} \gg J_s$, so that the electrical output power depends only weakly on J_s. The electrical output power is given by

$$P_{el} = J_m V_m A_{pv} \qquad [19]$$

where J_m and V_m are the current density and voltage that yield that maximum output power. These are obtained from the following:

$$\frac{dP_{el}}{dV} = 0 \Longrightarrow e^{X_m}(X_m + 1) - \frac{J_{sc}}{J_s} - 1 = 0 \qquad [20]$$

where

$$X_m = eV_m/kT_{pv}a_0 \qquad [21]$$

$$\frac{P_{el}}{P_{pv}} = X_m^2 \left(\frac{a_0 k T_{pv}}{E}\right) \frac{J_{sc} + J_s}{X_m + 1} \qquad [22]$$

$$J_{sc} = \int_c^{\lambda_g} S_r (1 - \rho_{pv}) q_{ie} d\lambda \qquad [23]$$

where S_r is the PV cell spectral response and ρ_{pv} is the cell back surface reflectance. The spectral response is calculated using

$$S_r = \frac{e\eta_Q}{E} = \frac{e\lambda}{hc_0}\eta_Q \qquad [24]$$

in which η_Q is the cell quantum efficiency.

The rare earth-YAG emitter is characterized by a single strong emission band with greatly reduced emission outside the band. In addition, at large wavelengths, an increased emittance is sometimes observed. In order to allow for such possibilities, we divide the emittances of the emitter into four bands:

$$0 \leq \lambda \leq \lambda_u, \epsilon_e = \epsilon_{ue} \qquad [25a]$$

$$\lambda_u \leq \lambda \leq \lambda_l, \epsilon_e = \epsilon_{be} \qquad [25b]$$

$$\lambda_l \leq \lambda \leq \lambda_c, \epsilon_e = \epsilon_{le} \qquad [25c]$$

$$\lambda_c \leq \lambda \leq \infty, \epsilon_e = \epsilon_{\infty e} \qquad [24d]$$

in which the strong emission band is given by [25b]. In addition, it is clear that an emitter and PV cell designed to work together would have the emission band of the emitter well-coupled to the bandgap of the cell. For this reason, we define the emission band limits in terms of the wavelength that corresponds to the PV cell bandgap energy [26] and a dimensionless bandwidth [27].

$$\lambda_l = \lambda_g = hc_0/eE_g = hc_0/sE_l \qquad [26]$$

$$\frac{\Delta E_b}{E_b} = (E_u - E_l)/\frac{E_u + E_l}{2} \qquad [27]$$

The optical properties of the other components are assumed to be adequately modeled by a similar four-band description.

Results and Discussion

Because this study aims only to investigate the performance of emitter and filter systems, the computational procedure is modified somewhat from that described above for the general TPV system. In the following, the emitter temperature is fixed at a chosen value, which determines the flux out of the emitter, and hence the fluxes into the components "downstream" of the emitter. In essence, this is equivalent to requiring that the components upstream have values that would yield the desired emitter temperature, and those component values are fixed in all the work described here. In addition, the window is assumed to be lossless and perfectly transmitting, with a transmittance of 1.0 for all wavelengths.

Seven sets of model data were investigated, representing seven different configurations of the system. Three configurations represent filter systems, and four represent selective emitter systems. The seven configurations are described briefly below.

Lossless Resonant Array Filter (RAF)

The dimensionless transmission bandwidth is 0.4. The filter is lossless, so all band absorptances (and hence emittances) are zero. The band limits are tied to the PV cell bandgap as described above. The band reflectances are 0.95, 0.5, 0.95 and 0.97. The band transmittances are 0.05, 0.5, 0.05 and 0.03.

Ideal Lossless Resonant Array Filter (IRAF)

Similar to the RAF system, but with an ideal transmission-band transmittance. The band reflectances are 0.9, 0.2, 0.98 and 0.98. The band transmittances are 0.1, 0.8, 0.02 and 0.02.

Lossless Plasma/Interference Filter (PIF)

The dimensionless transmission bandwidth is 0.6. The filter is lossless, so all band absorptances (and hence emittances) are zero. The band limits are tied to the PV cell bandgap as described above. The band reflectances are 0.8, 0.3, 0.9 and 0.8. The band transmittances are 0.2, 0.7, 0.1 and 0.2.

For all the above filter systems, the PV cell reflectance is zero, and the back surface of the receiver, which radiates through the window and onto the filter, is assumed to have an emissivity of 0.9.

Baseline Selective Emitter (SE)

The dimensionless emission bandwidth is 0.15. The band limits are tied to the PV cell bandgap as described above. The band reflectances are 0.8, 0.25, 0.8 and 0.25. The band emittances are 0.2, 0.75, 0.2 and 0.75. The band transmittances are zero, as are the PV cell reflectances.

Selective Emitter with Reflective Cell (SER)

Similar to the SE system except that the PV cell band reflectances are 0.03, 0.03, 0.8 and 0.8.

Selective Emitter with Wide Emission Band (SEW)

Similar to the SE system except that the dimensionless emission bandwidth is 0.3, representing a doubly-doped emitter.

Selective Emitter with Reflective Cell and Wide Emission Band (SERW)

Similar to the SE system except that the PV cell band reflectances are 0.03, 0.03, 0.8 and 0.8, and the dimensionless emission bandwidth is 0.3.

The principal results of the work are presented in Figs. 3-6.

Fig. 3(a) and (b) show the emitter/filter efficiency as a function of PV cell bandgap energy, for the seven system configurations, at emitter temperatures of 1500K and 2000K. The efficiency is seen to increase with increasing emitter temperature, and to decrease with increasing bandgap energy. At both temperatures, the efficiencies in increasing order are: SE, SEW, SER, PIF and SEWR, RAF, and IRAF. For the selective emitter systems, it can be seen that adding back-surface reflectance to the PV cell increases the efficiency somewhat more than does doubling the emitter emission bandwidth.

Also of interest is the fact that the ideal resonant array filter configuration shows relatively little degradation of efficiency with increasing bandgap energy, especially at high temperature. There are two distinct effects that cause this particular behavior. First, recall that the filter transmission band limits are tied to the bandgap energy of the cell, so that as the bandgap energy increases (at fixed temperature), the fraction of photons in the transmission band of the filter decreases. This explains the decrease in efficiency of the filter (and the emitter as well) with increasing bandgap. Second, as the temperature increases (for fixed bandgap), the radiation more effectively couples to the passband of the filter. In the case of the ideal resonant array filter, the emission band emissivity is large enough that the filter efficiency degrades only weakly with increasing bandgap energy.

Fig. 4 shows the PV cell efficiency, at four emitter temperatures, as a function of PV cell bandgap energy, for the selective emitter baseline system. There is little variation in cell efficiency among the various configurations, and the curves shown may be taken as representative.

Fig. 5(a) and (b) show the product of the emitte/filter efficiency and the cell efficiency, as a function of PV cell bandgap energy, at emitter temperatures of 1500K and 2000K. The curves for all configurations at all temperatures show well-defined maxima, as expected, given the shapes of the PV and emitter/filter efficiency curves. Because there is only a small degree of variation in PV cell efficiency among the seven configurations, the behavior of the efficiency product is dominated by the emitter/filter efficiency, and the ordering of the product efficiencies for the seven configurations is the same as that of the emitter/filter efficiencies.

Fig. 6(a) and (b) show the PV cell output power density, as a function of PV cell bandgap energy, for the seven system configurations, at emitter temperatures of 1500K and 2000K. All configurations display maxima which occur at somewhat larger bandgap energies for higher temperatures. The maxima all occur for bandgap energies in the region of 0.4-0.7 eV. There is much more variation between the output power of the various configurations at low bandgap energies than at high energies. At these smaller bandgap energies, the configurations can be seen to fall into two groups–a higher-efficiency group, consisting of (in decreasing order) the PIF, IRAF, SEW and SEWR systems, and a lower-efficiency group, consisting of the SE, SER and RAF systems.

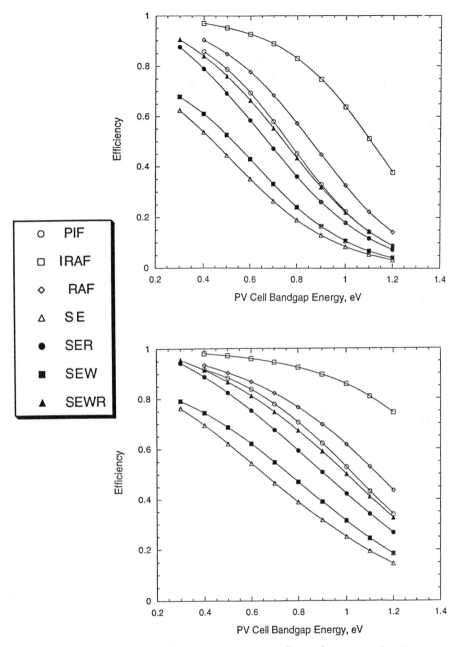

FIGURE 3. Emitter/Filter efficiencies for all configurations as a function of PV cell bandgap energy. Top: (a) Emitter temperature 1500K. Bottom: (b) Emitter temperature 2000K.

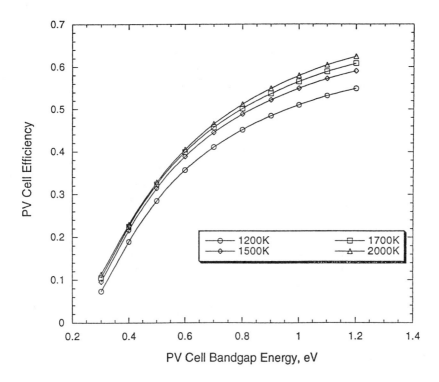

FIGURE 4. PV cell efficiency for selective emitter baseline (SE) configuration, emitter temperatures 1200K, 1500K, 1700K and 2000K.

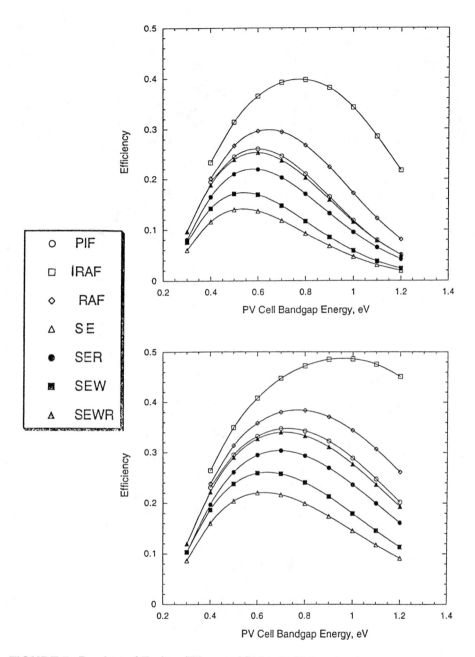

FIGURE 5. Product of Emitter/Filter and PV cell efficiencies for all configurations as a function of PV cell bandgap energy. Top: (a) Emitter temperature 1500K. Bottom: (b) Emitter temperature 2000K.

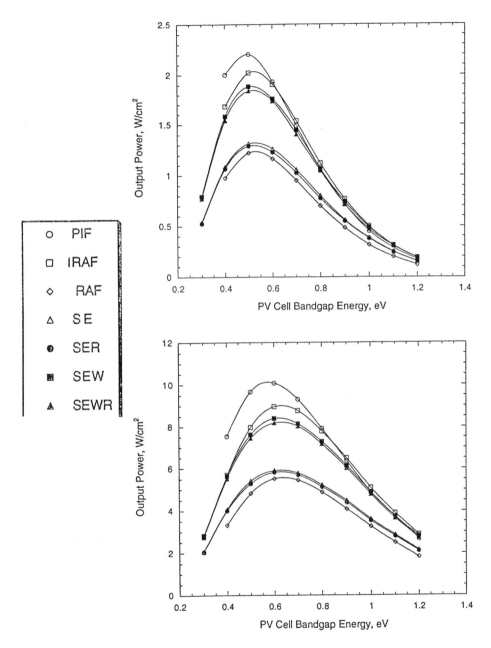

FIGURE 6. PV cell output power for all configurations as a function of PV cell bandgap energy. Top: (a) Emitter temperature 1500K. Bottom: (b) Emitter temperature 2000K.

Conclusions

The filter systems exhibited both the highest power and highest emitter/filter efficiency of the configurations investigated, although different filters gave the highest power and highest efficiency. It must be stressed that all filters were assumed to be lossless; experimental values for absorptance were unavailable, and it is unclear what degree of degradation in power and efficiency will be exhibited by these systems when losses are included.

The selective emitter systems showed considerable variation. The baseline system produced substantially lower efficiencies and output power than the best filter systems, but the best selective emitter systems produced power and efficiencies that were competitive with the best filter systems. Selective emitter technology is in its infancy, and significant future gains are expected.

No consideration was given to cost factors, as all of these technologies are sufficiently new that only very rough estimates of cost can presently be made. Based upon a crude assessment of emitter/filter complexity, it is expected that the two filter systems would be more expensive to manufacture than would the selective emitter systems. Any attempt to quantify the cost per watt of these systems is probably premature, although it is precisely these comparisons that will be the final determinant of the commercial potential of these technologies.

References

1. Good, Brian S., Chubb, Donald L., and Lowe, Roland A., "Temperature-Dependent Efficiency Calculations for a Thin-Film Selective Emitter," in AIP Conference Proceedings 321, The First NREL Conference on Thermophotovoltaic Generation of Electricity, 1994, pp. 263-275.

2. Stone, K. W., Kusek, S. M., Drubka, R. E., and Fay, T. D., "Analysis of Solar Thermophotovoltaic Test Data From Experiments Performed at McDonnell Douglas," in AIP Conference Proceedings 321, The First NREL Conference on Thermophotovoltaic Generation of Electricity, 1994, pp. 153-162.

3. Lowe, Roland A., Chubb, Donald L., and Good, Brian S., "Radiative Performance of Rare Earth Garnet Thin Film Selective Emitters," in AIP Conference Proceedings 321, The First NREL Conference on Thermophotovoltaic Generation of Electricity, 1994, pp. 291-297.

4. See, for example, Schock, A., and Kumar, V., "Radioisotope Thermophotovoltaic System Design and its Application for an Illustrative Space Mission," in AIP Conference Proceedings 321, The First NREL Conference on Thermophotovoltaic Generation of Electricity, 1994, pp. 139-152.

5. Baldasaro, Paul F., Brown, Edward J., Depoy, David M., Campbell, Brian C., and Parrington, Josef R., "Experimental Assessment of Low Temperature Voltaic Energy Conversion," in AIP Conference Proceedings 321, The First NREL Conference on Thermophotovoltaic Generation of Electricity, 1994, pp. 29-43.

6. Sze, S. M., "Semiconductor Devices: Physics and Technology," John Wiley and Sons, New York, 1985, pg. 290.

7. Olsen, L.C., Dunham, G., Huber, D. A., Addis, F. W., Anheier, N., and Coomes, E. P., "GaAs Solar Cells for Laser Power Beaming," Space Photovoltaic Research and Technology, 1991, NASA Conf. Pub. 3121.

IR Filters for TPV Converter Modules

W.E. Horne, M. D. Morgan, and V. S. Sundaram

The concept of thermophotovoltaic, TPV, energy conversion has been researched since the early 1960 time period. The TPV concept utilizes a semiconductor photovoltaic cell to convert radiant energy from a thermal source to electricity. In the most familiar application, a photovoltaic cell (or solar cell) is exposed to sunlight to produce electricity. However, the term TPV is usually reserved for applications involving nonsolar radiant heat sources ranging in temperature from about 1000K to 2000K. The TPV concept takes advantage of the fact that the photovoltaic conversion process is most efficient near the bandgap of the semiconductor material from which the PV cell is made. Figure 1 illustrates this phenomenon. Photons interact with the electrons in the semiconductor by imparting their energy to the electron as kinetic energy. This kinetic energy raises the electron from the valence band to the conduction band creating a mobile charge that can be collected across a PN junction to produce a current. Generally, each incident photon produces only one collectable charge; therefore, excess energy beyond that required to raise the electron to the conduction band cannot produce electric charge and is absorbed in the crystal lattice of the semiconductors as heat. Thus, theoretically as shown by the identified curve in figure 2, photons are most efficient at producing electric current when their energy is equal to the bandgap energy of the semiconductor. In practice the efficiency of collecting charge across a junction falls off very near the bandgap as shown by the actual cell response curve in figure 2. As the photon energy increases the photon-to-electric charge conversion reaches a maximum and then efficiency begins to drop again due to the excess kinetic energy transferred to each charge raised to the conduction band. The maximum photon-to-electric charge efficiency usually occurs at a photon energy slightly greater than the bandgap.

From the preceding discussion it can be seen that there is an optimum photon energy, or wavelength, for the conversion of radiant energy to electrical energy by a given semiconductor material as illustrated in figure 2. Thus (for maximum conversion efficiency) the optimum source would be monochromatic with a wavelength corresponding to the maximum photon to electric efficiency for the PV cell being used. Unfortunately most thermal energy sources have broad energy spectra as illustrated by the 1300K blackbody spectrum shown in figure 2. In order to make the TPV concept viable one must find a way to either suppress the emission of long wavelength photons or to return them back to the thermal source for reabsorption. Since the early 1960s workers have researched concepts to minimize the long wavelength, out of band thermal losses in TPV systems. These efforts have generally centered around the application of rare earth oxide materials as nonblackbody line emitters or the use of either wavelength selective first or second surface mirrors to reflect the unusable thermal energy back to the thermal source for reabsorption. Both of these approaches have met with some degree of success; however, this paper will discuss only the progress that has been made in the latter approach. Data will be presented on a recently developed new type of optical filter that appears to be almost ideally suited to the TPV application. It will also discuss a combination of the line emitter, optical filter approach to achieve maximum TPV efficiency.

The requirements for the reflective filter are quite stringent since the bulk of the available energy from low temperature sources is outside of the response range of the PV cells in a given TPV application. Figure 2 illustrates the problem graphically. Mathematically the TPV efficiency of a cell filter combination can be expressed by

$$\text{Eff} = \frac{V_{oc} \, FF \, \int_{\lambda=0}^{\lambda_{BG}} I(\lambda) \, d\lambda}{\int_{\lambda=0}^{\lambda=\alpha} [T_{filter}\alpha_{cell}(\lambda) + \alpha_{filter}(\lambda)] \, d\lambda}$$

where
- V_{oc} = cell open circuit voltage
- FF = cell fill factor
- $I(\lambda)$ = cell current as a function of wavelength
- $\alpha(\lambda)$ = cell/filter absorption as a function of wavelength
- λ = photon wavelength
- T = transmittance

Figure 3 shows the results of the above equation for a GaSb cell having a V_{oc} of 0.4 volts and FF = 0.75, a filter with various levels of long wavelength absorption, and a cavity having varying areas of nonactive reflective walls. From these calculations it can be seen that the role of the filter in minimizing the long wavelength absorption is very critical since the efficiency drops dramatically as the reflectance varies between 100% to 90%.

The most common type of optical filter is made by depositing alternating layers of two or more dielectric materials having different indices of refraction. If the optical thicknesses of the layers are carefully chosen to be the proper fraction of a wavelength, then their reflectances can interfere constructively or destructively to produce either a bandpass or a band rejection filter. The simplest form of the bandpass filter is a single layer anti-reflection coating placed on the PV cell surface to couple light from a low index air or vacuum ambient into the high index semiconductor. In the 1960 timeframe researchers at General Motors, Allison Division[1] developed a reflective filter for use with a germanium cell operating in the TPV mode. These coatings were fabricated using five alternating layers of high index silicon and low index silicon dioxide. The thicknesses of the layers were calculated using a computer model of the interference phenomena within the stack. Typical results reported for this approach are shown in figure 4. As can be seen from figure 4, the broadband parasitic losses of this filter were substantial when compared to the requirements indicated in figure 3.

In the late 1970 and early 1980 timeframe researchers at the Boeing Company looked at both interference filters as first surface mirrors, reflective back contacts on the PV cell as second surface mirrors, and combinations of the preceding approaches.[2] Figures 5, 6 and 7 illustrate typical results from these efforts. The long wavelength parasitic losses still substantial and the in-band transmittance was reduced by harmonic reflections in the interference stack filter which was used to enhance the second surface mirror formed by the cell back contact.

During the 1980s considerable work was done on the development of indium tin oxide coatings for various applications. These coatings have some of the attributes desired for a TPV first surface reflective filter. They show reasonably good transmittance at short wavelengths and relatively high, broad-band reflectance at long wavelengths. Figure 8 illustrates the optical characteristics of a good ITO film.[3] However, for the TPV applications there are problems with the ITO film. Its reflective cut-on (known as the plasma edge) is gradual and the cut on is accompanied by considerable absorption. Further, the ultimate reflectance of ITO appears to be limited to about 90%.

More recently, efforts have been made to enhance the ITO, or plasma edge, filter with a dielectric stack interference filter.[4] Figure 9 shows results reported for this kind of filter. The long wavelength absorption is still about 20% and the in-band transmission about 75%. However, based on the best results reported for ITO films, it appears that the long-wavelength reflection losses may ultimately be reduced to about 10% for this type of filter. As can be seen in figure 3, 10% absorption is still substantial in a TPV system.

Currently researchers at EDTEK are utilizing advances in the capabilities of e-beam and masked ion-beam lithography, to adapt a new type of front surface reflective filter to the TPV application. These new filters appear to be almost ideally suited for the TPV application, particularly for applications which have low energy density heat sources such as radioisotopes or solar concentrators. The EDTEK TPV filter (patent pending) is based on a high density array of antenna elements. Figure 10 illustrates the two basic types of antenna arrays that can be made. If slotted elements are etched into a metal film then an inductive resonance is achieved producing a bandpass filter. If metal elements are etched onto a dielectric substrate then a capacitive resonance is achieved producing a band reject filter. The operation of the filters depends on the interaction of electric and magnetic fields with the tiny, submicron conductive elements of the mesh. A primarily inductive mesh is formed if the field-induced currents can flow between filter elements, such as the array of crossed slots patterned in a continuous metal film. If the dimensions of the elements are comparable to the wavelength of the interacting electromagnetic field, then resonant conditions can be set up. The inductive resonant mesh configuration produces a bandpass filter. The degree of transmittance and reflectance of the filter is a function of the size and shape of the elements, the resistivity of the metal film, and the dielectric and optical properties of the substrate. The resonant wavelength is also a function of the dielectric properties of the media surrounding the film. This can be understood by noting that resonance involves oscillations between energy flowing (current) and energy storage (field strength in the dielectric). Changing the dielectric constant is thus equivalent to changing the spring constant of a mechanical oscillator system. The resonant frequency is independent of incident angle with the off normal transmittance obeying the cosine function with angle of incidence.

Peak transmittance is determined primarily by the density of filter elements etched into the metal film. The higher the density, the higher the peak transmittance. There is a similar relationship between element density and bandwidth. Given a PV cell quantum efficiency performance, the filter performance can be adjusted via element length, density and linewidth to maximize conversion efficiency. To the first order, filter peak transmission wavelength is determined by the length of the element and the refractive index on either side of the metal film. This is expressed below:

$$\lambda = 2 \cdot L \, [(n_1^2 + n_2^2)/2]^{1/2}$$

where λ is the peak wavelength, n_1 and n_2 are refractive indices of the medium above (air) and below (substrate) the film, respectively, and L is the filter element length.

As the light frequency moves away from the resonance band, the filter behaves optically like the parent metal film. Thus the filter exhibits efficient broadband, angular insensitive reflectance to out of band energy. For a gold film the out of band reflectance is on the order of 98% to 99%. No other approach to date has been able to achieve this level of broadband parasitic loss suppression.

The filters are ideal for the TPV application for the following reasons:

a. They accept in-band energy useful to the PV cells and transmits it into the PV cells.
b. The in-band energy not transmitted to the cells is reflected back to the source.
c. The filters exhibit the broadband reflective properties of gold film (>98%) in the out-of-band regions of the spectrum.
d. The resonant frequency of the filters is independent of angle of incidence and the transmittance falls off only as the cosine of the angle of incidence of the light.

The performance of the EDTEK TPV filter in combination with GaSb PV cells has been demonstrated[5]. Figure 11 shows the configuration of this test schematically and figure 12 shows a photograph of the test hardware. We measured performance of 13.3% filtered cell array efficiency with a clear path of engineering improvements to >20% indicated[5,6]. These improvements have now been achieved as will be discussed later.

For the TPV application we have developed the bandpass filter in a gold film. Figure 13 shows a typical antenna pattern used in our filters. Figures 14 and 15 show the optical characteristics of such a filter. One of the features of these filters is that they either transmit or reflect having the absorption properties of the parent metal film. Figure 16 compares the long wavelength reflectance of the filter compared to the reflectance of a gold film. As can be seen, the two curves are indistinguishable. Thus, since gold films can be made routinely with 98% to 99% IR reflectance, these filters do indeed minimize the long wavelength losses. Another feature is the insensitivity of the resonant frequency to angle of incidence of the thermal radiation which is usually omnidirectional. Figure 17 presents measured data for the transmittance of a filter at varying angles of incidence showing that the resonant frequency is fixed with the transmittance falling off approximately as the cosine of the incident angle.

Thus, a TPV filter has been developed which can be tuned to one's particular application. The transmission band can be tuned to the desired frequency by adjusting either the physical dimensions of the antenna element or the index of refraction of the substrate. Overall transmittance and band width are also adjustable parameters. Figures 18 19, and 20 show the optical characteristics of three different geometrical filter configurations. The different transmission, bandwidth, and long wavelength cut off characteristics are achieved simply by varying the geometry of the antenna elements. Thus, there is considerable latitude for tailoring the filters to particular TPV applications.

The parasitic loss in a TPV system can be further suppressed by using the EDTEK TPV filter in combination with a line emitting source. Figure 21 shows the effective energy spectrum from an Auburn University Erbia composite emitter [7] modified by an EDTEK TPV filter. Calculations indicate that such a system can exceed 30% efficiency with GaSb PV cells.

The fabrication of the filters from a continuous gold film facilitates system fabrication as illustrated by the test array shown in the photograph of figure 22. The filters are fabricated a "windows" in the gold film. The windows are positioned to correspond to the locations of the PV cell active areas underneath the filter substrate. Thus a smooth, continuous gold film covers the interconnects and inactive areas between cells. This avoids the absorption that occurs wherever cracks or discontinuities are present in the TPV cavity walls. Figure 23 shows a detailed drawing of a more practical array which maximizes the active area to inactive area ratio commonly referred to as the "packing factor". The continuous gold film framing the active areas minimizes packing factor losses in the TPV system. This minimization is particularly important for radioisotope or solar concentrator heat systems where the source energy density is low relative to the emitting surface area.

Considerable progress has been made in tailoring the filters for the radioisotope fueled TPV application. Figure 24 shows the transmission characteristic of filters used in the early feasibility test[5] compared to a more recently fabricated filter. The improvements realized in system efficiency due to these filter improvements raises the system efficiency from 13.3% to > 20%. These system efficiency improvements are discussed in detail in another paper[8].

Progress has also been made in lowering the cost of production for the filters. The early filters and most of the optimization study samples to date have been fabricated by direct-write e-beam lithography, DEBL. This process, illustrated in figure 25, requires two write strokes by the e-beam for each antenna element. Since there are billions of elements per square centimeter, the DEBL process is very time consuming and expensive (about $4,000.00/cm^2$). For this reason, EDTEK has collaborated with the Cornell University National Nanofabrication Facility (NNF) and the University of Houston to develop a masked ion beam lithography (MIBL) production process. The MIBL process is also illustrated in figure 22. In the first step of the MIBL process, the NNF DEBL process is used to expose the desired pattern onto an e-beam-resist-coated silicon membrane. An etching process developed at the University of Houston generates open windows in the silicon membrane corresponding to the antenna elements imprinted in the photoresist. Thus, a silicon stencil is obtained that can be 1 cm^2 or more in area. A collimated proton beam is then directed onto the stencil. The energy of the protons are selected so that they are stopped by the stencil body but transmitted through the windows to expose a photoresist placed on the prepared filter substrate underneath. Thus, a 1 cm^2 or more filter area can be exposed in about 10 seconds. By moving the substrate underneath the stencil and repeating the exposures, a very large filter area can be built-up in a very short time. EDTEK, NNF, and the University of Houston recently demonstrated the feasibility of this process during a Department of Energy (DOE) sponsored SBIR Phase I project. Figure 26 compares the characteristics of filters fabricated by MIBL to those fabricated by DEBL holding everything in the process constant except the method of photoresist exposure. The MIBL process produces filters equivalent to the DEBL process and will improve repeatability and quality control.

Under a DOE SBIR Phase II effort, EDTEK is currently transferring (under license) the MIBL process from the University of Houston to a production pilot line located at the EDTEK laboratory. It is expected that in full production this method will produce the filters for less than $1.00/cm^2$.

In summary, a brief review of the reflective method of reducing parasitic losses in TPV systems has been presented and a new type filter has been discussed that promises to raise TPV conversion efficiencies to the 20% to 30% range and to greater than 30% when used in combination with composite line emitters depending on the heat source, PV cell type, and system configuration.

REFERENCES

1. Allison Division of General Motors, "Radiant Energy Conversion," Final Report, U.S. Atomic Energy Commission Contract AT(30-1)-3587, GMAO 3587-23, 3/10/1967.

2. Horne, W.E. and Day, A.C., "Thermal Photovoltaic Space Power System, Phase 3," NASA Contract NAS8-33436 Interim Report, Submitted to Marshall Space Flight Center July 11, 1983.

3. American Vacuum Society Short Course, Transparent Conducting Oxides, Chicago 1992.

4. Baldasaro, Paul F., et al., "Experimental Assessment of Low Temperature Voltaic Energy Conversion," 1st NREL Conference on Thermophotovoltaic Generation of Electricity, AIP Conf. Proceedings 321, Copper Mountain, CO, 1994.

5. Horne, W. E., et al., "Thermophotovoltaic Thermal-to-Electric Conversion Systems Report," Contract No. 959595, Final Report, prepared by The Boeing Company for Jet Propulsion Laboratory, December 1993.

6. Schock, A., et al., "Radioisotope Thermophotovoltaic (RTPV) Generator and Its Application to the PLUTO Fast Flyby Missions, Fairchild Space & Defense Corporation, FSC-ESC-217-93-519.

7. Adair, Peter L. and Rose, Frank M., "Composite Emitters for TPV Systems," 1st NREL Conf. on Thermophotovoltaics Generation of Electricity, AIP Conf. Proceedings 321, Copper Mountain, CO, 1994.

8. Schock, A., C. Or, and V. Kumar, "Small Radioisotope Thermophotovoltaic Generator" and "Effect of Measured Infrared Filter Improvements on Performance of Radioisotope Thermophotovoltaic System". 2nd NREL Conference on TPV Electricity Generation, Colorado Springs, CO, July 17-19, 1995.

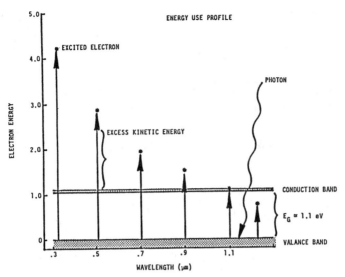

Figure 1. Photon Efficiency for Producing Electron-Hole Pairs in Silicon

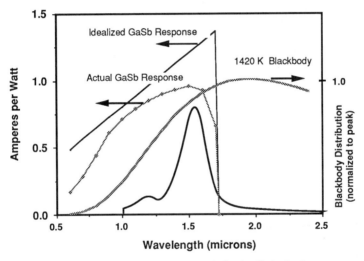

Figure 2. Comparison of Key TPV System Parameters Indicating Motivation for Applying a Line Filter or Source to the TPV System

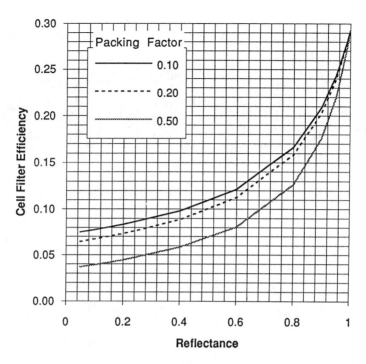

Figure 3. Typical Dependence of PV Cell/Filter Efficiency on Long Wavelength Reflectance for a 1473K Blackbody Source

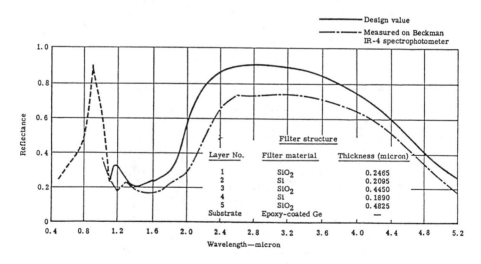

Figure 4. Comparison of The Design and Measured Reflectivity for Allison Fine-Layer Interference Filter (ref. 1)

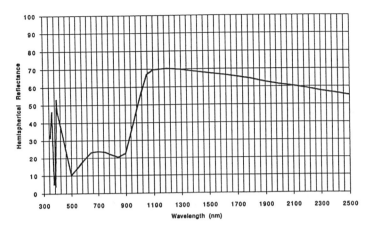

Figure 5. Spectral Reflectance of Silicon TPV Cell With Reflective Back Contact (ref. 2)

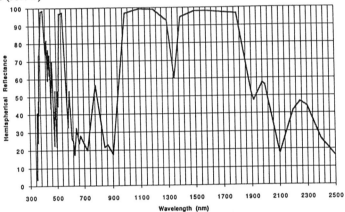

Figure 6. Spectral Reflectance of Broadband Dielectric Stock Filter (ref. 2)

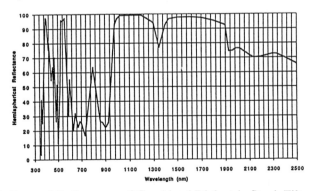

Figure 7. Spectral Reflectance of Combined Dielectric Stock Filter and Reflective Silicon Cell (ref. 2)

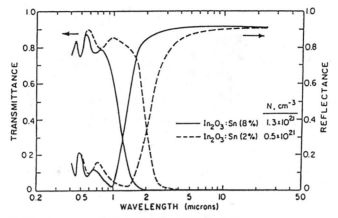

Figure 8. Reflectance of High Quality ITO (ref. 3)

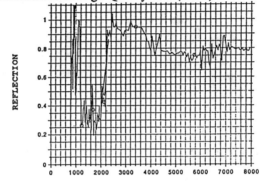

Figure 9. Reflectance of Interference Filter and Plasma Filter Combined (ref. 4)

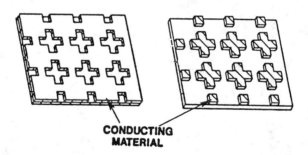

Figure 10. Illustration of Two Basic Types of Antenna Filter Elements. Slot Elements in a Metal Film Produce a Bandpass Filter. Dipole Elements on a Dielectric Produce a Band Reject Filter

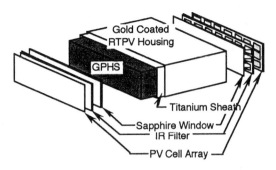

Figure 11. Schematic Diagram of RTPV Demonstration Experiment

Figure 12. Photograph of Experimental RTPV Demonstration Cavity

Figure 13. Photomicrograph of Typical TPV Filter Pattern

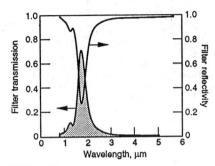

Figure 14. Typical Filter Characteristic

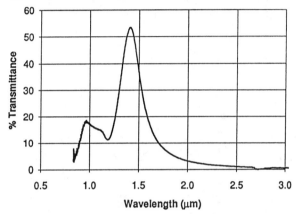

Figure 15. Typical Filter Transmittance Optimized for Use in a Radioisotope Fueled TPV System

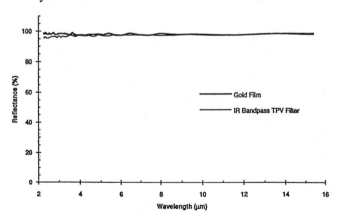

Figure 16. Comparison of Long Wavelength Reflectance of TPV Filter and Gold Film

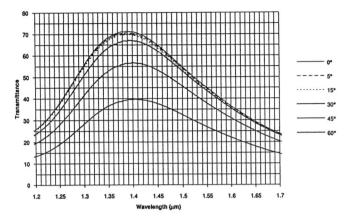

Figure 17. EDTEK TPV Filter Transmittance as a Function of Incidence Angle

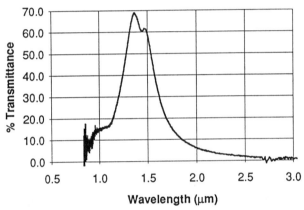

Figure 18. Transmittance of TPV Filter Configuration Alternate 1

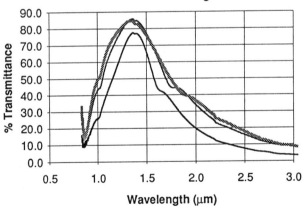

Figure 19. Transmittance Characteristic of TPV Filter Configuration Alternate 2

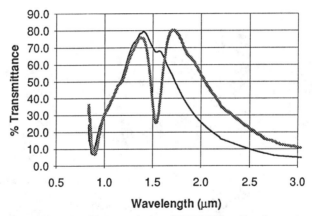

Figure 20. Transmittance Characteristic of TPV Filter Configuration Alternate 3

GaSb PV cell Efficiency
(Line Emitter only) = 17.61%

GaSb PV cell Efficiency (With Filtered
Line Emitter) = 39.98%

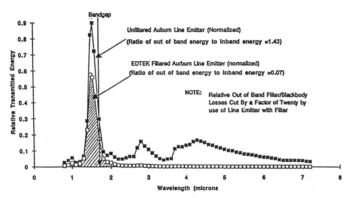

Figure 21. Comparison of Filtered and Unfiltered Composite Line Emitter Spectra and the Resultant GaSb PV Cell Efficiencies (composite line emitter data from ref. 7)

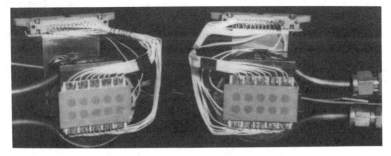

Figure 22. Photograph of Filter Arrays Showing Filter "Windows" Corresponding to active PV Cell Area Underneath

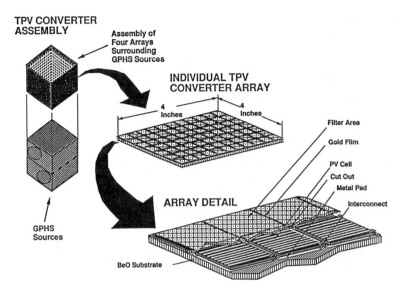

Figure 23. 5K Etch Illustrating a Practical TPV Array Utilizing to Continuous Gold Film Feature of the EDTEK TPV Filter to Minimize Packing Factor Losses

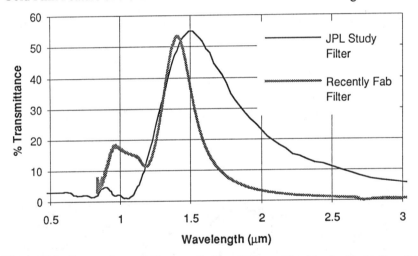

Figure 24. Comparison of Characteristics of Filters Used in the Early Feasibility Test (13.3% eff ref. 5) and Improved Filters (>20% eff ref. 8)

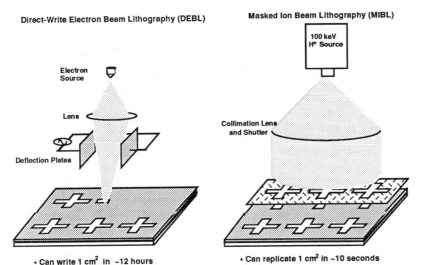

Figure 25. Illustration of the Direct Write E-Beam Exposure Process Compared to the Masked Ion Beam Exposure Process

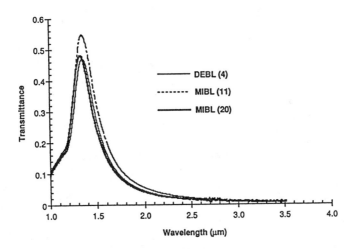

Figure 26. Comparison of Transmittance Characteristics of Filters Fabricated by the Direct Write E-Beam Lithography Process, DEBL, and the Masked Ion Beam Lithography Process, MIBL

SESSION I:
THERMOPHOTOVOLTAIC
SYSTEMS DESIGN AND PERFORMANCE

Effect of Expanded Integration Limits and of Measured Infrared Filter Improvements on Performance of RTPV System

A. Schock, C. Or, and M. Mukunda

Orbital Sciences Corporation
Germantown, MD 20874 U.S.A.

Abstract. In papers presented at last year's conference, the authors described an integrated Radioisotope Thermophotovoltaic (RTPV) power system design study they had conducted for the U.S. Department of Energy, and examined the system's applicability to an illustrative space mission to Pluto. The power system employed previously flown and safety-qualified radioisotope heat source modules, radiating their heat to gallium antimonide photovoltaic cells covered with spectrally selective infrared filters.

Analyses of the system required tabular data for the spectral reflectivity of the resonant gold filters and the quantum efficiency of the GaSb PV cells. These data had to be integrated over the range of wavelengths from 0 to ∞ to compute the converter heat flux and power density. In performing these integrations, we had erroneously truncated the upper and lower integration limits. When that was corrected in the present study, it was found that, for a given set of hot and cold temperatures, expanding the integration limits to their full range had almost no effect on the converter's power output, but significantly increased its heat input, which led to a corresponding decrease in conversion efficiency. The results reported in the present paper are all based on the expanded integration limits.

The previous OSC study had employed two alternative filter and cell data sets furnished by Boeing (now EDTEK) personnel: a conservative set, based on actual Boeing measurements; and an optimistic set, based on their projected improvements. OSC's system design studies indicated that the use of an RTPV instead of an RTG (Radioisotope Thermoelectric Generator) for the Pluto mission would substantially reduce the generator's mass and required fuel inventory, and would greatly increase its specific power and system efficiency, particularly for the projected filter and cell performance models.

In view of these encouraging results for the projected performance models, OSC extended a subcontract to EDTEK to develop improved filters and cells, and to demonstrate what performance improvements can in fact be achieved. The present paper reports the spectral reflectivities of improved filters produced under that subcontract, and their effect on the predicted performance of OSC's RTPV system. As shown in the paper, the filter improvements demonstrated to date have yielded about three quarters of the previously projected RTPV system performance improvement. Experiments are continuing to determine the effect of additional filter improvements and planned cell improvements on system performance.

© 1996 American Institute of Physics

INTRODUCTION

Last year the authors presented papers [1,2] describing their DOE-sponsored design, analysis, and optimization of a Radioisotope Thermophotovoltaic (RTPV) Power System for possible use on a NASA mission to Pluto, the only unexplored planet in the solar system. The present paper is based on the identical design, but its analysis differs in two major respects.

The previous analysis had truncated the range of wavelengths over which the numerical integrations were carried out, based on the conclusion that the truncated region had only a minor effect on the converter's performance. When that conclusion was reexamined recently, it was found to be incorrect, which changes the previously reported results. As shown in the present paper, expanding the wavelength limits to extend from zero to infinity has virtually no effect on the predicted power output, but significantly raises the predicted heat absorption, which lowers the predicted conversion efficiency. Except where otherwise noted, all results in the present paper are for the full integration limits.

The previous OSC analyses had been based on two sets of wavelength-dependent filter reflectivities and PV cell quantum efficiencies. Both sets had been supplied to us by Boeing (now EDTEK) personnel. One set was based on actual filter and cell measurements, and the other on their predictions of potential performance improvements. Since the projected filter and cell models yielded substantially better system performance than the measured models, OSC extended a subcontract to EDTEK in the fall of 1994 to develop improved filters and cells, to demonstrate what performance improvements can in fact be achieved. Except where otherwise noted, all results presented in this paper are based on the measured spectral reflectivities of the best improved filters produced to date by EDTEK under the OSC subcontract. As will be shown, OSC system analyses indicate that the initial filter improvements, for the same PV cells, have yielded about three quarters of the projected filter and system performance improvements.

SYSTEM DESIGN

This section briefly reviews the design of the power system (consisting of a radioisotope heat source, TPV converter, heat rejection radiator) and its integration with the Pluto spacecraft. These were described more fully in last year's papers [1, 2]. Optical, thermal and electrical analyses and optimization of the integrated system are discussed in subsequent sections.

Radioisotope Heat Source

The radioisotope heat source for the RTPV design, like that for the RTG options analyzed in earlier OSC studies [3], is based on General Purpose Heat Source (GPHS) modules [4]. These are the same modules that were used in the RTGs flown on the Galileo and Ulysses missions after very extensive safety analyses and tests and after passing stringent safety reviews, and which are slated for launch on the upcoming Cassini mission.

As shown, each GPHS module has a maximum thermal power of 250 watts, and contains four ^{238}PuO$_2$ fuel pellets encapsulated in iridium-alloy clads. The remaining module components are graphitic and are designed to protect the integrity of the iridium clads in case of accidents before, during, and after launch. There are two impact shells and one aeroshell made of fine-weave pierced fabric (FWPF), a very tough high-temperature three-dimensional carbon-carbon composite. The aeroshell would serve as an ablator in the unlikely event of inadvertent atmospheric reentry, and the impact shells would help to prevent breach of the clads during subsequent Earth impact. Between the impact shells and the aeroshell is a high-temperature thermal insulator consisting of a low-density composite of carbon-bonded carbon fibers (CBCF), to prevent overheating of the clads during the reentry heat pulse and overcooling and embrittlement of the clads during the subsequent subsonic atmospheric descent before earth impact.

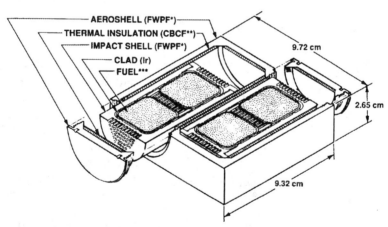

*Fine-Weave Pierced Fabric, a 90%-dense 3D carbon-carbon composite
**Carbon-Bonded Carbon Fibers, a 10%-dense high-temperature insulator
***62.5-watt ^{238}PuO$_2$ pellet

FIGURE 1. GPHS-General Purpose Heat Source Module (250 Watts) Sectioned at Mid-Plane.

As in last year's study [1,2], the RTPV generator design is based on the use of two GPHS modules, which would be a two-thirds reduction from the six modules used in the RTG design for the Pluto mission [3].

Thermophotovoltaic Converter Assembly

As shown in Figure 2, the two GPHS modules are enclosed in a common molybdenum canister. The inside of the canister's end caps is lined with iridium as a reaction barrier between the graphite and molybdenum, and the outside of its side walls is coated with tungsten to minimize sublimation. To reduce the temperatures in the heat source, the inside of the canister's side walls are roughened to raise their effective total emissivity to 0.60. So is the walls' outside, for reasons explained later in this paper.

Figure 2 presents an exploded view of the two-module heat source and of the thermophotovoltaic converter, and lists their components' construction materials. As indicated, the canister's two end faces are thermally insulated by a multifoil assembly, similar to those used in previous thermoelectric converters, and consisting of 60 layers of 0.008 mm Mo foils separated by ZrO_2 spacer particles. Our analysis showed that only 1.8% of the thermal power is lost through the multifoil insulation.

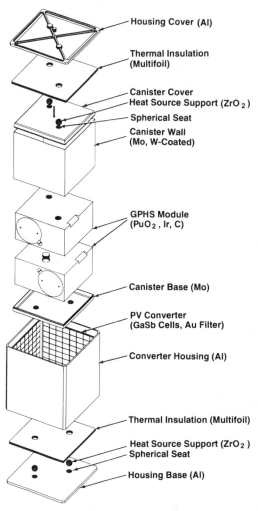

FIGURE 2. Exploded View of Converter.

As indicated in Figure 2, each of the canister's four side faces radiates heat to a photovoltaic array of 8x8 closely spaced gallium antimonide cells, slightly over 1 cm^2 each. The canister side walls and planar arrays of filtered PV cells are parallel to each other and closely spaced to provide a good view factor for radiative interchange. The spacing between the PV cells is the minimum necessary for the desired series-parallel connections. The cells on each face are bonded to a BeO substrate, which in turn is bonded to the generator's aluminum side wall.

Each of the four cell arrays is covered with a spectrally selective infra-red filter [5,6], to permit the transmission of those wavelengths which can be efficiently converted to electricity by the PV cells, and the reflection of those wavelengths that cannot. Much of the reflected radiation is absorbed by the heat source canister, which then re-emits it with a full spectral distribution. Thus, the unused reflected energy is conserved, which reduces the energy to be supplied by the radioisotope heat source and greatly increases the efficiency of the generator. Hence, the selective filter is a vital element of this RTPV system.

Each of the generator's four PV cell arrays is covered with a continuous resonant gold filter. The filter consists of a thin gold film deposited on a transparent substrate (e.g., sapphire), containing over two hundred million submicronic holes per cm^2 in its active regions (i.e., opposite each PV cells). The size, spacing, and geometry of the hole pattern determines the performance of the resonant filters.

Since the heat source modules are contained in a monolithic canister, unlike the stack of unsupported modules used in preceding RTGs [7], there is no need for a complex axial preload mechanism to hold the stack together during launch vibration. As indicated in Figure 2, each of the canister's two end faces is supported by a pair of diagonally opposed low-conductivity zirconia balls. The diagonal locations at the two end faces are orthogonal to each other. The zirconia balls penetrate through the multifoil thermal insulation, and are seated in spherical indentations on the outside of the canister end caps, and on the inside of the generator housing end caps. Thus, they provide both axial and lateral support to the heat source. For the dimensions shown, our analysis showed that only 0.5% of the heat source's thermal power is lost through the zirconia support balls. Hence, over 97% of the generated heat arrives at the converter.

The converter's 256 PV cells are arrayed in a series-parallel matrix. At each horizontal level, the cells are parallel-connected in groups of four, and these groups of parallel cells are series-connected to groups in adjacent horizontal levels. Thus, each generator side has two series-parallel networks of 8x4 cells, and the generator's eight networks are connected in series with each other to form a 64x4 series-parallel network, for a total output of approximately 28 volts.

Heat Rejection System

The RTPV needs much larger radiator fins than typical RTGs, because - as will be shown - they must operate at much lower heat rejection temperatures to achieve their high efficiencies. The optimum dimensions, i.e., the dimensions that maximize the system's specific power, were determined by detailed analyses described later. Detailed analyses are warranted because the radiators are the biggest mass component of the RTPV system.

Figure 3 shows an exploded view of the generator and of one of the four radiator fins, with typical dimensions. As shown, there is a large trapezoidal fin bonded to each side of the converter housing. The exploded fin view shows a central core consisting of an aluminum honeycomb with two embedded Al/NH_3 heat pipes. To each face of the honeycomb core, two skins are bonded: an inner skin of aluminum which has minimal thickness over most of its length, but is thickened near the fin root to provide increased structural strength for resisting bending moments during launch; and an outer skin consisting of a graphitized carbon-carbon composite to provide high thermal conductance in the fiber direction. The graphite fibers are oriented in the vertical direction, normal to the heat pipes' axes. In that direction they have a thermal conductivity twice that of copper, at about one fourth its density [8]. They serve to distribute the heat from the heat pipes over the width of the fin. They also provide the fin with high-emissivity surfaces.

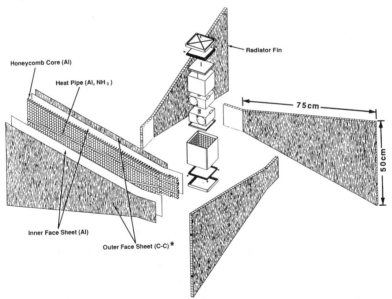

FIGURE 3. Exploded View of RTPV Generator and of Radiator Fin.

Finally, we need to examine whether these large fin sizes could really be accommodated on JPL's Pluto spacecraft design, which is shown in Figure 4. An earlier design had located the high-gain antenna at the top of the spacecraft, but this has now been moved to the side as shown. This frees up the space on top for mounting the optimized RTPV generator with its 75 cm-long fins. As shown, when the generator is rotated 45 degrees about its axis, its fins clear the antenna.

Alternative Arrangement

An alternative design arrangement to minimize system mass, suggested by A. Newhouse, is depicted in Figure 5. It calls for merging the RTPV radiator with the parabolic high-gain antenna of the Pluto spacecraft. Both of these components employ carbon-carbon face sheets, stiffened respectively by aluminum heat-pipes and ribs. As indicated, the radiator's high-conductivity fibers would be oriented in a circumferential direction, normal to its eight heat pipes. Merging the radiator and antenna would eliminate the need for the radiator's aluminum honeycomb and face sheets, since the radiator would be supported by the antenna's stiffening ribs, and since a parabolic dish is inherently much stiffer than a cantilevered planar fin. As will be shown, this arrangement - if feasible - would substantially raise the specific power of the RTPV system.

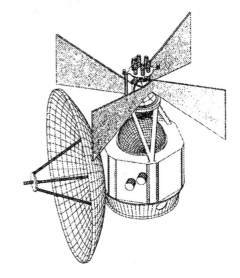

FIGURE 4. Optimized RTPV Generator Mounted on Top of Pluto Spacecraft.

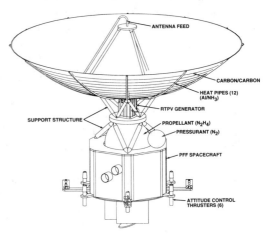

FIGURE 5. RTPV Generator and Pluto Spacecraft with Common Radiator/Antenna.

Finally, Figure 6 shows how an RTPV-powered Pluto spacecraft with a combined radiator/antenna could fit into a Proton launch vehicle.

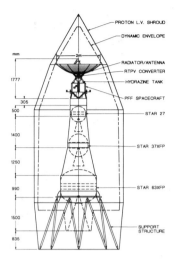

FIGURE 6. Fit of RTPV-Powered Pluto Spacecraft with Combined Radiator/Antenna into Proton Launch Vehicle.

EFFECT OF INTEGRATION LIMITS ON CONVERTER PERFORMANCE

As explained more fully in the authors' earlier papers [1,2], the net heat flux absorbed by the filtered PV cells is given by

$$q_{net} = \int_o^\infty \frac{2\pi h c^2 \lambda^{-5}[\exp(hc/\lambda k T_s)-1]^{-1}}{[\varepsilon_s(\lambda)]^{-1}+\{[R_c(\lambda)]^{-1}-1\}^{-1}} d\lambda, \qquad (1)$$

where **h** and **k** are Planck's and Boltzmann's constants, **c** is the speed of light, T_s is the absolute surface temperature of the heat source canister, and ε_s (λ) and R_c (λ) are the canister emissivity and filtered cell reflectivity at wavelength λ.

The converter's resultant short-circuit current density J_{sc} is given by

$$J_{sc} = 2\pi\alpha c e \int_0^\infty \frac{\lambda^{-4} Q(\lambda)[\exp(hc/\lambda k T_s)-1]^{-1}}{[\varepsilon_s(\lambda)]^{-1}+\{[R_c(\lambda)]^{-1}-1\}^{-1}} d\lambda, \qquad (2)$$

where α is the active (*i.e.*, unobstructed) area fraction of the PV cell array, e is the electronic charge, and $Q(\lambda)$ is the cell's quantum efficiency at wavelength λ. The cell's maximum output power density is then given by

$$P_{max} = J_{sc}V_{oc}F. \tag{3}$$

Equations for the open-circuit voltage V_{oc} and fill factor F were given in the previous paper [2]. Finally, the converter's efficiency is given by

$$\eta = P_{max}/q_{net} \tag{4}$$

In carrying out the numerical computation of Eqs. (1) and (2), the previously reported analyses had truncated the integration at wavelengths below 1 micron and above 4 micron. This was based on the mistaken assumption that outside those limits the integrands become negligible. When this assumption was reexamined in May 1995, it was found that it was correct for the short-circuit current density J_{sc} and power density P_{max}, but that it was not correct for the input heat flux q.

As shown in Figure 7, expanding the integration limits to embrace all wavelengths from zero to infinity results in a significant increase in the predicted heat flux and a corresponding drop in conversion efficiency. As seen, this is true both for the same heat flux and for the same source temperature.

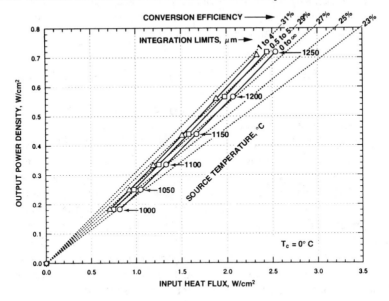

FIGURE 7. Effect of Integration Limits on Performance of TPV Converter Based on Projected Filter and Cell Models.

This conclusion is reconfirmed by Figure 8, which shows the effect of the integration limits on the computed performance of the RTPV system with the same filter and cell performance and with identical fin dimensions and range of graphite skin thicknesses.

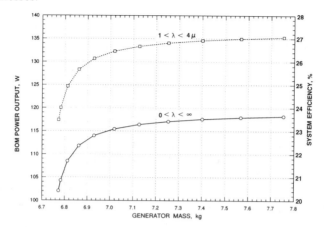

FIGURE 8. Effect of Integration Limits on Performance of RTPV System with Projected Cell and Filter, 75 cm Fin Length, and 50 cm Tip Height.

Clearly, the previously published analyses overpredicted the conversion efficiencies, typically by about 3 percentage points, and must be recomputed by integrating over the full range of wavelengths, as is done in the balance of the present paper and in all our subsequent presentations.

EFFECT OF FILTER AND CELL IMPROVEMENTS

Application of the Equations (1) through (4) requires three sets of wavelength-dependent data: the emissivity of the heat source canister $\varepsilon_s(\lambda)$, the quantum efficiency $Q(\lambda)$ of the PV cell, and the reflectivity $R_c(\lambda)$ of the filtered cell. For the latter two, OSC's previous study employed measured and projected performance models supplied to us by Boeing (now EDTEK) personnel in 1993. The effect of those models is depicted in Figure 9, which presents plots of the converter's output power density versus input heat flux for the projected and measured filter and cell performance models (for an illustrative 0°C cell temperature). The figure also shows lines of constant conversion efficiency and the effect of heat flux on source temperature. As can be seen, for a given heat flux the projected filter and cell models yield much higher conversion efficiencies than the measured models. Therefore, OSC initiated a subcontract at EDTEK to develop improved filters and cells, and to demonstrate what performance improvements can actually be achieved.

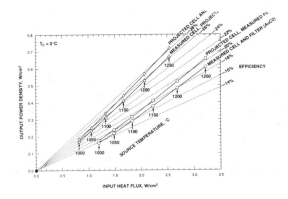

FIGURE 9. Effect of Projected Cell and Filter Improvements on Conversion Efficiency for Same Heat Flux.

In Figure 9 the differences between the first and second curves and between the third and fourth curves represent the effect of projected cell improvements; and the differences between the first and third curves and between the second and fourth curves represent the effect of projected filter improvements. Clearly, the efficiency gain produced by the projected filter improvement is much greater than that yielded by the projected cell improvement. Therefore, OSC requested that EDTEK give higher priority to filter improvement than to cell improvement.

The EDTEK subcontract started in the fall of 1994, and Figure 10 displays the spectral transmittance of the first three filters (AuR 1, 2, 4) produced under that subcontract, and compares them with the measured and projected transmittance of the filters used in last year's study.

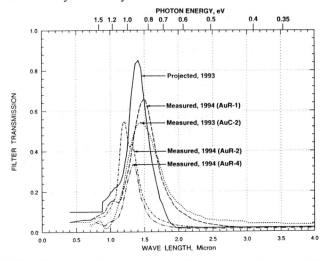

FIGURE 10. Spectral Transmittance of Projected and Measured Filters.

Two of the new filters, AuR2 and AuR4, were analyzed by OSC to generate curves of power output versus heat input. Figure 11 compares those curves with similar curves from the previously used projected and measured filters. As can be seen, for the same heat flux both of the new filters yield much higher efficiencies than the previously measured filter (AuC2). In fact, the best of the lot (AuR4) appears to have achieved more than 75% of the initially projected efficiency improvement. For example, for a typical input heat flux of 1.1 watt per cm^2, the previous measured filter (AuC2) yielded an efficiency of only 14%, but the best of the new batch of measured filters (AuR4) yielded a conversion efficiency of 22%, an improvement of eight percentage points. In fact, the efficiency of AuR4 is not far below the efficiency which EDTEK had projected for the improved filter (24%).

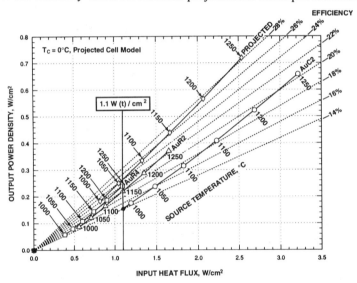

FIGURE 11. Efficiency Comparison of New Filters (AuR2, AuR4) with Previous Measured (AuC2) and Projected Filters.

While the filter improvement demonstrated so early in the EDTEK program was quite encouraging, it should be noted that - for the same 1.1-watt/cm^2 heat flux - the improved performance of AuR4 occurs at a much higher source temperature (1250°C) than AuC2 (970°C). This is because AuR4 has a much lower transmission peak than AuC2, as was shown in Figure 10. A temperature of 1250°C would lead to an iridium clad temperature that is still below the established limit of 1330°C for the GPHS clad. Temperatures above that limit would result in iridium grain growth leading to reduced impact resistance, which would only be acceptable if it occurred after the generator can no longer return to Earth. Figure 11 also shows that replacement of AuR4 with AuR2 would lower the source temperature from 1250°C to 1150°C, with only a 2% loss in conversion efficiency.

All of the above results were for filters (AuR2 and AuR4) made in 1994. More recently (in May 1995) EDTEK sent us measured performance data on a large selection of subsequently made filters [9]. The spectral transmittance of the ones yielding the best performance (SCAN A, H, and K) is compared with the previously measured (AuC2, AuR4) and projected filters in Figure 12. The figure also displays results for a filter (SCAN-E) made by Masked Ion Beam Lithography (MIBL). The MIBL process, under development by EDTEK under an SBIR contract, yields dramatic increases in filter production rates and corresponding decreases in production cost. Thus, it enables affordability of large-scale TPV applications.

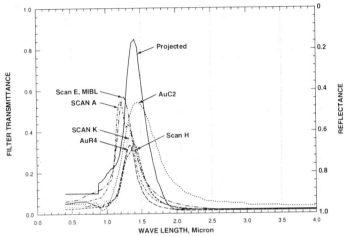

FIGURE 12. Spectral Transmittance of Recent and Previous Filters.

The filter transmissions displayed in Figure 12 were analyzed by OSC, with the results shown in Figure 13. As can be seen, there are several filters (AuR4, SCAN A, K, H) with almost identical performance curves. At a heat flux of 1.1 W/cm^2 all of these are seen to be more than three quarters of the way between the previous measured filter (AuC2) and the filter projected in 1993. This clustering is important, because it shows that AuR4 was not a singular fluke, but its performance is closely matched by a number of other filter designs. Once the process settles down, excellent performance reproducibility is anticipated.

Figure 13 also shows that the efficiency of the MIBL filter (SCAN E) is only a little lower than that of the best four filters made to date (20 versus 22% at 1.1 W/cm^2). This is very encouraging, because SCAN E is only an early attempt at making a MIBL filter, and ongoing investigation may yield even further performance improvements for these low-cost devices.

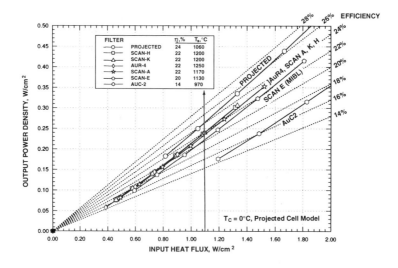

FIGURE 13. Comparison of Converters with Recent and Previous Filters.

Although the four filters (AuR4, SCAN A, H, K) yield very similar efficiencies, the Figure 13 insert table shows that at the same 1.1 W/cm² heat flux they produce significantly different source temperatures. Since SCAN A results in the lowest source temperature (1170°C versus 1250°C for AuR4), it was selected as the basis of the rest of the analyses in this report.

PARAMETRIC CONVERTER ANALYSIS

For a parametric range of heat source temperatures T_s and cell temperatures T_c, integration of Eqs. (1) and (2) were carried out over the full range of wavelengths, from 0 to ∞, and the converter's power density and efficiency were computed from Eqs. (3) to (4). The parametric results are displayed in Figures 14 through 17.

Figure 14 shows that the net heat flux is only a function of the source temperature, and is essentially independent of the cell tempera-ture in the range of interest.

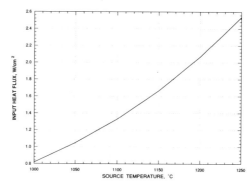

FIGURE 14. Effect of Source and Cell Temperatures on Net Heat Flux Absorbed by Converter.

Figures 15 and 16 show that the output power density and the converter efficiency are sensitive functions of the cell temperature. Lowering that temperature leads to significant performance improvements, albeit at the cost of increased radiator mass. Trade-offs between those parameters to maximize the system's specific power output are described in a later section.

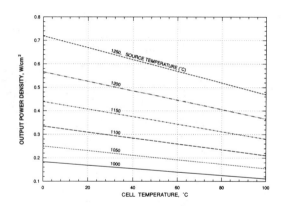

FIGURE 15. Effect of Source and Cell Temperatures on Maximum Output Power Density of Converter.

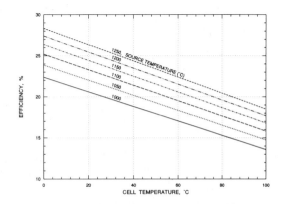

FIGURE 16. Effect of Source and Cell Temperatures on Efficiency of Converter.

The results of Figures 12, 13, and 14 are combined in Figure 17, which presents cross-plots showing the effect of q_{net} on T_s, P_{max}, and η, for a range of cell temperatures T_c. It again shows the performance improvement obtainable by lowering the cell temperature.

Figure 17 also shows that, for a given heat flux, higher source temperatures lead to higher power densities and efficiencies. From that, one might infer that these parameters can be significantly increased by lowering the heat source emissivity, which raises the source temperature for a given heat flux. But quite the opposite was found to be the case. This was discovered in the OSC study when the effect of roughening the tungsten surface on the converter's performance was analyzed.

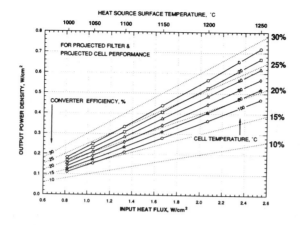

FIGURE 17. Effect of Net Heat Flux and Cell Temperature on Source Temperature, Power Density, and Efficiency.

The effective total emissivity, $\overline{\varepsilon}_s$ of the heat source canister is obtained by a weighted average of its spectral emissivity $\varepsilon_s(\lambda)$:

$$\overline{\varepsilon}_s = \frac{\int_0^\infty \lambda^{-5}[\exp(hc/\lambda kT_s)-1]^{-1}\varepsilon_s(\lambda)d\lambda}{\int_0^\infty \lambda^{-5}[\exp(hc/\lambda kT_s)-1]^{-1}d\lambda} \quad (5)$$

Applying the spectral emissivities for smooth tungsten shown by the solid curve in Figure 18 gives a value of 0.21 for $\overline{\varepsilon}_s$. It was assumed that roughening the tungsten (e.g., by grit blasting) would raise its effective total emissivity to 0.60. This corresponds to a 45% reduction of the smooth-tungsten spectral reflectivity [1-$\varepsilon_s(\lambda)$], resulting in the spectral emissivity for roughened tungsten shown by the dashed curve of Figure 18.

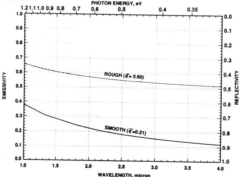

FIGURE 18. Effect of Surface Roughness on Spectral Emissivity of Tungsten Canister.

The effect of that emissivity increase is shown in Figure 19, which compares the converter's computed performance for smooth and roughened canisters at a 0°C cell temperature. As expected, for the same heat flux the roughened canister yields a significantly lower temperature, by about 100°C. But contrary to our expectations, the resultant power density and conversion efficiency was not lower but was in fact somewhat higher for the roughened canister.

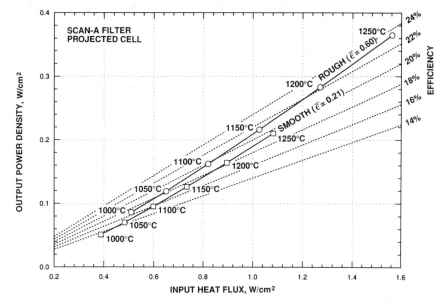

FIGURE 19. Effect of Canister Roughness on its Temperature, Converter Efficiency, and Output Power.

After some reflection, the author suggests the following explanation for this surprising result. The efficiency of the TPV system depends on the canister's efficient absorption and re-emission of the energy reflected by the spectrally selective filter. That absorption is facilitated by roughening the tungsten surface. With a smooth canister, much of the radiation reflected by the filter is re-reflected by the canister, and continues to bounce back and forth until it is finally absorbed. Since the filter is less than 100% efficient and the cells' active area fraction is less than 100%, these multiple reflections are necessarily somewhat lossy. This would explain why the smooth canister yields a somewhat lower efficiency, in spite of its much higher temperature. Since lowering the heat source temperature without loss of performance is a desirable goal, our RTPV design study assumed the use of a roughened tungsten canister.

PARAMETRIC SYSTEM ANALYSIS AND OPTIMIZATION

Analysis of the power system, including its heat rejection system, requires a coupled thermal and electrical analysis. The coupled analysis was carried out by means of a thermal analysis code (SINDA [10]) that had been modified by OSC, and by a standard thermal radiation code (SSPTA [11]). For the former we constructed a 197-node model, and for the latter a model consisting of 496 surfaces.

OSC made two major modifications in the thermal analysis code. The net heat flux q_{net} from the sides of the heat source to the converter cells at each iteration was computed by integration of Eq. (1), with appropriate corrections for gaps between cells and obstruction by the electrical grid; and the waste heat flowing to the radiator fins was computed by subtracting the converter's electrical power generation rate from the heat generation rate of the heat source. The power generation rate was computed by multiplying the total cell area of the generator by the power density P_{max} obtained from Eq. (2).

In the coupled analysis, the heat generation rate is known, but the heat source surface temperature T_s and cell temperature T_c are not. Therefore, the analysis must be carried out iteratively. In each iteration, the two thermal codes compute a new set of canister and cell temperatures, which are used as inputs in the next iteration. This iterative procedure is repeated until the modified code converges on a consistent solution.

For a fixed heat source thermal power and converter design, the only other system design parameters are the size of the radiator fins and the thickness of the high-conductivity carbon-carbon face sheets (see Figure 3). Increasing either of those parameters will lower the cell temperature, which increases the power output and efficiency, as shown in Figures 15 and 16. But this benefit is obtained at the cost of increased system mass. Thus, the system design requires a trade-off between system mass and performance (efficiency, power output). The goal of the optimization study is to determine the radiator dimensions which maximize the power system's specific power. As was explained in detail in the previous papers [1,2], the mass of the radiator fins includes an allowance for thickening the aluminum honeycomb skins near the fin roots, to enable the cantilevered fins to withstand the bending moments during launch vibration.

Effect of Graphite Skin Thickness

Let us first examine the effect of varying the graphite skin thickness on system characteristics for a set of illustrative radiator fin dimensions. For a 75 cm root-to-tip fin length, a 50 cm tip height, a 9.5 mm honeycomb thickness, a 0.076 mm aluminum skin thickness, and a converter with 90% active cell area, the effect of varying the graphite skin thickness from 0 to 0.76 mm is illustrated in Figure 20. The figure shows the effect of graphite skin thickness on system mass, cell temperature, output power, system efficiency, and specific power. In each of the three figures, the solid curve represents results based on measured values of SCAN-A filter transmittance and cell quantum efficiency, the dotted curve is based on SCAN-A filter and projected cell performance, and the dashed curve is based on projected vaules of the two.

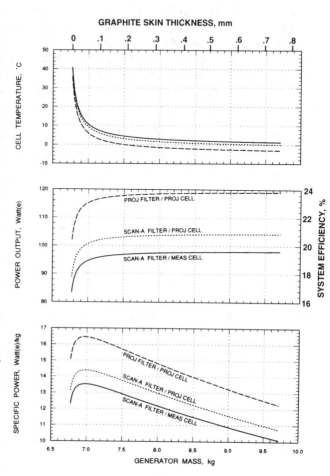

FIGURE 20. Effect of Graphite Skin Thickness.

As shown in Figure 20, the projected properties (primarily the filter reflectance) have a pronounced effect on system performance. It is also noteworthy that initially the addition of the graphite skins lowers the cell temperature which increases the output power and efficiency significantly, but after adding a surprisingly small thickness (typically 0.15 mm) further additions of graphite only increase the mass with little further increase of power or efficiency.

Effect of Filter and Cell Models on System Performance

Curves similar to Figure 20b were generated by OSC for most of the filters produced by EDTEK to date (May '95), with the results displayed in Figure 21. All except the top curve are for the cell performance measured in 1993. As can be seen, most of the gold filters are seen to be substantially better than the earlier AuC2. As shown, for the same cell performance the best of the gold filters (A, H, K, and AuR4) yield about 75% of the system performance improvement projected in 1993. By contrast, the aluminum filters are generally much poorer than the recent gold filters, because the reflectivity of aluminum is much lower than that of gold. In fact, the performance of the three aluminum filters is only a little better than that of the 1993 gold filter (AuC2). The difference between the top two curves in Figure 21 shows how much more the performance of the system with the projected filters would be improved if the projected cell performance were realized.

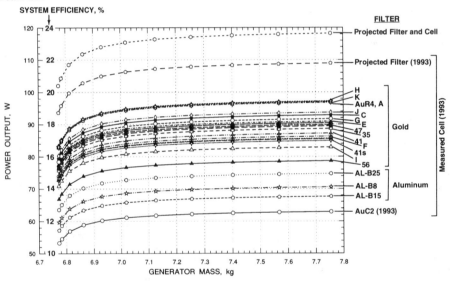

FIGURE 21. Effect of Filters on Performance of RTPV System with 75 cm Fin Length and 50 cm Tip Height.

Figure 22 compares the effect of the best filters on the power output, system efficiency, system mass, and specific power of the system with projected cell improvements. Again we see that the best filters yield almost 75% of the predicted system improvement from the 1993 measured and projected filters. The figure also shows that even the early MIBL filter (SCAN-E) yields encouraging system improvements.

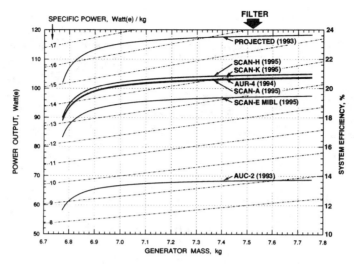

FIGURE 22. Effect of Measured and Projected Filters on Performance of RTPV System with Projected Cell.

EFFECT OF RADIATOR FIN DIMENSIONS

Similar analyses were carried out for fin lengths ranging from 50 to 100 cm and for fin tip heights of 30 and 50 cm. The results for all cases showed similar trends, confirming the previous conclusion that the specific power of the design is maximized at a graphite skin thickness around 0.15 mm. With thicker graphite skin, the increase in power output is quite small and is outweighed by the increased graphite mass.

The results of the parametric design studies are displayed in Figure 23 and 24. Both figures show curves representing the results of thermal, electrical, and mass analyses for fin lengths ranging from 50 to 100 cm and fin tip heights ranging from 30 to 50 cm, with the graphite skin thickness as the implicit variable within each curve. Each point on each curve is the result of an iterative solution of the coupled thermal and electrical analyses, using the modified thermal analysis code described earlier. All curves assume aluminum skins varying from 0.08 mm at the fin tip to whatever is needed near the fin root to survive a 40-g launch load without exceeding the allowable stress limit.

Figure 23 shows plots of cell temperature versus generator mass. For each fin size, the upper curve is based on the measured (SCAN-A) filter transmittance and the projected PV quantum efficiency model, and the lower curve is for the projected filter and cell characteristics. As can be seen, the larger fins lead to very low cell temperatures, but at substantially higher masses.

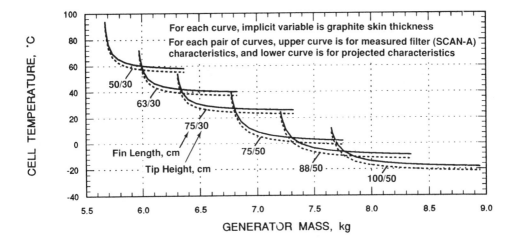

FIGURE 23. Effect of Fin Dimensions on Cell Temperature.

The trade-offs between mass and performance for the measured and projected filters are summarized in Figure 24. For each fin size, it presents curves of output power and system efficiency versus generator mass, with graphite skin thickness as the implicit variable. It also shows diagonal lines of constant specific power, which identify the fin dimensions that maximize the generator's specific power.

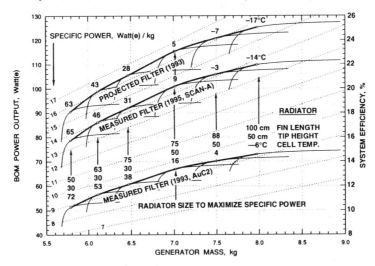

FIGURE 24. Effect of Filter on System Performance for Various Radiator Sizes.

For each filter and cell performance model, the figure shows a family of performance curves for different fin dimensions and a tangent envelope curve, with indicated cell temperatures at their point of tangency. For each performance model, the corresponding envelope curve represents the highest specific power that can be achieved by optimizing the system's radiator geometry. For every point on the envelope, there is some combination of fin length, fin tip height, and graphite skin thickness that will achieve the indicated performance

As can be seen, for all three performance models the system's specific power is maximized with a 75 cm fin length and 50 cm fin height. But note that this optimum is quite broad. As illustrated in Table 1, major deviations from the optimum design result in only modest reductions in specific power. Thus, the designer has wide latitude in trading off power and efficiency versus mass and size to meet specific mission goals. For example, for the projected filter and cell performance, the BOM power could be raised from 115 watts to 125 watts by lengthening the fins from 75 cm to 100 cm. As shown, this would increase the generator's mass from 7.0 to 8.0 kg, but would only decrease its specific power from 16.4 to 15.7 w/kg. Conversely, if desired the generator mass could be reduced from 7.0 kg to 5.8 kg by reducing the fin size. As shown, this would reduce the BOM power from 115 watts to 90 watts, but would only lower the generator's specific power from 16.4 to 15.6 w/kg.

TABLE 1: Effect of Off-Optimum Design on RTPV Performance.

Goal	Low Mass	Max Sp Power	High Power
Fin Length, cm	50	75	100
Fin Tip Height, cm	30	50	50
Cell Temperature, °C	63	5	-17
Power (BOM), Watt	90	115	125
Efficiency (BOM), %	18.1	23.1	25.0
System Mass, kg	5.8	7.0	8.0
Specific Power, W/kg	15.6	16.4	15.7

Figure 25 compares the performance envelope for the RTPV system with trapezoidal fins from Figure 24 with the corresponding envelope for the system with parabolic fins that was previously displayed in Figure 5.

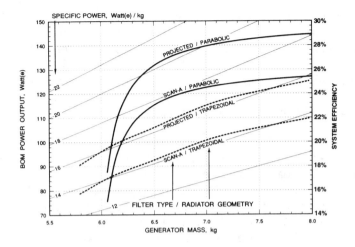

FIGURE 25. Effect of Radiator Geometry on BOM RTPV Performance.

As can be seen, for both the measured and projected filter and cell performance models, the parabolic fins in which the spacecraft's antenna is combined with the power system's radiator yields a substantially higher maximum specific power and efficiency than the system with trapezoidal fins.

Table 2 compares the BOM operating temperatures and system performance characteristics of a Pluto RTG with those of the RTPV with trapezoidal fins, for both the Measured and Projected filter and cell performance models.

TABLE 2: RTG/RTPV BOM Performance Comparison.

Generator	RTG		RTPV			
Radiator Fins	Rectangular		Trapezoidal		Parabolic	
Performance Model	Unicouple		Measured	Projected	Measured	Projected
Generator Mass, kg	15.8	18.1	7.0	7.0	6.7	6.7
No. of Heat Source Modules	5	6	2	2	2	2
Thermal Power, watts	1250	1500	500	500	500	500
Operating Temp, °C:						
Clad	1304	1315	1278	1190	1279	1191
Aeroshell	1053	1064	1205	1105	1207	1106
Canister	-	none	1183	1079	1185	1080
Converter	982/252	993/259	10	-5	-33	-39
Radiator Heatpipe	-	none	-9	-13	-58	-62
Output Voltage	19	23	26.8	27.5	31.0	31.8
Output Current, amps	4.6	4.6	3.5	4.2	3.8	4.3
Output Power, watts	87	106	95	115	118	136
System Efficiency, %	7.0	7.1	19.0	23.1	22.3	27.2
Specific Power, watts/kg	5.7	5.9	13.5	16.5	16.7	20.3

CONCLUSION

Finally, Figure 26 compares the predicted BOM performance of the RTPV system options, if successfully developed, with the results of recently published [12] system studies based on three point designs for the much more mature RTGs with SiGe thermoelectric unicouples.

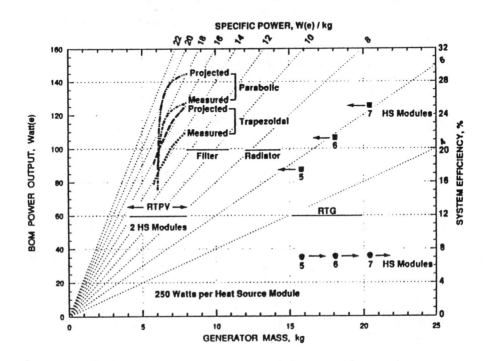

FIGURE 26. RTPV/RTG Performance Comparison.

As can be seen, for the same 250-watt heat source modules and the same output power, replacement of an RTG with a trapezoidal-fin RTPV would reduce the fuel requirement and the generator mass by approximately a factor of three, and would roughly triple the power system's efficiency and specific power. Even greater performance gains could be achieved by combining the RTPV's radiator with the spacecraft's parabolic antenna, which could quadruple the system's efficiency and specific power. This is why DOE is pursuing the development of RTPV technology, in spite of the RTG's much greater maturity.

REFERENCES

[1] Schock, A., and V. Kumar, "Radioisotope Thermophotovoltaic System Design and its Application to an Illustrative Space Mission," in *Proceedings of the NREL Conference on Thermophotovoltaic Generation of Electricity*, held at Copper Mountain, CO, July 1994, published by the American Institute of Physics.

[2] Schock, A., M. Mukunda, T. Or, and G. Summers, "Analysis, Optimization, and Assessment of Radioisotope Thermophotovoltaic System Design for an Illustrative Space Mission," in *Proceedings of the NREL Conference on Thermophotovoltaic Generation of Electricity*, held at Copper Mountain, CO, July 1994, published by the American Institute of Physics.

[3] Schock, A., "RTG Options for Pluto Fast Flyby Mission," IAF-93-R.1.425a, Presented at the 44th Congress of the International Astronautical Federation, Graz, Austria, October 1993.

[4] Schock, A., "Design Evolution and Verification of the General-Purpose Heat Source," #809203 in *Proceedings of 15th Intersociety Energy Conversion Engineering Conference*, held in Seattle, WA, 1980.

[5] Chase, S.T. and R.D. Joseph, "Resonant Array Bandpass Filters for the Far Infrared," *Applied Optics*, Vol. 22, No. 11, 1 June 1983, pp 1775-1779.

[6] Kogler, Kent J. and Rickey G. Pastor, "Infrared Filters Fabricated from Submicron Loop Antenna Arrays," *Applied Optics*, Vol. 27, No. 1, 1 January 1988, pp 18-19.

[7] Schock, A., "Use of Modular Heat Source Stack in RTGs," #799305 in *Proceedings of 14th Intersociety Energy Conversion Engineering Conference*, held in Boston, MA, 1979.

[8] Denham, H.B., et. al. "NASA Advanced Radiator C-C Fin Development," in *Proceedings of the 11th Symposium on Space Nuclear Power Systems*, CONF-940101, M.S. El-Genk and M.D. Hoover, eds., American Institute of Physics, New York, AIP Conf. No. 301, 3:1119-1127, 1994.

[9] Horne, W.E., M.D. Morgan, and V.S. Sundaram, "IR Filters for TPV Converter Module," to be presented at the 2nd NREL Conference on Thermophotovoltaic Generation of Electricity, in Colorado Springs, CO, July 1995.

[10] Gaski, J., *SINDA (System Improved Numerical Differencing Analyzed)*, version 1.315 from Network Analysis Associate, Fountain Valley, CA, 1987.

[11] Little, A.D., *SSPTA (Simplified Space Payload Thermal Analyzer)*, version 3.0/VAX, for NASA/Goddard, by Arthur D. Little Inc., Cambridge, MA, 1986.

[12] Schock, A., and C. T. Or, "Effect of Fuel and Design Options on RTG Performance Versus PFF Power Demand," in *Proceedings of the 29th Intersociety Energy Conversion Engineering Conference*, held in Monterey, CA, 7-12 August 1994.

Small Radioisotope Thermophotovoltaic (RTPV) Generators

A. Schock, C. Or, and V. Kumar

Orbital Sciences Corporation
Germantown, MD 20874 U.S.A.

Abstract. The National Aeronautics and Space Administration's recently inaugurated New Millennium program, with its emphasis on miniaturized spacecraft, has generated interest in a low-power (10- to 30-watt), low-mass, high-efficiency RTPV power system. This led to a Department of Energy (DOE)-sponsored design study by OSC (formerly Fairchild) personnel, who had previously conducted very encouraging studies of 75-watt RTPV systems based on two 250-watt General Purpose Heat Source (GPHS) modules. Since these modules were too large for the small RTPVs described in this paper, OSC generated derivative designs for 125-watt and 62.5-watt heat source modules. To minimize the need for new development and safety verification studies, these contained identical fuel pellets, clads, impact shell, and thermal insulation as the previously developed and safety-qualified 250-watt units. OSC also generated a novel heat source support scheme to reduce the heat losses through the structural supports, and a new and much simpler radiator structure, employing no honeycombs or heat pipes. OSC's previous RTPV study had been based on the use of GaSb PV cells and spectrally selective Infra-Red (IR) filters that had been partially developed and characterized by Boeing (now EDTEK) personnel. The present study was based on greatly improved selective filters developed and performance-mapped by EDTEK under an OSC-initiated subcontract. The paper describes illustrative small-RTPV designs and analyzes their mass, size, power output, system efficiency, and specific power, and illustrates their integration with a miniaturized New Millennium spacecraft.

INTRODUCTION

Revolutionary developments are occurring in the U.S. space program to enable the launchings of many small and inexpensive science missions instead of the infrequent and very expensive missions launched in recent decades. This requires drastic reductions in the mass and cost of spacecraft and payloads through use of advanced technologies, which are the goals of NASA's recently initiated New Millennium program. It also requires corresponding reductions of the payload's power demand and increases in power system efficiency and specific power, since power systems have traditionally constituted about one third the mass of the spacecraft.

© 1996 American Institute of Physics

Those space missions that require too much energy for reasonable sized batteries and that must operate where sunlight is inadequate for practical solar power systems can utilize radioisotope power systems. Such systems, employing thermoelectric converters, have been successfully used on dozens of space missions, where they have demonstrated excellent reliability and durability. But their efficiencies are too low to meet the objectives of the New Millennium program. This has led to NASA interest in much more efficient radioisotope power systems. There are three prime contenders that offer the possibility of tripling the efficiency of thermoelectric conversion systems at low power outputs: thermophotovoltaic (RTPV) systems, free-piston Stirling engines with linear alternators, and Alkali Metal Thermal to Electric Conversion (AMTEC) systems. The authors have conducted detailed studies [1,2,3,4,5,6] of 75-watt radioisotope power systems of the first two options for the Department of Energy, and have recently initiated a study of the third option. The studies of the first two options showed that both of the advanced systems would yield much higher efficiencies (and therefore reduced fuel requirements) than present Radioisotope Thermoelectric Generators (RTGs [7,8,9,10]), but that TPV systems promise much lower weights and higher specific powers than present-technology Stirling systems [5,6,10]. Recently, the NASA Administrator requested that DOE prepare a much smaller (20-watt) RTPV system design. DOE assigned that task to OSC, with the results reported in the present paper.

SMALL RTPV GENERATOR DESIGN

The previous study of a 75-watt(e) RTPV had employed a 500-watt radioisotope heat source consisting of two 250-watt General Purpose Heat Source (GPHS) modules [11] which DOE had developed and safety-qualified for various RTG missions (Galileo, Ulysses, Cassini). As shown in Figure 1, each GPHS module contains four 62.5 watt(t) PuO_2 fuel pellets encapsulated in iridium-alloy clads. The remaining module components are graphitic and are designed to protect the integrity of the iridium clads in case of accidents before, during, and after launch.

There are two impact shells and one aeroshell made of fine-weave pierced fabric (FWPF), a very tough high-temperature three-dimensional carbon-carbon composite. The aeroshell would serve as an ablator in the unlikely event of inadvertent atmospheric reentry, and the impact shells would help to prevent breach of the clads during subsequent Earth impact. Between the impact shells and the aeroshell is a high-temperature thermal insulator consisting of carbon-bonded carbon fibers (CBCF), to prevent overheating of the clads during the reentry heat pulse and overcooling and embrittlement of the clads during the subsequent subsonic atmospheric descent prior to earth impact.

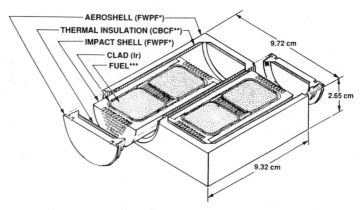

*Fine-Weave Pierced Fabric, a 90%-dense 3D carbon-carbon composite
**Carbon-Bonded Carbon Fibers, a 10%-dense high-temperature insulator
***62.5-watt ^{238}PuO$_2$ pellet

FIGURE 1. GPHS-General Purpose Heat Source Module (250 Watts) Sectioned at Mid-Plane.

The previously flown 250-watt GPHS modules are too big for a 20-watt RTPV generator. To minimize the need for new development and safety tests, OSC elected to base its design on a 125-watt heat source, containing identical fuel pellets, clads, impact shell, and thermal insulator, and employing the same aeroshell material and wall thickness as the 250-watt units, as shown in Figure 2.

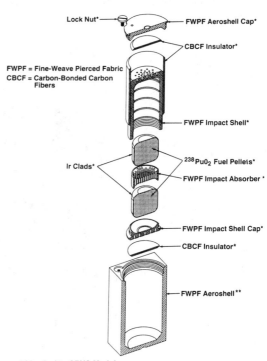

* Identical to GPHS Module
** Identical Material and Wall Thickness

FIGURE 2. Exploded View of 125-watt Heat Source (Sectioned at Midplane).

HEAT SOURCE CANISTER AND SUPPORT SCHEME

As shown in Figure 3, the heat source is enclosed in a molybdenum canister. This is to prevent contaminants released by the hot heat source from reaching the gold filter and the gallium antimonide photovoltaic (PV) cells, which could degrade their performance. The outside of the canister is coated with tungsten, to minimize sublimation (only 10^{-6} monolayers in ten years) and the tungsten coating is roughened to enhance the generator's conversion efficiency, for reasons explained in previous reports [2,3,4]. Where the inside of the canister is in contact with the aeroshell it is lined with iridium, to prevent reaction between molybdenum and graphite.

The support scheme previously proposed for the 500-watt heat source would have led to excessive heat losses from the small (125-watt) heat source. To avoid this, OSC devised a heat source support scheme in which each end of the canister is supported by a single zirconia ball seated in spherical indentations on the outside of the high-temperature canister end caps and on the inside of the low-temperature aluminum housing of the generator. Thus, the heat source is supported both axially and laterally.

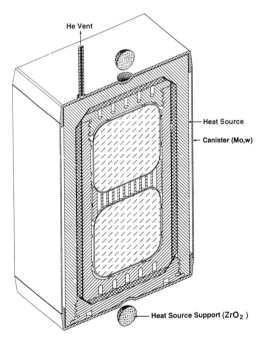

FIGURE 3. 125-Watt Heat Source Enclosed in Molybdenum Canister.

Figure 4 shows the high-temperature canned heat source inside the generator's low-temperature aluminum housing, and depicts the heat source support scheme. The canister's two end faces and its two inactive side faces are thermally insulated by a multifoil assembly (not shown), consisting of 60 layers of 0.008 mm W foils separated by ZrO_2 spacer particles, similar to those successfully used in previous long-duration thermoelectric converter tests. As shown, the helium generated by the fuel's alpha decay is vented to space through a semi-permeable vent plug.

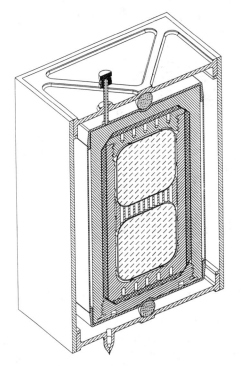
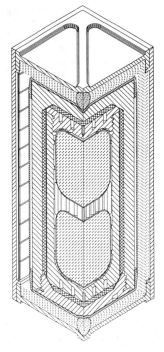

FIGURE 4. 125-Watt Heat Source in Al Converter Housing, Showing Low-Loss Support Scheme.

FIGURE 5. Quarter Section of 20-Watt RTPV Converter. (Showing Multifoil Thermal Insulation and PV Cells)

As indicated in Figure 5, the canister's two end faces and two of its four side faces are covered with the multifoil thermal insulation, and the other two side faces radiate heat to photovoltaic cells covered with spectrally selective infra-red filters (not shown), which are bonded to the inside of the aluminum housing.

Our analysis showed that 8.6% of the heat source's thermal power is lost through the multifoil insulation, and only 1.0 % passes through the ZrO_2 support balls. Thus, over 90 % of the thermal power is transferred to the PV cells.

As shown in Figure 6, each of the canister's two active faces radiates heat to a photovoltaic array of 6x11 closely spaced gallium antimonide cells, 7.9 x 8.4 mm each. The canister side walls and planar arrays of filtered PV cells are parallel to each other and closely spaced to provide a good view factor for radiative interchange. The spacing between the PV cells is the minimum necessary for the desired series-parallel connections. The cells on each face are bonded to a BeO substrate, which in turn is bonded to the generator's aluminum side wall.

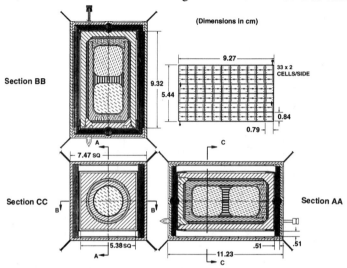

FIGURE 6. Sectioned Views of RTPV Heat Source and Converter, with 66 x 2 Series-Parallel Cell Network to Produce 28-Volt Output.

Each of the two cell arrays is covered with a spectrally selective infra-red filter [12,13], to permit the transmission of those wavelengths which can be efficiently converted to electricity by the PV cells, and the reflection of those wavelengths that cannot. Much of the reflected radiation is absorbed by the heat source canister, which then re-emits it with a full spectral distribution. Thus, the unused reflected energy is conserved, which reduces the energy to be supplied by the radioisotope heat source and greatly increases the efficiency of the generator. Hence, the selective filter is a vital element of this RTPV system.

Each of the generator's two PV cell arrays is covered with a resonant gold filter. The filter consists of a continuous thin gold film deposited on a transparent substrate (e.g., sapphire), about 54 x 93 mm, containing over two hundred million submicronic holes per cm^2 in its active regions (i.e., opposite each PV cell). The size, spacing, and geometry of the hole pattern determines the performance of the resonant filters.

The converter's 132 PV cells are arrayed in a series-parallel matrix. At each horizontal level, the cells are parallel-connected in groups of two, and these groups of parallel cells are series-connected to groups in adjacent horizontal levels. Thus, the generator consists of a 66 x 2 series-parallel network, for a total output of approximately 28 volts.

Heat Rejection System

The RTPV System needs much larger radiator fins than typical RTGs, because the PV cells must operate at much lower heat rejection temperatures to achieve their high efficiencies. For the previously published 75-watt design, the optimum radiator dimensions, i.e., the dimensions that maximize the system's specific power, were determined by detailed analyses.

Figure 7 shows an exploded view of the previous 75-watt generator [1, 3, 4] and of one of its four radiator fins, with typical dimensions. As shown, it had a large trapezoidal fin bonded to each side of the converter housing. The exploded fin view shows a central core consisting of an aluminum honeycomb with two embedded aluminum/ammonia heat pipes. To each face of the honeycomb core, two skins were bonded: an inner skin of aluminum which had minimal thickness over most of its length, but was thickened near the fin root to provide increased structural strength for resisting bending moments during launch; and an outer skin consisting of a graphitized carbon-carbon composite to provide high thermal conductance in the fiber direction. The graphite fibers were oriented in the vertical direction, normal to the heat pipes' axes. In that direction they have a thermal conductivity twice that of copper, at about one fourth its density [14]. The graphite skins served to distribute the heat from the heat pipes over the width of the fin, and also provided the fin with high-emissivity surfaces

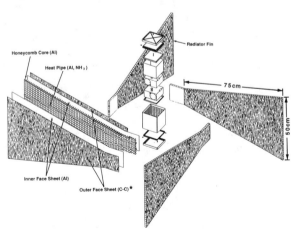

FIGURE 7. Exploded View of 75-Watt RTPV Generator and of Radiator Fin.

For the small RTPV generator, OSC was able to devise a much simpler radiator, with no honeycombs or heatpipes. As indicated by the cut-away view shown in Figure 8, it consists of 12 mutually stiffened trapezoidal fins, emanating at 45 degrees from the 12 edges of the converter housing. The mutual stiffening provided by neighboring fins greatly reduces the bending moments during launch vibration, and makes it possible to dispense with the previously shown honeycombs and with the embedded heat pipes. Each fin consists of a 0.020" aluminum core, bonded to high conductivity carbon-carbon face sheets. The dimensions shown are illustrative. Their optimization will be discussed shortly.

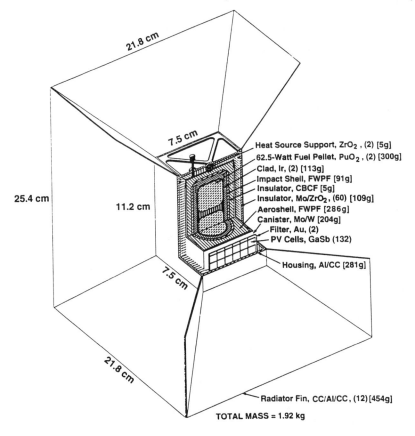

FIGURE 8. Small RTPV Radiator Configuration, Showing Mass Breakdown of Generator with Typical Radiator Height (25 cm) and Graphite Thickness (0.18 mm)

As can be seen for the illustrated dimensions the heat source and the radiator respectively comprise 52% and 24% of the generator's 1.92 kg mass.

Effect of Filter and Cell Improvements

As described in detail in previous publications [2,3,4], analysis of the RTPV generator requires three sets of wavelength-dependent data: the emissivity of the heat source canister, the quantum efficiency of the PV cell, and the reflectivity of the IR filter. For the latter two, OSC's previous study employed measured and projected performance models supplied to us by Boeing (now EDTEK) personnel in 1993 [2]. The effect of those models is depicted in Figure 9, which presents plots of the TPV converter's output power density versus input heat flux for the projected and measured filter and cell performance models (for an illustrative 0°C cell temperature). The figure also shows lines of constant conversion efficiency and the effect of heat flux on source temperature. As can be seen, for a given heat flux the projected filter and cell models yield much higher conversion efficiencies than the measured models. Therefore, OSC initiated a subcontract at EDTEK to develop improved filters and cells, and to demonstrate what performance improvements can actually be achieved.

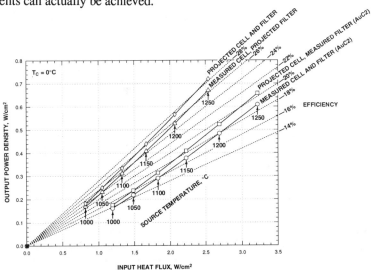

FIGURE 9. Effect of Projected Cell and Filter Improvements on Conversion Efficiency for Same Heat Flux.

In Figure 9 the differences between the first and second curves and between the third and fourth curves represent the effect of projected cell improvements; and the differences between the first and third curves and between the second and fourth curves represent the effect of projected filter improvements. Clearly, the efficiency gain produced by the projected filter improvement is much greater than that yielded by the projected cell improvement. Therefore, OSC requested that EDTEK give higher priority to filter improvement than to cell improvement.

The EDTEK subcontract started in the fall of 1994, and several of the filters made to date (May 1995) demonstrate very encouraging improvement over the filter measured in 1993 (AuC2). As shown in Figure 10, the performance of each of several new filters has already closed 75% of the gap between the 1993 measured and projected filter performances. This enhances our confidence that EDTEK's continuing filter and cell improvement studies [15] will suceed in achieving their projected performance goals.

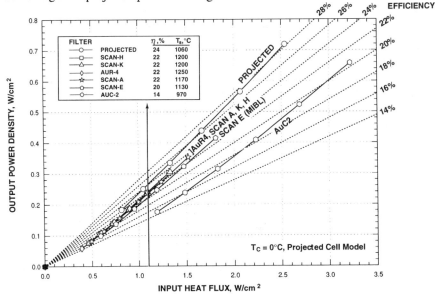

FIGURE 10. Comparison of Converters with Recent and Previous Filters.

Parametric System Analysis and Optimization

Analysis of the RTPV power system, including its heat rejection system, requires coupled optical, thermal, and electrical analyses. As described in detail in previous reports [2,3,4], these analyses were carried out by means of a thermal analysis code (SINDA [16]) that had been modified by OSC, and by a standard thermal radiation code (SSPTA [17]).

In the coupled analysis, the heat generation rate is known, but the heat source surface temperature and cell temperature are not. Therefore, the analysis must be carried out iteratively. In each iteration, the two thermal codes compute a new set of canister and cell temperatures, which are used as inputs in the next iteration. This iterative procedure is repeated until the modified code converges on a consistent solution.

For a fixed heat source thermal power and converter design, the only other system design parameters are the size of the radiator and the thickness of the high-conductivity carbon-carbon face sheets (see Figure 8). Increasing either of those parameters will lower the cell temperature, which increases the power output and efficiency. But this benefit is obtained at the cost of increased system mass. Thus, the system design requires a trade-off between system mass and performance (efficiency, power output). The goal of the optimization study is to determine the radiator dimensions which maximize the power system's specific power.

Effect of Graphite Skin Thickness

Let us first examine the effect of varying the graphite skin thickness on system characteristics for an illustrative 25 cm generator height and a converter with 90% active cell area. The effect of varying the graphite skin thickness from 0 to 1.0 mm is illustrated in Figure 11. The figure shows the effect of graphite skin thickness on system mass, cell temperature, output power, system efficiency, and specific power. In each of the three plots, the solid curve represents results based on measured values of SCAN-A filter transmittance and cell quantum efficiency, the dotted curve is based on SCAN-A filter and projected cell performance, and the dashed curve is based on projected values of the two.

As shown in Figure 11, for each performance model the initial addition of the graphite skins lowers the cell temperature which increases the output power and efficiency significantly, but after adding a surprisingly small thickness (typically 0.15 mm) further additions of graphite only increase the mass with little further increase of power or efficiency.

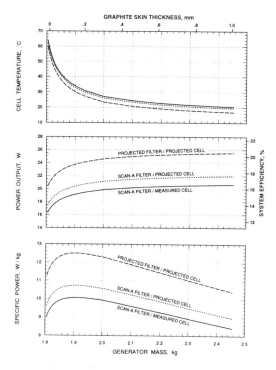

FIGURE 11. Effect of Graphite Skin Thickness.

Effect of Radiator Fin Dimensions

The results of parametric design studies are displayed in Figures 12 and 13. Both figures show curves representing the results of thermal, electrical, and mass analyses for generator heights ranging from 20 to 36 cm, with the graphite skin thickness as the implicit variable within each curve. Each point on each curve is the result of an iterative solution of the coupled thermal and electrical analyses, using the modified thermal analysis code described earlier.

Figure 12 shows plots of cell temperature versus generator mass. For each radiator size, the upper curve is based on the measured (SCAN-A) filter transmittance and the projected PV quantum efficiency model, and the lower curve is for the projected filter and cell characteristics. As can be seen, the larger fins lead to very low cell temperatures, but at substantially higher masses.

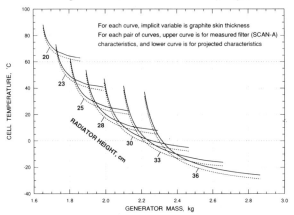

FIGURE 12. Effect of Radiator Height on Cell Temperature

The trade-offs between mass and performance for the measured and projected filters are summarized in Figure 13. For each radiator size, it presents curves of output power and system efficiency versus generator mass, with graphite skin thickness as the implicit variable. It also shows diagonal lines of constant specific power, which identify the radiator height that maximizes the generator's specific power.

For each filter and cell performance model, the figure shows a family of performance curves for different radiator dimensions and a tangent envelope curve, with indicated cell temperatures at their point of tangency. For each performance model, the corresponding envelope curve represents the highest specific power that can be achieved by optimizing the system's radiator geometry. For every point on the envelope, there is some combination of radiator height and graphite skin thickness that will achieve the indicated performance.

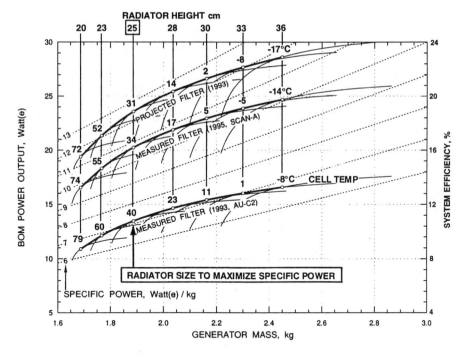

FIGURE 13. Effect of Filter on System Performance for Various Radiator Sizes.

Comparison of the three envelopes shows that EDTEK's improved filter (SCAN-A) has already succeeded in closing ~75% of the performance gap between their 1993 measured and projected filters.

As seen in Figure 13, for all three performance models the system's specific power is maximized with a 25 cm radiator height. But note that this optimum is quite broad. As illustrated in Table 1, major deviations from the optimum design result in only modest reductions in specific power. Thus, the designer has wide latitude in trading off power and efficiency versus mass and size to meet specific mission goals.

Table 1. Effect of Off-Optimum Design on RTPV Performance.

Goal	Low Mass	Max Sp Power	High Power
Radiator Height, cm	20	25	36
Cell Temperature, °C	72	31	-17
Power (BOM), Watt	19	24	29
Efficiency (BOM), %	15.5	18.9	22.9
System Mass, kg	1.7	1.9	2.5
Specific Power, W/kg	11.5	12.5	11.7

ADDENDUM

After completing the 20-watt RTPV generator study requested by the NASA Administrator, OSC decided to supplement it with a similar study of an even smaller generator, based on a 62.5-watt (1-fuel-capsule) derivative of the previously flown (4-capsule) General Purpose Heat Source (GPHS) module. As before, the small heat source employs a fuel pellet, clad, impact shell, and thermal insulation identical to the GPHS, and uses the same aeroshell material and wall thickness to minimize the need for new development.

As shown in Figures 14 and 15, the 62.5-watt heat source is inserted into a TPV converter that is smaller but otherwise identical to the previious designs, using 6.2 x 6.8 mm PV cells. The converter contains 128 filtered PV cells, connected in a 64 x 2 series-parallel network to yield a 28-volt output.

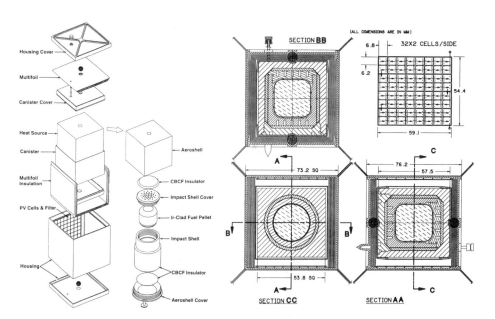

FIGURE 14. Exploded View of 62.5-Watt Heat Source and Thermophotovoltaic Converter

FIGURE 15. Sectioned View of 62.5-Watt Heat Source and Thermophotovoltaic Converter

As shown, the converter is roughly a 76 mm (3") cube. The combined mass of the heat source and converter is 0.94 kg.

The generator employing the 62.5-watt heat source uses the same radiator configuration as the previous design, and its size represents a trade-off between maximizing the power output and system efficiency versus minimizing the generator size and mass. As shown in Figure 16, for the illustrated radiator size the generator is a 18 cm (7") cube weighing 1.18 kg.

The analytical results for the smaller generator, for both the measured and projected filter and cell performance models, are displayed in Figure 17. As can be seen, the specific power of the generator maximizes with a 18 cm (7") radiator, at a system mass of 1.18 kg. For the best filter (SCAN-A) demonstrated to date and the measured cell performance, the generator yields a BOM power output of 8.8 watts(e), a system efficiency of 14%, and a 7.4 w/kg specific power. For the projected filter and cell performance, the generator's BOM power rises to 11 watts(e), the system efficiency to 18%, and the specific power to 9.4 w/kg. If desired, its BOM power and efficiency could be raised to 12.5 watts and 20% by employing a 23 cm radiator, which would raise the generator mass to 1.36 kg, but yield almost the same specific power (9.3 w/kg). Conversely, an 8.4-watt output could be achieved with a 13 cm cubical generator weighing 1.04 kg.

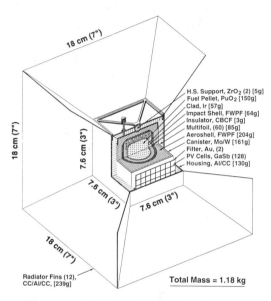

FIGURE 16. Cutaway View and Mass Breakdown of RTPV Generator with 62.5-Watt Heat Source

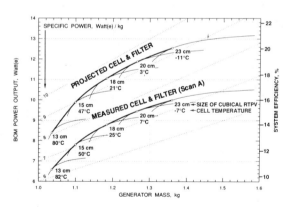

FIGURE 17. Effect of Filter and Cell Models on Performance of RTPV with 62.5-Watt Heat Source

Finally, Figure 18 compares the generator performance maps for RTPVs employing 125- and 62.5-watt heat sources based on projected filter and cell performance models. As can be seen, the smaller generator has about half the size and power output as the larger unit. The efficiencies of the two generators are about the same, but the specific power of the smaller unit is significantly lower.

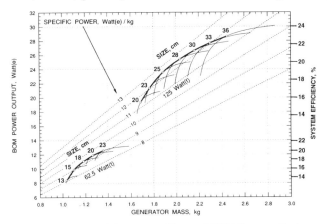

FIGURE 18. Effect of Generator Size on RTPV System Performance. Based on Projected Filter and Cell Characteristics

Because of its compact size and low mass and fuel inventory, either generator may be of considerable interest for powering the highly miniaturized spacecraft planned for NASA's New Millennium program. To illustrate this, Figure 19 depicts a possible arrangement for integrating two 10-watt RTPVs with JPL's preliminary design for a New Millennium spacecraft for missions to the outer solar system. This arrangement has not yet been analyzed, but it seems clear that even better performance could be achieved by utilizing the spacecraft's 46 cm diameter graphitic high-gain antenna as the RTPV's radiator

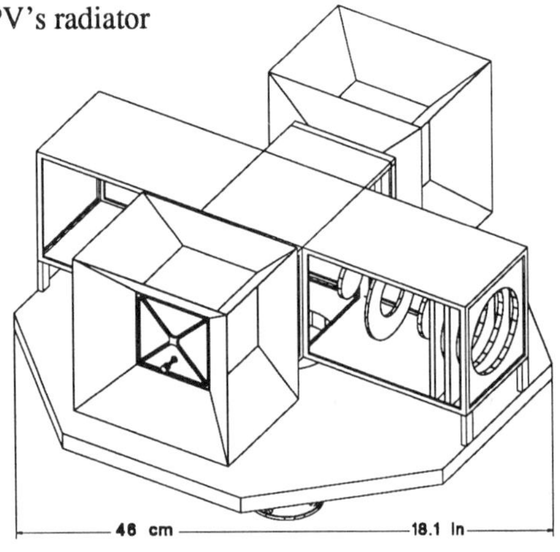

FIGURE 19. Illustrative Arrangement for Mounting Two 10-Watt RTPVs on JPL's Design for New Millennium Miniaturized Spacecraft for Outer Solar System Missions

References

[1] Schock, A. and V. Kumar, "Radioisotope Thermophotovoltaic System Design and its Application to an Illustrative Space Mission," *Proc. of the NREL Conference on Thermophotovoltaic Generation of Electricity*, held at Copper Mountain, Colorado, July 1994, published by the American Institute of Physics.

[2] Schock, A., M. Mukunda, T. Or, and G. Summers, "Analysis, Optimization, and Assessment of Radioisotope Thermophotovoltaic System Design for an Illustrative Space Mission," *Proc. of the NREL Conference on Thermophotovoltaic Generation of Electricity*, held at Copper Mountain, Colorado, July 1994, published by the American Institute of Physics.

[3] Schock, A., et. al. "Design, Analysis, and Optimization of a Radioisotope Thermophotovoltaic (RTPV) Generator, and its Applicability to an Illustrative Space Mission," 45th Congress of the International Astronautical Federation, Jerusalem, Israel, 9-14 October 1994.

[4] Schock, A., C. Or, and M. Mukunda, "Effect of Expanded Integration Limits and of Measured Infrared Filter Improvements on Performance of RTPV System," to be presented at the 2nd NREL Conference on Thermophotovoltaic Generation of Electricity, in Colorado Springs, Colorado, July 1995.

[5] Schock, A., "RSG Options for Pluto Fast Flyby Mission," IAF-93-R.1.425b, 44th Congress of the International Astronautical Federation, Graz, Austria, 16-22 Oct 1993.

[6] Schock, A. Or, C.T., and V. Kumar, "Radioisotope Power System Based on Derivative of Existing Stirling Engine," IECEC-95-159, to be presented at the 30th Intersociety Energy Conversion Engineering Conference, in Orlando, Florida, August 1995.

[7] Schock, A., and C. T. Or, "Effect of Fuel and Design Options on RTG Performance Versus PFF Power Demand," *Proc. of the 29th Intersociety Energy Conversion Engineering Conference*, held in Monterey, CA, 7-12 August 1994.

[8] Schock, A., C. T. Or, and V. Kumar, "Design Modifications for Increasing the BOM and EOM Power Output and Reducing the Size and Mass of RTG for the Pluto Mission," *Proc. of the 29th Intersociety Energy Conversion Engineering Conference*, held in Monterey, CA, 7-12 August 1994.

[9] Schock, A., "RTG Options for Pluto Fast Flyby Mission," IAF-93-R.1.425a, 44th Congress of the International Astronautical Federation, Graz, Austria, 16-22 Oct 1993.

[10] Schock, A., "Comparison of Thermoelectric Space Power System with Alternative Conversion Options", Proceedings of the 14th International Conference on Thermoelectrics, held in St. Petersburg, Russia, 7-10 June 1995.

[11] Schock, A., "Design Evolution and Verification of the General-Purpose Heat Source," #809203, *Proc. of 15th Intersociety Energy Conversion Engineering Conference*, held in Seattle, WA, 18-22 August 1980.

[12] Chase, S.T. and R.D. Joseph, "Resonant Array Bandpass Filters for the Far Infrared," Applied Optics, Vol. 22, No. 11, 1 June 1983, pp 1775-1779.

[13] Kogler, Kent J. and Rickey G. Pastor, "Infrared Filters Fabricated from Submicron Loop Antenna Arrays," Applied Optics, Vol. 27, No. 1, 1 Jan 1988, pp 18-19.

[14] Denham, H.B., et. al. "NASA Advanced Radiator C-C Thin Development," *Proc. of the 11th Symposium on Space Nuclear Power Systems*, CONF-940101, M.S. El-Genk and M.D. Hoover, eds., American Institute of Physics, New York, AIP Conference No. 301, 3: 1119-1127, 1994.

[15] Horne, W. E., M. D. Morgan, and V. S. Sundaram, "IR Filters for TPV Converter Module," to be presented at the 2nd NREL Conference on Thermophotovoltaic Generation of Electricity, in Colorado Springs, CO, July 1995.

[16] Gaski, J., SINDA (System Improved Numerical Differencing Analyzed), version 1.315 from Network Analysis Associate, Fountain Valley, CA, 1987.

[17] Little, A.D., SSPTA (Simplified Space Payload Thermal Analyzer), version 3.0/VAX, for NASA/Goddard, by Arthur D. Little Inc., Cambridge, MA, 1986.

Design of a TPV Generator with a Durable Selective Emitter and Spectrally Matched PV Cells

David B. Sarraf* and Theresa S. Mayer**

*Thermacore, Inc., 780 Eden Rd., Lancaster, PA 17601
** Department of Electrical Engineering, Penn State University, University Park, PA 16802

Abstract: This paper describes the conceptual design of a TPV system that uses a durable powder metal selective emitter surface and spectrally matched InGaAs PV cells. The materials and components have been selected to maximize output power density and system efficiency at an operating temperature of 1100°C. The proposed generator offers several notable features including a high system efficiency, an inherent resistance to thermal shock and vibration, and an ability to use a variety of fuels interchangeably.

INTRODUCTION

Thermophotovoltaics (TPV) was proposed over twenty years ago as a means to achieve efficient direct heat to electric energy conversion [1]. Initial research on these systems, however, was limited by relatively immature materials technologies. In particular, the two critical components in TPV systems, the thermal emitter and the photovoltaic cells, suffered from low efficiencies. Recent advances in material processing for both components have renewed interest in TPV as an efficient, quiet, and clean alternative to electromechanical energy conversion using heat engines and electrical generators.

In order to compete successfully with other direct energy conversion systems such as fuel cells and thermoelectric generators, a reliable TPV system that maximizes efficiency and power density is required [2]. Currently, two system configurations provide the most promise for meeting these objectives. The first consists of a filtered broadband system [3]. In this configuration, a characteristic greybody spectrum is emitted from a metallic or Silicon Carbide (SiC) emitter operating in the temperature range of 900°C–1500°C. The radiant energy with wavelengths that are close to the bandgap of the PV cells is passed to the PV array. The remaining radiant energy that cannot be converted efficiently to electrical energy by the PV array is reflected back to the emitter using passband filters (plasma or interference). The majority of the reflected energy is reabsorbed into the broadband emitter where it can be readmitted as useful energy rather than being lost to the system as waste heat. In many of these systems, the passband filters are also used to protect the PV cells from the products of combustion.

© 1996 American Institute of Physics

Although the filtered broadband systems are capable of producing relatively high output power densities even at low temperatures, losses in the filters and cells, and reabsorption/emission of the radiant energy by the broadband emitter limit overall system efficiencies. At typical operating temperatures of less than 1300°C, the metallic and SiC emitters are expected to be very durable and have a long lifetime. Moreover, because the PV cells are protected from the combustion products and are maintained at safe operating temperatures using external cooling, they are not expected to degrade during operation. There is, however, concern regarding the reliability of the passband filters due to their presence in a harsh combustion environment.

The second configuration that has been investigated is based on a selective emitter [4]. Here, rather than inserting passband filters between the emitter and PV arrays, the emitter is formed with a rare earth oxide that provides a selective line emission. In general, the full width at half maximum of the emission is on the order of 300 nm [4], which is very narrow as compared to the filtered broadband passband. By choosing the bandgap of the PV cells to match the selective emission of the emitter, the conversion of radiant to electrical energy is very efficient. In particular, losses due to reabsorption/emission by the emitter are eliminated. Moreover, improved spectral matching between the selective emitter and the PV cells reduces cell conversion losses that are inherent in broadband systems (due to the dependence of conversion efficiency on incident wavelength). Unfortunately, at a given operating temperature the amount of radiant energy incident on the PV array is reduced greatly. Therefore, improved system efficiency comes at the expense of a reduction in the output power density.

Currently, most selective emitter configurations that are described in the literature are based on fibrous ceramic emitters due to their excellent emission properties at wavelengths of 0.98, 1.55, 2.0, and 2.4 µm [4]. In order to achieve useful power densities at these wavelengths, however, emitter operating temperatures higher than those used in broadband systems are required. The fragile nature of the fibrous selective emitters coupled with the higher operating temperatures have caused much uncertainty regarding their durability and longevity. Moreover, because the fuel is actually burned in the fibrous emitter, these systems also require a transparent window that separates the cells from the combustion products.

From this brief overview of filtered broadband and selective emitter systems, there is a clear tradeoff between system efficiency and output power density. Moreover, there are potential concerns regarding the long-term reliability of many of the components including the spectral filters and fibrous emitters. In this work, a conceptual design is presented that is based on mechanically durable and rugged components that can be configured to maximize both the efficiency and the output power density of a TPV system operating at 1100°C. The following sections will describe the current efforts underway at Thermacore and Penn State to develop a holmia-based powder metal selective emitter and spectrally matched InGaAs PV cells. Finally, a conceptual system design will be outlined and the predicted system performance will be summarized.

COMPONENT SELECTION

In this work, a TPV system design is proposed that couples a holmia-based powder metal selective emitter to spectrally matched InGaAs PV cells in an inverted configuration. A diagram of the generator is shown in Fig. 1. As demonstrated in this figure, the cylindrical emitter surrounds the PV cells on the outside rather than the inside. Moreover, to provide higher output power densities, parabolic reflectors are used to concentrate the radiant energy on six narrow arrays of PV cells. Finally, a water cooled Porous Metal Heat eXchanger (PMHX) is placed behind the PV arrays to remove the waste heat and guarantee safe PV cell operating temperatures.

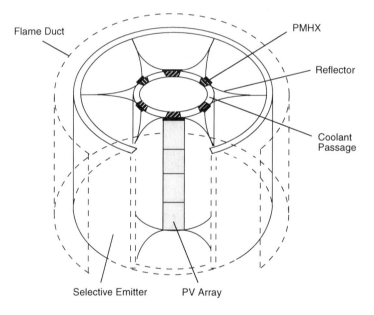

FIGURE 1. Proposed TPV system design based on a powder-metal selective emitter and spectrally matched InGaAs PV cells.

A holmia-based powder metal selective emitter was chosen because it offers several notable features including:
(1) A selective line emission that is centered around 2.0 μm which corresponds to the maximum in the output radiant energy available from a 1100°C greybody spectrum.
(2) Durability and ruggedness and the ability to withstand thermal and mechanical shock.
(3) The availability of simple joining techniques that allows the formation of an evacuated space between the emitter and the cells.
(4) Isolation of the combustion products without the need for additional transparent windows.

To provide a good spectral match to the 2.0 μm holmia emission peak, a PV cell with a cutoff wavelength of 2.1 μm is required (Eg = 0.6 eV). Of the more commonly studied semiconductor systems, InGaAs can be tailored to reach this bandgap by controlling the composition of indium in the active device layers. In particular, the cutoff wavelength of $In_{0.67}Ga_{0.33}As$ is approximately 2.1 μm.

The following subsections describe the fabrication of a powder metal selective emitter surface and the growth and fabrication of InGaAs PV cells.

Powder Metal Selective Emitter Surface

A cross section of the proposed powder metal selective emitter surface is shown in Fig. 2. It consists of a metallic superalloy base, a layer of sintered nickel powder, and a layer of a rare earth oxide. The superalloy base provides the emitters mechanical strength and ductility and resistance to oxidation and flames while the rare earth oxide layer provides its selective emittance property. The intermediate nickel layer is included to absorb the mismatch in the coefficient of thermal expansion between the base and the oxide layer and to promote adhesion of the thin rare earth oxide layer. The relatively low emissivity of this nickel layer should also minimize the broadband emission of the selective emitter surface. In the final design, the base will be constructed using Inconel 800 or Haynes Alloy A-214, the intermediate nickel layer will use -200 mesh nickel powder, and the rare earth oxide layer will be comprised of a holmia oxide.

Several powder metal selective emitter surfaces have been fabricated and tested for mechanical stability. A photograph of one such selective emitter surface is shown in Fig. 3. It is comprised of an Inconel 800 base, and a nickel powder metal layer that was divided into four sections. The leftmost section of the nickel powder metal layer has no coating, whereas of the remaining three sections, one is coated with thoria and two with neodymia (although holmia is specified in the system design, neodymia was used because it was readily available). One of the

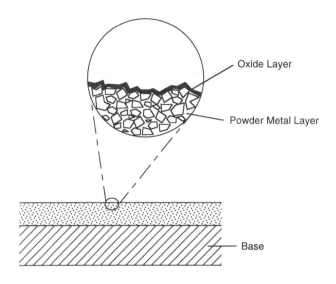

FIGURE 2. Cross section of the powder metal selective emitter surface.

FIGURE 3. Photograph of an emitter surface made of Inconel 800. From left to right, the sections are coated with nothing, thoria, neodymia oxide, and neodymia carbonate.

neodymia sections was coated directly with neodymium oxide while the other section was coated with neodymium carbonate that was reduced to its oxide during the firing process. Although the optical properties have not been measured, the samples have good mechanical properties. In particular, the samples have withstood thermal cycling and exposure to high temperatures without noticeable degradation of the base or the thin oxide layer.

Spectrally Matched InGaAs PV Cells

A typical epitaxial structure for an InGaAs PV cell operating at a cutoff wavelength of 2.1 µm is given in Table 1. The epitaxial layers of InGaAs and InAlAs are deposited on InP substrates using Molecular Beam Epitaxy (MBE). To permit the growth of high quality $In_{0.67}Ga_{0.33}As$ with such a large lattice mismatch to the InP substrate (1.2%), a thin linearly-graded buffer layer (LGBL) is inserted between the substrate and the active layers. Such buffer layers have been used to minimize the number of active layer threading dislocations [5] that often behave like electrically and optically active defects [6]. Moreover, in contrast to other buffer layers that have been used for the growth of lattice mismatched materials by Vapor Phase Epitaxy (VPE) and metalorganic chemical vapor deposition (MOCVD) [7], these LGBL's are typically less than 2 µm thick making them compatible with the relatively slow growth rate (1–1.5 µm/h) of MBE.

The active device layers consist of a 3000 nm thick n-$In_{0.67}Ga_{0.33}As$ base doped $5 \times 10^{17}/cm^3$ followed by a 200 nm thick p^+-$In_{0.67}Ga_{0.33}As$ emitter doped $5 \times 10^{18}/cm^3$. A 50 nm p^+–$In_{0.67}Ga_{0.33}As$ window layer is included to reduce the surface recombination velocity at the front surface of the device and a 50 nm p^+–$In_{0.67}Ga_{0.33}As$ cap layer facilitates the fabrication of low resistance, nonalloyed top ohmic contacts.

Following the growth of the epitaxial structure, the PV cells are fabricated using a conventional mesa process. This relatively simple process consists of three main steps: ohmic contact formation, mesa device isolation, and surface passivation/ antireflection deposition. The top-side ohmic contact is made by lifting-off Ti/Pt/Au. Next, in order to minimize absorption in the cap layer, the heavily doped p-type InGaAs cap layer is removed in the areas surrounding the fingers using a process that etches selectively InGaAs over InAlAs. The devices are then isolated electrically by chemically etching to a depth of 0.35 µm. Finally, the surface is passivated using Silicon Nitride which also serves as an antireflection coating at these wavelengths.

TABLE 1. A typical epitaxial structure for an InGaAs PV cell operating at a cutoff wavelength of 2.1 μm.

Layer	Material	Doping (cm^{-3})	Thickness (nm)
Contact	In$_{0.67}$Ga$_{0.33}$As:Be	1x10^{19}	50
Window	In$_{0.67}$Al$_{0.33}$As:Be	5x10^{18}	50
Emitter	In$_{0.67}$Ga$_{0.33}$As:Be	5x10^{18}	200
Base	In$_{0.67}$Ga$_{0.33}$As:Si	1x10^{17}	3000
n$^+$–Buffer	In$_{0.67}$Ga$_{0.33}$As:Si	5x10^{18}	50
Linearly Graded Buffer	In$_x$Ga$_{1-x}$As:Si (0.53 < x < 0.67)	5x10^{18}	2000
Lattice Matched Buffer	In$_{0.53}$Ga$_{0.47}$As:Si	5x10^{18}	300
n$^+$–Substrate	InP:S	2–5x10^{18}	

GENERATOR DESIGN AND PERFORMANCE

A conceptual design for a 250 Watt TPV generator based on a powder metal selective emitter and spectrally matched PV cells is presented in this section. A cross section and top view of the proposed generator are shown in Fig. 4. As described previously, the key components of the generator are the selective emitter surface, the parabolic reflectors, the InGaAs PV arrays, a pair of bellows, and a liquid cooled Porous Metal Heat eXchanger (PMHX). In this design, the powder metal emitter surface forms the outer cylinder of the annulus and serves as a duct for hot combustion gasses. The radiant energy being emitted from the powder metal surface is concentrated onto the six InGaAs PV arrays using parabolic reflectors as shown most clearly in the top view of Fig. 4. The InGaAs PV arrays are mounted on the inner surface that also serves as a water cooled PMHX. This PMHX forms an extended surface of high thermal conductivity and is used to dissipate the waste heat resulting from the cell conversion process. Finally, a pair of bellows that accommodates the differential thermal expansion between the emitter and the PV array is used to seal and evacuate the gap between the two components.

In addition to the features that have been mentioned previously, this combination of components and system configuration has other advantages. In particular, the geometry of the generator provides inherent isolation between the

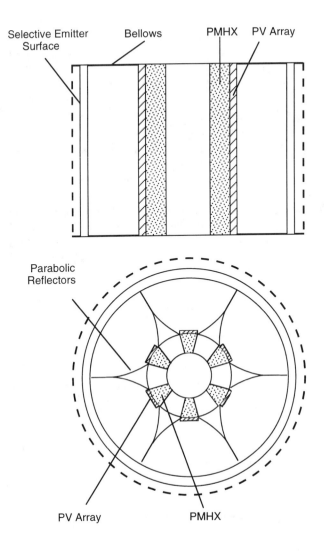

FIGURE 4. Cross section and top view of the proposed TPV generator based on a powder-metal selective emitter and spectrally matched InGaAs PV cells.

fuel/heat source and the cells without the need of additional transparent windows. This configuration can also be used with dirty fuels such as oil or kerosene with little to no effect on generator performance. The evacuated gap between the selective emitter and the PV cells provides several additional benefits. In an evacuated environment, the nickel powder will not oxide and degrade the performance of the emitter. Moreover, the parasitic heat losses and PV cell heating due to convection from the emitter will be minimized.

Specifications for a proposed 250 Watt TPV generator operating at 1100°C are given in Table 2. The selective emitter is 6" in diameter and 10" long and is made with an Inconel 800 base and a -200 mesh nickel powder intermediate layer that is coated with holmia oxide. Throughout the analysis, it is assumed that the emitter is isothermal. The PV cells are 1 cm x 1cm and are interconnected to form an array that is 1 cm wide and 10" long. Six arrays will be mounted on the 2" inner surface that also serves as a heat exchanger. The device active layers will be fabricated using $In_{0.67}Ga_{0.33}As$ that has a cutoff wavelength of 2.1 µm. The cells are maintained at 31°C using 20°C water with a flow rate of 1 liter/min.

These specifications were determined from an analysis that only considered the thermal to electrical energy conversion (ie. emitter to PV cell). First, the radiant energy as a function of wavelength was calculated for a holmia-based selective emitter assuming that the peak output power density was 85% of the ideal blackbody. From this calculation, it was determined that the integrated output

TABLE 2. TPV generator specifications and predicted performance

Emitter	
Size:	6" diameter, 10" long
Base material:	Inconel 800, 0.032" thick
Emitter oxide:	Holmia
Temperature:	1000°C, assumed isothermal
PV Cells	
Array size:	1 cm x 10" long (mounted on 2" diameter PMHX)
Number of arrays:	6
Active layer material:	$In_{0.67}Ga_{0.33}As$
Cooling System	
Fluid:	water
Inlet temperature:	20°C
Outlet temperature:	31°C
Average cell temperature:	25°C
Flowrate:	1 liter/min

power density leaving the emitter is 1.6 W/cm^2. For the specified reflector geometry, this results in an incident power density on the PV array of 6.4 W/cm^2. Approximately 10% of the incident energy will be reflected back to the emitter by the metallic grid on the PV cells (did not account for recuperation of the reflected energy and did not consider end losses in this calculation). Taking this into consideration, the incident power density available for conversion is approximately 5.75 W/cm^2. Using predicted PV cell conversion efficiencies that range from 25 – 30%, output power densities of 1.5 – 1.75 W/cm^2 can be achieved. These values of output power density are comparable to those predicted from filtered broadband systems operating at 1100°C. Finally, because the fuel to thermal losses are minimized with a selective emitter configuration, the overall system efficiencies are expected to be high.

CONCLUSIONS

In summary, a TPV system design that uses a durable powder metal selective emitter and spectrally matched InGaAs cells has been proposed. This combination of components coupled with an inverted configuration should result in a system that has a high conversion efficiency with an output power density (1.5 – 1.75 W/cm^2) comparable to filtered broadband systems. In addition, this design offers an inherent resistance to thermal shock and vibration and an ability to use a variety of fuels interchangeably.

ACKNOWLEDGMENTS

The work performed at Thermacore, Inc. by D.B.S. was sponsored by internal research and development funds while T.S.M. was supported, in part, by a subcontract to Sensors Unlimited, Inc. funded by the Navy STTR program Phase I (N00014-94-C00262).

REFERENCES

1. Schwartz, R. J., "A p-i-n thermo-photo-voltaic diode," *IEEE Transactions on Electron Devices*, Vol. 16, No. 7, pp. 657–663, 1969.
2. Rosenfeld, R. L., "An ARPA Perspective on TPV," in *Proc. of 1st NREL Conference on Thermophotovoltaic Generation of Electricity*, p. 301, 1994.

3. Baldasaro, P. F., E. J. Brown, D. M. Depoy, B. C. Campbell, J. R. Parrington, "Experimental Assessment of Low Temperature Voltaic Energy Conversion," *in Proc. of 1st NREL Conference on Thermophotovoltaic Generation of Electricity*, pp. 29–43, 1994.

4. Nelson, R. E., "Thermophotovoltaic Emitter Development," *in Proc. of 1st NREL Conference on Thermophotovoltaic Generation of Electricity*, pp. 80–95, 1994.

5. Lord, S. M., B. Pezeshki, and J. S. Harris, Jr., "Investigation of High In Content InGaAs Quantum Wells Grown on GaAs by Molecular Beam Epitaxy," *Electronics Letters*, Vol. 28, No. 13, pp. 1193-1195, 1992.

6. Martinelli, R. U., T. J. Zamerowski, and P. Longeway, "2.6 µm InGaAs photodiodes," *Applied Physics Letters*, Vol. 53, No. 11, p. 989, 1988.

7. Linga, K. R., G. H. Olsen, V. S. Ban, A. M. Joshi, W. F. Kosonocky, "Dark Current Analysis and Characterization of InGaAs/InAsP Graded Photodiodes with x>0.53 for Response to Longer Wavelengths," *IEEE Journal of Lightwave Technology*, Vol. 10, No. 8, pp. 1050-1054, 1992.

Thermophotovoltaic Energy Converters Based on Thin Film Selective Emitters and InGaAs Photovoltaic Cells

N. S. Fatemi[†], R. H. Hoffman[†], D. M. Wilt[*], R. A. Lowe[**], L. M. Garverick[†], and D. Scheiman[††]

[†]*Essential Research, Inc., Cleveland, OH*
[*]*NASA Lewis Research Center, Cleveland, OH*
[**]*Kent State University, Kent, OH*
[††]*NYMA, Inc. Brookpark, OH*

This paper presents the results of an investigation to demonstrate thermophotovoltaic energy conversion using InGaAs photovoltaic cells, yttrium-aluminum-garnet- (YAG-) based selective emitters, and bandpass/reflector filters, with the heat source operating at 1100 °C. InGaAs cells were grown on InP by organometallic vapor phase epitaxy with bandgaps of 0.60 and 0.75 eV and coupled to Ho-, Er-, and Er-Tm-doped YAG selective emitters. Infrared reflector and/or shortpass filters were also used to increase the ratio of in-band to out-of-band radiation from the selective emitters. Efficiencies as high as 13.2% were recorded for filtered converters.

INTRODUCTION

A thermophotovoltaic (TPV) energy conversion system is comprised of two main components: a heat source and a converter. The thermal energy from the heat source is converted to electricity by the converter. A converter consists of a photovoltaic (PV) cell, and either a blackbody (or more accurately, a graybody) emitter, or a spectrally selective emitter. The graybody emitter emits radiation in a broad spectrum, while the selective emitter emits in a narrow emission band. In this work, we have fabricated such converters by coupling low-bandgap PV cells to selective emitters based on rare-earth elements.

Since optimized TPV converters are projected to have total system efficiencies of greater than 20%,[1] one area of potential application for them is to replace the conventional thermoelectric converters. Specifically, it is estimated that by substituting TPV converters in a radioisotope thermoelectric generator (RTG), the system efficiency can be increased by as much as a factor of three or more.[1] The general purpose heat source modules are used in RTGs to provide thermal energy. These modules generate 250 W (thermal) at the beginning of life, and operate at 1100 °C.

Several selective emitters are suitable for operation at 1100 °C, based on rare-earth oxides or hosts doped with rare-earth ions. Such emitters generally exhibit one strong emission peak at a characteristic wavelength, and low emissivity elsewhere. The rare-earth elements of interest and their corresponding center of emission band wavelengths are Nd (2.4 µm), Ho (2.0 µm), Tm (1.8 µm), and Er (1.5 µm). These emitters are most efficiently coupled to PV cells having energy bandgaps (Eg) slightly below the energy of the center of their emission peak. For example, a Ho-based emitter is a suitable mate for a PV cell with an Eg of ~0.60 eV (2.07 µm).

The low-bandgap PV cells almost exclusively consist of III-V semiconductor compounds. The most commonly used cells for TPV applications are GaSb, InGaAs (on InP), and InGaAsSb (on GaSb). Whereas GaSb has a fixed composition bandgap (0.73 eV), the compositions of InGaAs and InGaAsSb cells can be varied over a wide range, enabling their bandgaps to be tuned to the spectral emission characteristics of desired selective emitters.

Ordinarily, a TPV converter consists of only emitters and PV cells. However, because the currently available selective emitters exhibit excessive emission at $\lambda > 5$ µm, interposing a reflector and/or bandpass filters between the cells and emitters can increase the converter efficiency by reflecting back some of the longer wavelength, out-of-band radiation from the emitters back to the heat source.

We have fabricated a number of converters using InGaAs on InP PV cells, rare-earth doped single-crystal yttrium aluminum garnet (YAG) thin-film selective emitters, and bandpass/reflector filters. These converters were tested with an electrically powered heat source operating at 1100 °C.

EXPERIMENT

A fast-switching horizontal low-pressure organometallic vapor phase epitaxy (OMVPE) reactor was employed for all semiconductor material growth. The reactant species were trimethylindium, trimethylgallium, diethylzinc, phosphine, arsine and silane. $In_xGa_{1-x}As$ layers were grown on p-type InP wafers as prepared by the substrate vendor. A thin layer of InP was first grown on all substrates to provide a clean surface to nucleate the InGaAs alloys. Two different cell compositions were grown. The lattice-matched, Eg = 0.75 eV

InGaAs alloy was grown directly on the InP layers. The mismatched alloy with Eg = 0.60 eV was grown on top of three 0.1 μm thick compositionally stepped buffer layers. All InGaAs cells were capped with an InP window layer; this layer was 0.1 μm thick on lattice-matched cells, and 0.05 μm thick on mismatched cells. Growth temperature and pressure were 620 °C and 190 torr, respectively, throughout all OMVPE runs.

AuGe[2] and AuZn[3] contact systems were used for the front grid and back metallization, respectively. Specific contact resistivities (ρ_c) for the as-fabricated AuGe system was in the low 10^{-6} Ω/cm^2 range. Heat treatment of the contacts in the 300 to 350 °C range, reduced ρ_c by more an order of magnitude. The grid shadowing (GS) was 20%. A cross-sectional view of the completed cell is shown in figure 1.

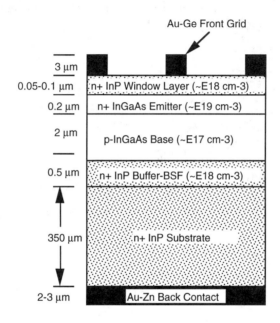

Figure 1. Cross-sectional view of InGaAs/InP PV cell.

The Ho-, Er-, and Er-Tm-doped YAG selective emitters were characterized using a custom-developed test bed designed for measuring thin-film selective emitters for thermophotovoltaic applications. Normal spectral emittance measurements were made from 1.2 to 3.2 μm with a spectroradiometer constructed from a 1/8 m monochrometer, a temperature-controlled PbS detector, and an

800 Hz chopper. The spectroradiometer was calibrated with a 1273 K blackbody reference.[4] Reflectivity and transmission data were obtained using a Perkin-Elmer Lambda 19 spectrophotometer.

The experimental configuration used to test the converters is shown in figure 2. The radiant emission from the selective emitter is guided and delivered to the cell test plane with a highly reflective platinum tube. The cell is mounted on a water-cooled heat sink to prevent thermal damage to the cell . If a filter was included in the converter, it was placed near the top surface of the PV cell and was not actively cooled. Radiant power delivered to the PV cells was measured with a calibrated thermopile detector placed at the plane of the cell.

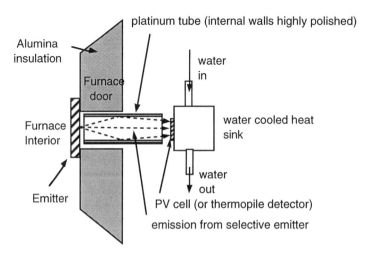

Figure 2. Schematic diagram of test apparatus used to test converters.

RESULTS AND DISCUSSION

Selective Emitters and Filters

Three different rare-earth doped YAG selective emitters were used in this work. One was doped to 25% with Ho, the second was doped to 40% with Er, and the third was co-doped with 25% Er and 25% Tm. Each emitter was backed with a Pt foil. Spectral emittance measurements for these emitters were made at an average temperature of 1100 °C. Figure 3 shows the spectral emittance of the Ho-doped selective emitter, which was 1.1 mm thick. Figure 4 shows the spectral emittance of the Er-doped emitter, which was 0.4 mm thick. Finally, figure 5 shows the spectral emittance of the Er Tm co-doped emitter, which was 0.9 mm thick.

Co-doped emitters are attractive because they offer the potential for higher power output under the selective emission band than what is possible with a singly doped emitter. However, as shown in figures 3 to 5, the co-doped emitter has a lower peak value of the emittance in the emission band than those for the singly doped emitters, i.e. ~0.4 versus ~0.6 to 0.7. This is possibly caused by a detrimental interaction occurring between the dopant species in the YAG lattice.

Similar to several other oxide ceramics, YAG has very high emittance (~0.95) at longer wavelengths (i.e., >5 μm) at high temperatures.[5] This has limited the radiative efficiency (defined as power emitted in the emission band divided by the total emitted power at all wavelengths) of these emitters to ≤30%.

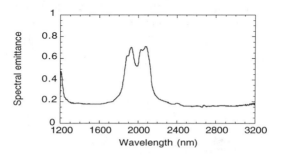

Figure 3. Spectral emittance of 25% Ho-doped YAG selective emitter operating at 1100 °C.

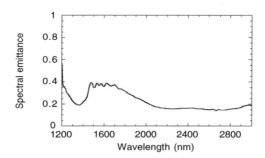

Figure 4. Spectral emittance of 25% Er/25% Tm-doped YAG selective emitter operating at 1100 °C.

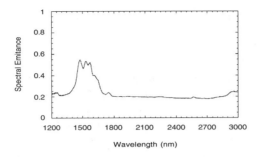

Figure 5. Spectral emittance of 40% Er-doped YAG selective emitter operating at 1100 °C.

To enhance the emitter radiative efficiency, we used a SiC disc mounted to the Pt substrate (shown in fig. 6). The SiC acts as a thermal absorber to improve the thermal coupling from the electric furnace to the selective emitter. As a result, a decrease in ΔT from 220 to 170 °C was observed across the emitter, increasing the peak value of the spectral emittance in the emission band from 0.65 to over 0.70 for the Ho-YAG emitters.

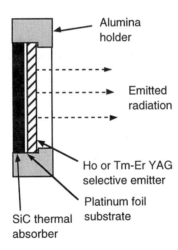

Figure 6. Cross-sectional view of thin-film selective emitter using YAG coupled with SiC thermal absorber.

To some extent, the unusable radiation energy emitted below the bandgap of the PV cell can be recycled in a closed TPV system, through the use of reflector and/or bandpass filters. A gold-on-sapphire infrared (IR) reflector filter was used to partly suppress the transmission of longer IR wavelengths (>2.5 µm) from the YAG, reflecting a portion of that radiation back to the heat source. This is illustrated in figure 7, where the transmission characteristics of the filter is shown. We also used a dielectric stack/gold on sapphire shortpass/IR reflector combination filter to further reduce the longer IR transmission from the YAG. The transmission cutoff wavelength for this filter was at ~1.66 µm, making it suitable for use with an Er-generated emission peak (see fig. 8). However, it should be noted that when using such filters, their absorptivity should also be taken into account. This is illustrated in figure 9, where the sum of the transmission and reflectance for the shortpass/reflector filter is shown for $\lambda \leq 2.5$ µm. As seen in the figure, this sum does not add up to 100%, indicating the absorptivity to average ~10% for $\lambda \leq 2.5$ µm.

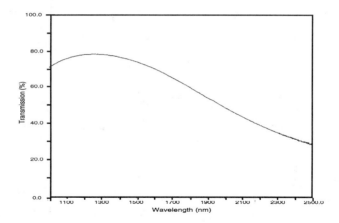

Figure 7. Transmission characteristics of gold-on-sapphire IR reflector filter.

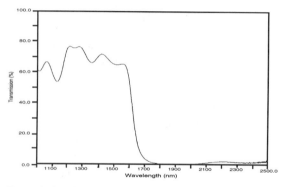

Figure 8. Transmission characteristics of combination shortpass/IR reflector filter.

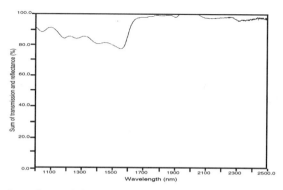

Figure 9. Sum of transmission and reflectance for combination shortpass/IR filter.

InGaAs PV Cells

InGaAs cell structures were grown via OMVPE on InP substrates. By varying the ratio of In to Ga, two different compositions of InGaAs were obtained with bandgaps of 0.60 and 0.75 eV. The cell with Eg = 0.75 eV is suitable for use with the Er-YAG emitter, whose center-of-emission peak is at 1.5 μm (0.827 eV). The cell with Eg = 0.60 eV is suitable for use with the Ho-YAG emitter, whose center-of-emission peak is at 2.0 μm (0.62 eV). However, unlike the cell with Eg = 0.75 eV, this cell is lattice-mismatched to the InP substrate. Consequently, a high concentration of threading dislocations is generated at the InGaAs/InP interface. These dislocations can propagate through the active regions of the cell, creating crystalline defects. In turn, the defects will reduce the minority carrier lifetime (τ) and increase the reverse saturation current (J_o) in the cell. As a result, open-circuit voltage (Voc) and fill-factor (FF) values suffer in these cells.

This effect is illustrated in table I, where the air-mass zero (AM0) results for 0.60 and 0.75 eV InGaAs cells are given. As shown in the table, all performance parameter values for the 0.60 eV cell, including the short-circuit current (Jsc), are below their theoretical potential. For our future work on these cells, we will modify the composition and thickness of the graded buffer layers grown at the cell/substrate interface to diminish the detrimental effects of lattice mismatch. On the other hand, the lattice-matched cell (Eg = 0.75 eV), performed closer to its theoretical potential. However, the Voc of this cell (349 mV) was still lower than what can be achieved in an optimized cell, i.e. in the 375 to 400 mV range.[6,7]

TABLE I. AIR-MASS ZERO RESULTS FOR InGaAs/InP PV CELLS WITHOUT ANTIREFLECTION (AR) COATING, GS = 20%.

Eg, eV	Voc, mV	FF, percent	Jsc, mA/cm^2	Efficiency, %
0.60	192	58.8	28.8	2.3
0.75	349	69.7	27.6	4.7

Converter Operation at 1100°C

Four different converter configurations consisting of various combinations of cells and emitters with and without filters were tested with the heat source operating at 1100 °C:

1. Ho-YAG emitter, 0.60 eV InGaAs PV cell, and reflector filter.
2. Ho-YAG emitter, 0.75 eV InGaAs PV cell, and reflector filter.
3. Er-Tm-YAG emitter, 0.75 eV InGaAs PV cell, and shortpass/reflector filter.
4. Er-YAG emitter, 0.75 eV PV InGaAs cell, and shortpass/reflector filter.

The results were as follows:

1. Ho-YAG emitter, 0.60 eV InGaAs PV cell, and reflector filter.

The center of the emission peak for the Ho-doped YAG selective emitter at 1100 °C occurs at $\lambda = 2.0$ μm, with the base of the emittance peak spanning between ~1.8 to 2.2 μm (see fig. 3). As a result, the 0.60 eV (2.07 μm) cell is almost ideally tuned to the spectral emittance characteristics of the Ho-YAG, with its bandgap cutoff wavelength missing slightly part of the emission band.

The total measured power output of the Ho-YAG at 1100 °C was 3.19 W/cm^2. The reflector filter (see fig. 7) allows the transmission of only 0.637 W/cm^2. Most of the radiation is reflected back to the source by the filter, but as discussed earlier, a small fraction of it is absorbed in the filter. This absorption will reduce the converter efficiency to some extent. On the other hand, part of the reflected radiation from the PV-cell metallization gridlines (GS = 20%) to the heat source may be recycled. At present, exact values for these losses and/or gains in efficiency are not known for the system described here.

The I-V curves for the most efficient cells tested without and with the reflector filter are shown in figures 10(a) and (b), respectively. Table II summarizes the results in tabular form. As shown in the table, using the reflector filter greatly increased the efficiency of the cell: from 1.3 to 4.4%, a factor of 3.4. Because the radiative efficiency of YAG-based emitters is ≤30%, the recycling of the out-of-band radiation via filtering is crucial in the fabrication of efficient converters.

Contrary to expectation, it should also be noted that the Voc values for unfiltered PV cells in table II are slightly lower than those for filtered cells. We believe that this is caused by the slightly higher cell temperatures for the unfiltered cells compared with the filtered cells.

2. Ho-YAG emitter, 0.75 eV InGaAs PV cell, and reflector filter.

Because of the performance-robbing lattice mismatch present in our 0.60 eV InGaAs cells, we decided to replace them with the lattice-matched 0.75 eV cells. The results for two of the best cells are shown in table III, and the I-V curves for the most efficient cell tested without and with the reflector filter are shown in figures 11(a) and (b) respectively.

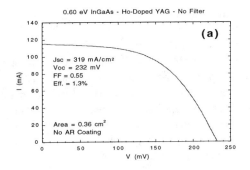

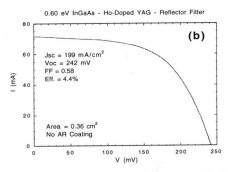

Figure 10. I-V curves for 0.60-eV InGaAs PV cell coupled to Ho-YAG (a) Without filter. (b) With filter with source temperature 1100 °C.

TABLE II. PERFORMANCE PARAMETERS FOR 0.60 eV InGaAs PV CELLS COUPLED TO Ho-YAG. MEASUREMENTS TAKEN WITH AND WITHOUT IR REFLECTOR FILTER, AT SOURCE TEMPERATURE 1100 °C.

I.D.	Power in, W/cm^2	Filter	Voc, mV	Jsc, mA/cm^2	FF, %	Efficiency (No AR) %	Efficiency (with AR*) %
256-D	3.19	No	226	315	54.0	1.3	1.7
	0.637	Yes	240	189	59.0	4.2	5.9
295-D	3.19	No	232	319	55.0	1.3	1.7
	0.637	Yes	242	199	58.0	4.4	6.1

*Efficiencies with AR coating are calculated.

A comparison of the results reported in tables II and III reveals that the 0.75 eV cells have the same efficiency as the 0.60 eV cells (1.3%). Moreover, the filtered 0.75 eV cells are more efficient than the 0.60 eV cells (5.2 versus 4.4%), despite the fact that the bandgap of the latter is better tuned to the spectral emission characteristics of the Ho-doped YAG emitter. This is mainly a result of the poor Voc and FF values obtained with 0.60 eV cells. The Voc and FF values for the 0.75 eV cells are much closer to their theoretical maximum; whereas, for the defect-ridden 0.60 eV lattice-mismatched cells, these parameters fall far short of their maximum potential. Also shown in table III, the use of the reflector filter enhances the cell efficiency by a factor of about four, from 1.3 to ~5%.

3. Er-Tm YAG emitter, 0.75 eV InGaAs PV cell, and shortpass/reflector filter.

As shown previously, the 0.75 eV cells perform as well as or better than the 0.60 eV cells, even when coupled to a Ho-based emitter. On the other hand, an Er-based selective emitter is ideally tuned to the bandgap of the 0.75 eV cells. We had two Er-based YAG emitters available: one singly doped with Er and the other co-doped with Er and Tm. The center of the emission peak for Er is at 1.5 µm (0.826 eV), and for Tm, 1.8 µm (0.689 eV). In this section, we will present the results of coupling the 0.75 eV cells with the Er-Tm co-doped YAG emitter.

As mentioned earlier, the peak value of the emittance in the emission band for the Er-Tm co-doped emitter was rather low, ~0.4 (see fig. 5). In addition, since the center of the emission peak for Tm is at 1.8 µm, its contribution to power conversion (coupled to a 0.75 eV cell) is negligible. Nonetheless, the Er-Tm YAG/0.75 eV cell converter proved to be more efficient than others described in previous sections.

The total measured power output of the Er-Tm YAG at 1100 °C was 3.82 W/cm^2. The shortpass/reflector combination filter (see fig. 8) reduced the power transmission by a factor of 10, to only 0.382 W/cm^2. Again, most of the radiation is reflected back to the source by the filter, but as shown in figure 10, a fraction of the radiation is absorbed in the filter.

The I-V curves for the most efficient cell tested without and with the shortpass/reflector filter are shown in figures 11(a) and (b), respectively. Table IV summarizes the results in tabular form. As indicated in the table, using the shortpass/reflector filter greatly improved the efficiency of the cell from 1.6 to 7.9-8.5% a factor of ~5. The performance of the 0.75 eV lattice-matched InGaAs on InP cell coupled to an Er-based emitter seems to be superior to that of a Ho-based emitter coupled to either lattice-matched or lattice-mismatched cells.

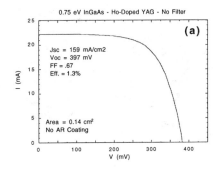

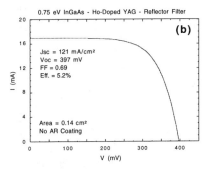

Figure 11. I-V curves for 0.75 eV InGaAs PV cell coupled to Ho-YAG (a) Without filter. (b) With filter with source temperature 1100 °C.

TABLE III. PERFORMANCE PARAMETERS FOR 0.75 eV InGaAs PV CELLS COUPLED TO Ho-YAG. MEASUREMENTS TAKEN WITH AND WITHOUT IR REFLECTOR FILTER, AT SOURCE TEMPERATURE 1100 °C.

I.D.	Power in W/cm^2	Filter	Voc, mV	Jsc, mA/cm^2	FF, %	Efficiency (No AR) %	Efficiency (with AR[*]) %
293-1C	3.19	No	383	153	68.0	1.3	1.7
	0.637	Yes	391	111	70.0	4.8	6.6
293-2D	3.19	No	383	159	67.0	1.3	1.7
	0.637	Yes	397	121	69.0	5.2	7.2

[*]Efficiencies with AR coating are calculated.

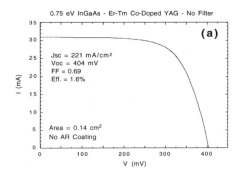

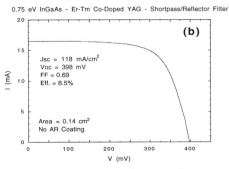

Figure 12. I-V curves for 0.75 eV InGaAs PV cell coupled to Er-Tm YAG (a) Without filter. (b) With filter with source temperature 1100 °C.

TABLE IV. PERFORMANCE PARAMETERS FOR 0.75 eV InGaAs PV CELLS COUPLED TO Er-Tm YAG. MEASUREMENTS TAKEN WITH AND WITHOUT SHORTPASS/REFLECTOR FILTER, AT SOURCE TEMPERATURE 1100 °C.

I.D.	Power in, W/cm^2	Filter	Voc, mV	Jsc, mA/cm^2	FF, %	Efficiency (No AR) %	Efficiency (with AR[*]) %
293-1C	3.82	No	400	231	62.0	1.6	2.2
	0.382	Yes	398	115	66.0	7.9	11.0
293-2D	3.82	No	404	221	69.0	1.6	2.2
	0.382	Yes	398	118	69.0	8.5	11.9

[*]Efficiencies with AR coating are calculated.

4. Er-YAG emitter, 0.75 eV InGaAs PV cell, and shortpass/reflector filter.

The peak value of the emittance in the emission band for the singly doped Er-YAG emitter (~0.6) was significantly better than that for the Er-Tm YAG (~0.4). However, the thickness of the Er-YAG emitter was thinner than the optimal, i.e., 0.4 mm versus ~0.9 mm[8]. Nevertheless, this converter configuration produced the highest efficiency of all the converters discussed so far.

The total measured power output of the Er-YAG at 1100 °C was 2.74 W/cm^2. Again, the shortpass/reflector combination filter (see fig. 8) reduced the power transmission by a factor of ~11, to 0.241 W/cm^2. The I-V curves for the most efficient cell tested without and with the shortpass/reflector filter, and without anti-reflection (AR) coating, are shown in figures 12(a) and (b), respectively. The I-V curves for the most efficient cell tested without and with the shortpass/reflector filter and with AR coating (Ta_2O_5), are shown in figures 13(a) and (b), respectively. It should be noted that the AR coating thickness of 2400 Å applied to this cell was thicker than optimum (i.e., 1800 to 1900 Å).

These results are also summarized in table IV. Again, as shown in the table, the use of the shortpass/reflector filter substantially improved the efficiency of the cell: from 2.3 to 13.2%—a factor of ~6.

Although, neither the InGaAs cells nor the selective emitters were fully optimized for this work, it seems clear that high conversion efficiencies with YAG-based emitters can only be obtained through the use of filters. In order to attain efficiencies ≥15%, The emitter radiative efficiency of the rare-earth doped YAG crystals has to increase substantially. The filters used in this work could accomplish this to some degree, but they also diminished the total transmitted power delivered to the PV cell by a factor of 5 to 10. Development of selective emitters with more ideal spectral emittance characteristics could eliminate the necessity to use such filters.

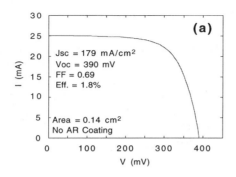

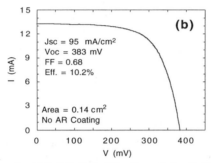

Figure 13. I-V curves for 0.75 eV InGaAs PV cell coupled to Er-YAG (a) Without filter. (b) With filter with source temperature 1100 °C.

TABLE III. PERFORMANCE PARAMETERS FOR 0.75 eV InGaAs PV CELLS COUPLED TO Er-YAG. MEASUREMENTS TAKEN WITH AND WITHOUT SHORTPASS/REFLECTOR FILTER, AT SOURCE TEMPERATURE 1100 °C.

I.D	Power in, W/cm^2	Filter	Voc, mV	Jsc, mA/cm^2	FF, %	Efficiency (No AR) %	Efficiency (with AR[*]) %
293-2D	2.74	No	390	179	69.0	1.8	2.5
	0.241	Yes	383	95	68.0	10.2	14.3
284-2	2.74	No	387	220	74.0	---	2.3[†]
	0.241	Yes	376	115	74.0	---	13.2[†]

[*]Efficiencies with AR coating are calculated.
[†]Ta_2O_5 AR coating (2400 Å).

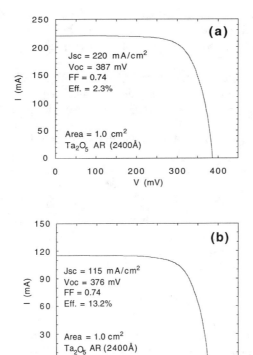

Figure 14. I-V curves for 0.75 eV InGaAs PV AR coated cell coupled to the Er-YAG (a) Without filter. (b) With filter with the source temperature at 1100 °C.

SUMMARY

The main conclusions drawn from this work are as follows:

1. The quality of the selective emitters and PV InGaAs cells are extremely important in the fabrication of high-efficiency radioisotope thermophotovoltaic (RTPV) converters.

2. To some extent, the use of bandpass/reflector filters can compensate for the unusable radiation energy emitted below the bandgap of the PV cell by the YAG by recycling a portion of this out-of-band radiation.

3. The performance of the lattice-matched InGaAs/InP cells (Eg = 0.75 eV) was as good or better than that of lattice-mismatched cells (Eg = 0.60 eV), even when coupled to a Ho-doped emitter.

4. Efficient converters can be fabricated with filtered Er-YAG selective emitters and lattice-matched InGaAs/InP PV cells, operating at lower wavelengths (1.5 μm) than theoretical optimum (>2 μm), for a source temperature of 1100 °C.

5. The best conversion efficiency (13.2%) was obtained with a 0.75 eV InGaAs/InP PV cell coupled to an Er-doped YAG selective emitter and a shortpass/reflector filter.

REFERENCES

1. A. Schock and V. Kumar, "Radioisotope thermophotovoltaic system design and its application to an illustrative space mission," First NREL TPV Conf. Proc., Copper Mountain, CO, July 1994, p.139.

2. N.S. Fatemi and V.G. Weizer, "On the electrical and metallurgical behavior of AuZn contacts to p-type InP." J. Appl. Phys. **77**, 5241 (1995).

3. V.G. Weizer and N.S. Fatemi, "A simple, extremely low resistance contact system to *n*-InP that does not exhibit metal-semiconductor intermixing during sintering," Appl. Phys. Lett. **62**, 2731 (1993).

4. D.L. Chubb and R.A. Lowe, "Thin film selective emitter," J. Appl. Phys. **74**, 529 (1993).

5. R.M. Sova, M.J. Linevsky, M.E. Thomas, and F.F. Mark, "High-temperature optical properties of oxide ceramics," Johns Hopkins APL Technical Digest, **13**, 375 (1992).

6. D.M. Wilt, N.S. Fatemi, R. Hoffman, P. Jenkins, D. Brinker, D. Scheiman, and R. Jain, "High efficiency InGaAs photovoltaic devices for thermophotovoltaic applications." Appl. Phys. Lett. **64**, 2415 (1994).

7. M.W. Wanlass, J.S. Ward, K.A. Emery, and T.J. Coutts, "$Ga_xIn_{1-x}As$ thermophotovoltaic converters," First World Conference on Photovoltaic Energy Conversion Proc. Waikoloa, HI, December 5-9, 1994, p.1685.

8. R.A. Lowe, B.S. Good, and D.L. Chubb, "The effect of thickness and temperature on the radiative performance of thin film selective emitters for thermophotovoltaic applications," 30th Intersociety Energy Conversion Engineering Conf. (IECEC), July 30-August 4, 1995.

Development of a Small Air-Cooled "Midnight Sun" Thermophotovoltaic Electric Generator

Lewis M. Fraas, Huang Han Xiang, She Hui, Luke Ferguson,
John Samaras, Russ Ballantyne, Michael Seal[†] and Ed West[†]

JX Crystals Inc., 1105 12th Ave NW, Suite A2, Issaquah, WA 98027
[†] *Western Washington University, Bellingham, WA 98225*

Abstract. A natural gas fired thermophotovoltaic generator using infrared-sensitive GaSb cells and a silicon carbide emitter is described. The emitter is designed to operate at 1400°C. Twelve GaSb receivers surround the emitter. Each receiver contains a string of series connected cells. Special infrared filters are bonded to each cell. These filters transmit short wavelength useful IR to the cells while reflecting longer wavelength IR back to the emitter. Combustion air is supplied to the burner through a counterflow heat exchanger where the air is preheated by the exhaust from the burner. The unit is air cooled and designed to produce approximately 100 Watts of electric power.

Introduction

JX Crystals is developing a small, portable 100 Watt thermophotovoltaic cogenerator designed to provide heat and electricity to a single room off-grid dwelling such as a mountain cabin. Our TPV unit can also be adapted for use in a recreational vehicle or a sail boat. Our design concept is shown schematically in figure 1.

The horizontal cross section at the top of this figure shows that the core of the generator is cylindrically symmetric. The central SiC ceramic emitter is surrounded by twelve TPV receiver assemblies. Each receiver assembly contains a series string of GaSb cells facing inward toward the infrared (IR) emitter. Special IR filters are bonded to each cell. These filters transmit short wavelength useful IR to the cells while reflecting longer wavelength IR back to the emitter. Heat is removed from the cell string by fins located on the back of each receiver.

Referring to the vertical cross section in figure 1, the fins and SiC burner are again evident. A blower is provided on one side of the central cylinder to provide both combustion air and cooling air. Combustion air is provided to a counterflow heat exchanger located on the top of the central cylinder. The combustion air is preheated by the hot exhaust gases rising from the top of the burner. A large fraction of the blower air is used for forced air cooling of the TPV receivers. The cooling air is fed to a plenum located below the generator cylinder and thence

© 1996 American Institute of Physics

upward through an array of holes in the generator base vertically along the receiver back fins. This clean air heated by the receivers back fins can then be used to heat the room. The smaller quantity of exhaust gas from the combustion process exits the top of the generator and can be vented outside the dwelling.

The DC electric power produced by the generator can be used to charge a battery or to run appliances. Our demonstration unit is provided with a small 12 Volt battery which we use to start the blower. However in actual commercial application, a larger car-size or marine-size battery would probably be available. For a mountain cabin in northern climates, the unit would probably be used for heating over a 24 hour period and hence, it would produce 100 Watts x 24 hrs = 2.4 kWh of electricity daily. During most of the time, this energy would be delivered to a battery. This 2.4 kWh of energy would then probably be used mainly in the evening hours by the owner, for example, for 4 hours from 6 pm to 10 pm. The result would be the availability of 600 Watts for 4 evening hours daily.

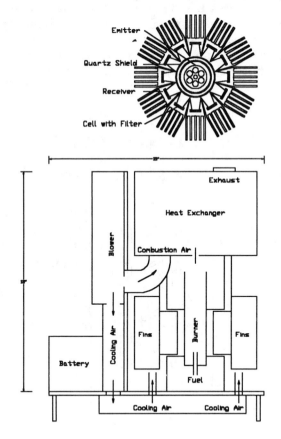

FIGURE 1. Prototype 100 Watt air-cooled TPV electric generator

Measurements

The challenge in developing the "Midnight Sun" generator is not just the development of the key components such as the TPV receivers and the SiC emitter but the integration of these key components into a working, optimized generator.

We have previously reported on the fabrication and measurement of GaSb TPV cells and circuits. Flash test measurements on 1 cm^2 cells at currents up to 6 Amps show V_{oc} as high as 0.5 V and fill factors as high as 0.8. Flash test measurements on circuits have shown power densities up to 2.2 Watts/cm^2. These measurements show that we can successfully fabricate GaSb TPV cells and circuits.

Herein, we report on the stable performance of circuits in combination with a SiC emitter as a function of emitter temperature. In the present experiment, IR filters are attached to each cell and the circuit is attached to a water cooled heat sink. In order to collect this systematic receiver data, the receiver test station shown conceptually in figure 2 was fabricated. It consist of an electrically heated SiC globar surrounded by a SiC IR emitter tube identical in dimensions to the SiC emitter in our natural gas fired TPV generator. The insulating cuffs at the top and bottom and the fused silica heat shields are also identical in dimensions to those in our generator.

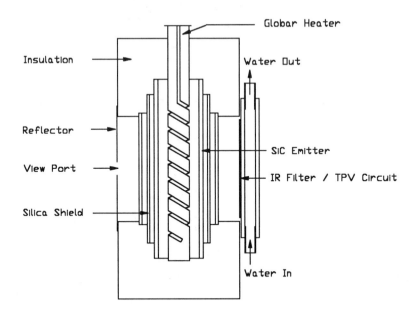

FIGURE 2. TPV receiver test station

A reflecting cylinder consisting of "everbright" aluminum surrounds the emitter. A water cooled TPV test receiver is placed into a vertical slot in this reflecting cylinder while the temperature of the emitter can be monitored through a hole in the cylinder with an optical pyrometer. The optical pyrometer used responds to IR wavelengths between 1.2 and 1.6 microns.

Why build a receiver test station? Why not use the generator itself? There are four significant advantages associated with this test station. First, indivual receivers can be tested. Second, the emitter temperature can easily be measured simultaneously with the receiver test. Third, the power input can be easily measured by recording the current and voltage supplied to the glowbar. And finally fourth, the heat load on the receiver can be measured by recording the input and output water temperatures and the water flow.

Figure 3 shows the short circuit current and the open circuit voltage vs. emitter temperature for a 7 cell circuit as recorded in our receiver test station. The most significant result is that the receiver current rises to 3.4 Amps at 1400°C at which time the open circuit voltage was recorded at 3.12 V. The result is significant because, while we have recorded currents and voltages in generator operation previously, we have never had a correlation with emitter temperature. Flash test measurements on the figure 3 test circuit were consistent with the DC data recorded in figure 3 except that the flash test showed a higher V_{oc} of 3.3 V as expected for a circuit at room temperature.

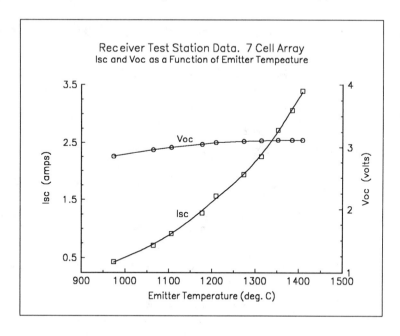

FIGURE 3. current & voltage vs emitter temperature

Having established individual 7 cell circuit performance, we then proceeded to fabricate 12 similar circuits and mount them on air cooled finned receivers and to combine them with a natural gas fired SiC emitter. IR filters were bonded onto the cells in these 12 circuits. Table 1 shows the flash test parameters for these circuits.

The resultant "Midnight Sun" generator is shown in figure 4. The SiC burner is described in a separate paper. At the present time, our prototype TPV generator routinely produces enough power to operate a DC color television.

TABLE 1. Performance Parameter Summary for the 12 Circuits Incorporated in "Midnight Sun" Generator

Circuit #	FF	Voc (V)	Isc (A)	Imax (A)	Vmax (V)	Pmax(W)*
7c2f	0.735	3.267	2.790	2.593	2.583	6.697
7c4f	0.709	3.286	2.807	2.470	2.646	6.536
7c5f	0.757	3.334	2.841	2.628	2.729	7.173
7c6f	0.774	3.320	2.806	2.684	2.686	7.207
7c7f	0.761	3.330	2.784	2.590	2.725	7.056
7c8f	0.740	3.315	2.774	2.452	2.773	6.801
7c9f	0.753	3.311	2.778	2.593	2.671	6.928
7c10f	0.738	3.305	2.764	2.484	2.715	6.745
7c11f	0732	3.319	2.792	2.449	2.769	6.781
7c12f	0.757	3.310	2.814	2.602	2.710	7.050
7c13f	0.728	3.290	2.756	2.464	2.681	6.605
7c16f	0.761	3.310	2.810	2.599	2.725	7.080

* The sum of Pmax for the 12 circuits is 82.7 Watts

Conclusions

JX Crystals is now fabricating IR cells, IR filters, and cooled TPV receiver assemblies and we have developed a test station to allow TPV receiver qualification to customer specifications. In addition, JX Crystals now has a working prototype for a gas-fired portable off-grid TPV generator.

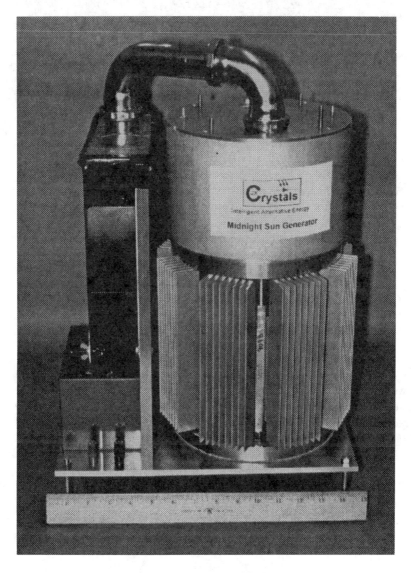

FIGURE 4. Photograph of prototype TPV generator

Demonstration of a Candle Powered Radio Using GaSb Thermophotovoltaic Cells

Douglas J. Williams and Lewis M. Fraas

JX Crystals Inc., 1105 12th Ave NW, Suite A2, Issaquah, WA 98027

Abstract. We have discovered that the flame from a single candle when surrounded by a bracelet-sized ring of GaSb cells can provide enough power to operate a transistor radio. Spectral measurements of the flame show that a large fraction of the flame energy is in the infrared in a blackbody-like emission band centered at 1.3 microns. GaSb cells capture infrared energy out to 1.7 microns, providing enough energy to power the radio. Traditional silicon solar cells only capture energy out to 1.1 microns and produce less than 30% of the power of GaSb cells, which is not enough to operate the radio.

Introduction

We have discovered that the flame from a single candle when surrounded by a bracelet sized ring of GaSb cells can provide enough power to operate a transistor radio. Specifically, the thermophotovoltaic (TPV) bracelet is an eight sided ring containing two rows of eight series connected 0.75 cm^2 active area GaSb cells. The ring is approximately 1.6" in diameter and 1" tall and produces an open circuit voltage of 4.3 V and a short circuit current of 24 mA. Figure 1 shows photographs of the TPV bracelet / radio combination with the flame energy required for operation provided by either (a) a small oil lamp, or (b) a candle.

Measurements

We find this candle demonstration of thermophotovoltaics fascinating. However, two questions immediately arise: why does it work and will it work with traditional silicon solar cells? To answer these questions, we measured the spectral output of the oil-lamp flame and we purchased some silicon cells and fabricated a TPV bracelet of similar dimensions for a direct comparison.

Figure 2 shows the spectral output of the oil-lamp flame over the range from 0.5 to 5 microns. First, note that most of the energy emitted is in the infrared. Unlike sunlight, only a small fraction of the emitted energy is visible. Second, note that the emitted energy can be characterized as being associated with blackbody like emission or OCO and HOH combustion product gas line emissions. Most of the energy is contained in the blackbody like emission band centered at approximately 1.3 microns. This band can be attributed to small glowing carbon

FIGURE 1. Photographs of the TPV bracelet / radio combination with the flame energy provided by either (a) a small oil lamp or (b) a candle

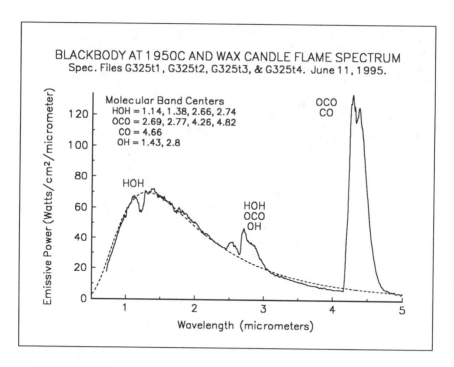

FIGURE 2. Blackbody at 1950°C and wax candle / oil lamp flame spectrum

particles. When the flame is adjusted up too high, these carbon particles become evident as black smoke rising from the tip of the flame. When the flame is adjusted properly, however, these particles are completely consumed in the yellow body of the flame. From the fit of the flame spectrum to a blackbody curve, it is evident that the characteristic temperature of the glowing carbon particles is approximately 1950°C. Finally, note that a low band gap photovoltaic cell like GaSb with infrared response extending out to 1.7 microns will capture a large fraction of the radiated energy from a flame. However, a traditional silicon solar cell with response out only to 1.1 microns will not do as well.

We calculated that the current produced from a similar sized silicon cell near a 1950°C emitter should be 5 to 6 times less than for GaSb. In order to check this calculation, we purchased 1.8 cm^2 active area silicon cells from Edmund Scientific. We then connected eight silicon cells together in a bracelet with identical dimensions and placed the silicon cell bracelet around a candle. The current produced was only 10 mA while the open circuit voltage was 3.6 V. This was not enough to operate the radio. Table 1 summarizes our measurement results for both GaSb and silicon bracelets operating in combination with the same candle flame. The current density ratio is then 24/(0.75) over 10/1.8. Experimentally, the

TABLE 1. Comparison of GaSb and Si TPV bracelet performance

Bracelet Type	Cell Area (cm^2)	I$_{sc}$ (mA)	V$_{oc}$ (V)	I$_{sc}$ x V$_{oc}$ (mW)
GaSb	16 x 0.75 = 12	24	4.3	103
Silicon	8 x 1.8 = 14.4	10	3.6	36

GaSb cells produced 5.8 times more current per unit area than the silicon cells. The power density advantage is 4.2*24/(16*0.75) over 3.6*10/(8*1.8). Experimentally, the GaSb cells produced 3.4 times more power per unit area than the silicon cells.

Interestingly, the current produced by a GaSb cell near the candle flame is approximately the same as the current produced by a GaSb cell in outdoor sunlight. However, we calculate that when GaSb cells are placed 1" away from a 1950°C blackbody with an emissivity of unity, the output current should be equivalent to several hundred suns. This discrepancy can be explained by noting that the carbon particles in a flame occupy a very small fraction of the flame volume. Referring to the spectrum shown in figure 2, the OCO line at 4.3 microns is much higher than the blackbody-like emission at 1.3 microns because of a higher volume density for OCO. In other words, although an open flame demonstrates the TPV effect, much higher power levels are possible by heating blackbody ceramic materials to high temperatures with a flame.

Conclusions

We have found that a candle powered radio is a simple desk top demonstration of the TPV concept, i.e. fuel to electricity with no moving parts. We have also found that a candle flame provides a simple demonstration of the value of low bandgap TPV cells.

Laboratory Development TPV Generator

Glenn A. Holmquist,* Eva M. Wong,* and Cye H. Waldman**

*Quantum Group, Inc., 11211 Sorrento Valley Road, San Diego, CA 92121
**Consultant, P.O. Box 231157, Encinitas, CA 92023-1157

Abstract. A laboratory model of a TPV generator in the kilowatt range was developed and tested. It was based on methane/oxygen combustion and a spectrally matched selective emitter/collector pair (ytterbia emitter-silicon PV cell). The system demonstrated a power output of 2.4 kilowatts at an overall efficiency of 4.5% without recuperation of heat from the exhaust gases. Key aspects of the effort include: 1) process development and fabrication of mechanically strong selective emitter ceramic textile materials; 2) design of a stirred reactor emitter/burner capable of handling up to 175,000 Btu/hr fuel flows; 3) support to the developer of the production silicon concentrator cells capable of withstanding TPV environments; 4), assessing the apparent temperature exponent of selective emitters; and 5) determining that the remaining generator efficiency improvements are readily defined combustion engineering problems that do not necessitate breakthrough technology.

The fiber matrix selective emitter ceramic textile (felt) was fabricated by a relic process with the final heat-treatment controlling the grain growth in the porous ceramic fiber matrix. This textile formed a cylindrical cavity for a stirred reactor. The ideal stirred reactor is characterized by constant temperature combustion resulting in a uniform reactor temperature. This results in a uniform radiant emission from the emitter. As a result of significant developments in the porous emitter matrix technology, a TPV generator burner/emitter was developed that produced kilowatts of radiant energy.

SUMMARY-OVERVIEW

The objective of this effort was to demonstrate that TPV technology is practical for multi-kilowatt power generation. The emitter/burner design/development effort met this goal. A Laboratory TPV generator based on an ytterbia selective emitter spectrally matched with silicon PV cells demonstrated 2.4 kilowatts. The key was successful fabrication of ceramic ytterbia material having the strength to operate in the emitter/burner configuration. As a result of the developmental work for

© 1996 American Institute of Physics

processing the ceramic fiber matrix, a well-stirred reactor structure was produced which exhibited excellent strength and thermal cycling durability.

Also, of importance was the achievement of the thermal management of the combustion system. The generator design configuration was predicated on its use as a power source for an undersea vehicle. Accordingly, the stirred-reactor design adopted was a right circular cylinder configuration with a combustion-heated emitter radiating outward through a heat shield to the surrounding PV cells. This design offers features that made it attractive for the originally intended use.

The efficiency of the generator can be described as the product of three sub-efficiencies:
- the efficiency of converting the fuel's Btu/hr energy content to photons ($\eta_{Btu/hr-rad}$.)
- the efficiency with which the PV cells collect the photons ($\eta_{photon-collection}$).
- the efficiency of converting the photons into PV cell power output ($\eta_{rad-electric}$).

The first sub-efficiency, converting the fuel's Btu/hr content to photon efficiency, was most affected by the design/development efforts. Concerning the second sub-efficiency the concentric emitter and PV cell configuration lends itself to achieving up to 90% photon collection efficiency.

Finally, the strong temperature dependence of the selective emitters was developed with the resulting capability to model radiant emittance changes with increased emitter temperature. Quantum, not a PV cell manufacturer, determined the best PV cell vendor with regard to material, cost, design maturity and deliverability.

As a result of this development effort, it is possible to demonstrate future improvements through such refinements as optimized recuperation, improved burner design and better photon management. These improvements can be benchmarked against the demonstrated performance of the basic system.

The balance of this paper details the approach taken to develop the apparatus. It includes the results obtained and discussion thereof. Finally, conclusions and recommendations are made.

APPROACH

The requirements for the originally intended underwater vehicle application resulted in the general configuration of the generator. A power analysis identified the needed emitter surface area and temperature requirements and the methane flow requirements. A laboratory bench model (Figure 1) was developed to conform to these parameters. The original application also dictated the use of oxygen as the fuel oxidizer. This choice remained throughout the development effort.

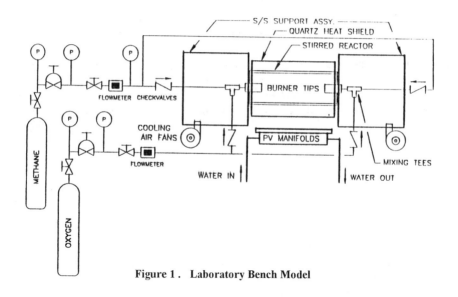

Figure 1. Laboratory Bench Model

Design Parameter Analysis

The generator design configuration was predicated on the end use as a power source for a submarine launched underwater vehicle. Thus it would be horizontally operational and an essentially infinite heat sink would be available. The design developed was a right circular cylinder which operates as a stirred reactor with the emitter forming the reactor wall surrounded by a fused silica heat shield which provides an annulus for passage of the hot exhaust gases. The PV cells are arrayed in strips a short distance from the heat shield. Figure 2 illustrates the positioning of the torch or burner and recuperator configuration, in addition to the reactor.

The development of the operating parameters of the fuel-oxygen reactor proceeded from two directions: 1) the fuel flow necessary to give kilowatts of energy after preliminary efficiencies were estimated; and 2) the calculation of the necessary temperature of a radiant emitter of given surface area to illuminate a PV cell with sufficient energy to produce the desired output. Figure 3 shows a schematic of the TPV generator which is useful in assessing the values of the key parameters. The maximum efficiencies noted in Figure 3 are shown as they were initially estimated and were later substantiated, with interpretation, from reference 1.

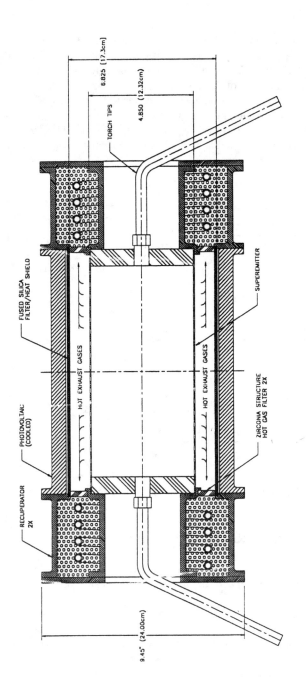

Figure 2. Preliminary Burner Design

In the generator, the ceramic emitter is heated to incandescence by methane/oxygen combustion exhaust gases passing through the porous material. The primary mechanism for heat loss from the emitter is radiant emittance. The efficiency of interest, $\eta_{Btu/hr\text{-}rad}$, is the ratio of the selective emitter radiance with wavelengths below 1.07 micrometers divided by the total energy content of methane gas flow. To a first approximation, all the useful radiant emittance of an ideal ytterbia selective emitter is less than 1.07 micrometers in wavelength. The Stefan-Boltzman equation and the average emissivity for ytterbia at elevated temperatures yield the radiant emittance of interest.

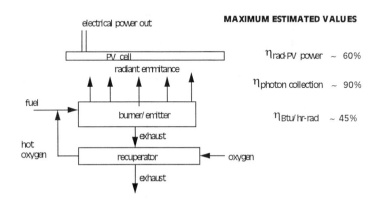

Figure 3. TPV Generator Schematic Diagram

Preliminary Radiant Emittance Analysis

- The selective or superemitter is a right circular cylinder 12 cm in diameter and 24 cm long. Thus the burner or reactor surface area is 905 cm^2

$$\text{area} = 12 \times 24 \times 3.142 = 904.8 \text{ cm}^2 \qquad (1)$$

- The cylinder is formed from a sheet of selective emitter material wrapped around stabilized zirconia end caps and sealed with sol-gel ytterbia cement along the length of the cylinder.

- For the analysis, the selective emitter is assumed to be heated to between 1900 and 2000°C
- The total radiant emittance per cm^2, w, will follow the Stefan-Boltzman law using the <u>average</u> emissivity (ϵ) for the entire band at that temperature and the Stefan-Boltzman constant (σ).
- At 1900°C $\quad w = \epsilon \sigma T^4 = \epsilon \times 5.6686 \ 10^{-12} \times (1900 + 273)^4 \ watts/cm^2 \quad$ (2)
  $\quad\quad\quad\quad = \epsilon \times 126.4 \ watts/cm^2$
- At 2000°C $\quad w = \epsilon \sigma T^4 = \epsilon \times 5.6686 \ 10^{-12} \times (2000 + 273)^4 \ watts/cm^2 \quad$ (3)
  $\quad\quad\quad\quad = \epsilon \times 151.3 \ watts/cm^2$

A porous fiber matrix of the material can have an emissivity greater than the solid material due to the emittance of radiation from the cavities or pores, and the measured ytterbia porous matrix emissivities were as high as 0.2 at 1900 degree temperatures, thus:

- $w = 25.28 \ w/cm^2$ at 1900°C

- $w = 30.26 \ w/cm^2$ at 2000°C

Continuing the preliminary analysis, the 905 cm^2 surface area at 1900°C would radiate 22,880 watts. The estimated possible maximum efficiency of the silicon PV cell ($\eta_{rad-elect}$) was 60%. White & Hottel (ref. 1) reported a range of 40% to 60%, and that 40% should be available for a silicon PV cell coupled with an ytterbia emitter. Assuming 40% the generator output would then be 9,150 watts. This would be further reduced to 8,240 watts assuming a 90% efficient photon collection efficiency ($\eta_{photon \ collection}$).

Preliminary Methane Flow Analysis

- The PV cell electrical power output goal is 10,000 watts. As in Figure 3 assume:

$$\eta_{overall} \equiv \eta_{Btu/hr-rad} \bullet \eta_{photon-collection} \bullet \eta_{rad-electric} \quad (4)$$

- Assuming a PV cell 40% efficiency, the emittance from the superemitter must be 25,000 watts to generate 10 kilowatts.
- Assuming the photon collection efficiency is 90%, the necessary emittance is increased to 27,777 watts.
- 27,777 divided by the methane flow power content to radiated energy efficiency ($\eta_{Btu/hr-rad}$) results in the necessary input power.
- Using the estimated maximum value of 45% estimated for $\eta_{Btu/hr-rad}$ the resulting methane requirement is 61,726 watts.
- Assume the energy content of the methane is 1024 Btu/hr per standard cubic foot per hour (SCFH), therefore,

$$1 \, SCFH \bullet 1024 \frac{Btu/hr}{SCFH} \bullet 0.293 \frac{watts}{Btu/hr} = 300 \, watts \qquad (5)$$

- The resulting methane flow must be 205.6 SCFH (5.82 m³/hour).

These parameters identified realizable goals for the approach. For oxidation of methane, twice the volume of oxygen is required for stoichiometric combustion, yielding three volumes of products per volume of methane input. The ceramic fiber matrix would have to show excellent mechanical strength to withstand gas flows of 617 SCFH (standard cubic feet per hour) of gas at the operating temperature and pressure of the burner. A strong porous ceramic fiber matrix is key to withstanding this flow and the pressure from flow resistance at operating temperatures.

Strength Improvement of the Ceramic Fiber Matrix

During operation an emitter temperature of 2000°C is desired. The emitter material must have mechanical integrity and tolerance for thermal cycling. A fiber matrix with fiber diameters on the order of 10 micrometers was the approach to develop a material to withstand such thermal cycling and to maintain interlocked fiber mechanical strength as well as show enhanced radiation properties. To attain this structure, the microstructure of a textile material, (felt) must be mimicked by the ceramic matrix. To achieve these properties, Quantum utilizes a relic process with an organic felt textile as the precursor material.

The relic process has been exploited to produce a variety of different material forms such as fiber, fibrous mat, or reticulated ceramic shapes. The inherent flexibility of the process also enables the composition of the final product to be tailored by control of the precursor solution. The process does not require specialized equipment, chemicals or precursors, and the physical size of the product is limited only by the size of the processing equipment. The process lends itself well to an inexpensive batch type manufacturing operation.

The relic process is a means of arriving at an oxide of nearly any form by the use of an inexpensive solution, such as a nitrate, which is inbibed into a porous, sacrificial organic medium. The basic process consists of soaking the preform in the concentrated salt solution for a prespecified period of time. Excess solution is then removed, and the preform is dried at a low temperature followed by conversion of the metal salt to a metal hydroxide. The preform is then rinsed to remove any residual nitrates and dried before the final heat treatment, during which complete organic removal and conversion of the hydroxide to an oxide occur. A flow chart illustrating the process is shown in Figure 4, and the basic chemical conversion reactions are shown below, where M represents a trivalent rare earth metal.

$$M(NO_3)_3 + 5NH_3 = M(OH)_3 + 4N_2 + 6H_2O \qquad (6)$$

$$2M(OH)_3 + O_2 = M_2O_3 + 3H_2O + O_2 \qquad (7)$$

Complete pyrolysis of the preform is typically controlled in two stages. The first stage of the final heat treatment is performed by slowly heating from room temperature to an intermediate temperature in an inert atmosphere to prevent rapid charring of the structure (which will result in structural damage). Shrinkage of the preform during the first stage is on the order of 40%. The second stage of heat treatment is performed in the presence of oxygen to convert the hydroxide to the oxide and control the final microstructure. At this stage, the material may be more rapidly heated to high temperatures to promote grain growth and sintering if required. Further shrinkage during the second stage is less than 5%.

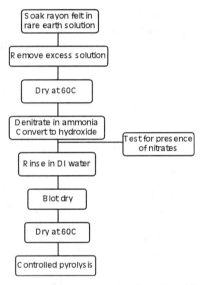

Figure 4. Flowchart illustrating relic process used to form fiber felt.

The resulting emitter felt is a continuous structure of discontinuous, mechanically interlocked ceramic fibers with fiber diameter on the order of 10 micrometers. Microporosity exists between the fibers in the as-processed form. A certain amount of porosity is required for optical efficiency. This must be balanced, however, with mechanical strength and material formability. As the fibrous mat becomes more open, it also becomes less formable due to the discontinuous nature of the fibrous structure. Conversely, as the fibrous mat becomes more closed, optical efficiency is reduced when the mat can not be evenly heated and therefore will not evenly illuminate and radiate. The balance required between these properties can be controlled by selection of the optimal fiber size and density. The theory of fracture mechanics, ceramic fibers and grain growth is discussed in Appendix B.

RESULTS AND DISCUSSION

Power Generation

2,400 watts of power were generated at an input methane flow equivalent to 52,650 watts (179,700 Btu/hr from 175.5 SCFH methane). Burner failure occurred at 210 SCFH methane, the operation at the 175.5 point was the demonstrated maximum safe operation point. The burner efficiency was not augmented with recuperation, and this efficiency (much less than the 45% used in the above calculation) can be improved upon with straightforward combustion engineering techniques. In the test, one of twelve PV cell strips (6 cm by 24 cm) recorded open circuit voltage and short circuit current values for the four individual 6 cm by 6 cm arrays. The fill factor for these arrays has been measured to be 75%. The resultant product (Isc x Voc x ff) yielded the power from each of four arrays which summed to 200 watts for the strip. The power levels were greater for the two center arrays. Twelve times the sum gave the 2,400 watt value.

To provide insight into the balance between the combustion and radiation temperature a rough estimate of the energy flow was conducted for the 2,400 watt operating parameters:

- the mass flow of methane was at 175.5 SCFH and the heat flow from combustion was approximately 179,700 Btu/hr at a mass flow of methane and oxygen of 39.5#/hr.
  Ignoring disassociation this results in 17.75 #/hr of H_2O and 21.7#/hr of CO_2
- Approximate values of C_p at 2000°C are, for H_2O - 1.25 and for CO_2 - 3.1 Btu/#-°K
- Assuming 10 % losses and 20% of the remainder converted to radiation, the temperature gradient across the emitter thickness is

$$\Delta t_e = \frac{dQ/dt(1\text{-losses}) \times (\text{percentage radiated})}{[dm/dt \times C_p]_{water} + [dm/dt \times C_p]_{CO2}} \approx 344°C \qquad (8)$$

The temperature in the reactor has an upper limit of 2200°C, as the melting point of the emitter is 2250°C. Therefore if these assumptions are correct the emitter radiating surface temperature is limited to approximately 1850°C from this rough estimate. This temperature is approximately the preliminary estimate of 1900 to 2000°C.

Ceramic Fiber Matrix Strength Improvements

A fiber matrix with fiber diameters on the order of 10 micrometers did, in fact, demonstrate the strength to withstand TPV generator operation, with respect to both thermal cycling and to maintaining interlocked fiber mechanical strength. Continuing the discussion presented in the Approach, and now introducing our results, it follows that tailoring the strength of the ceramic fibers necessitated careful control of the processing temperature and time of heat treatment. The heating rate at low temperatures is most critical during the first stage due to the large volume shrinkage that occurs when the organic preform is pyrolized and the hydroxide is converted to an oxide. During the second stage, the material is dimensionally stable, and the heating rate is less critical. Control of the final time and temperature, however, ultimately determines the strength of the material by altering the microstructure. The material may be kept amorphous at low temperatures, but a microcrystalline grain structure is preferred for increased strength. Recent laboratory measurements of relative strength indicate that higher temperatures and longer times increase the ultimate tensile strength of the material, as shown in Table I.

High temperatures initiate crystalline grain growth, which directly influences mechanical strength. Besides heat treatment, there are many other processing variables which can affect the strength of the material, and the continued improvement of material strength over development time can be seen in Figure 5. The increased strength enabled the forming of self-supporting emitter cylinders 12 cm in diameter and 24 to 32 cm long. These maintained their integrity until the initial high temperature burner operation, when this extreme temperature resulted in further sintering the ceramic matrix into its final shape and strength.

Table I. Measured Average σ_{UTS} as a Function of Heat Treatment Conditions

Temperature	Time	Measured Average σ_{UTS}, kPa
450°C	1 hour	18.80
600°C	2 hours	31.32
	6 hours	43.68
800°C	2 hours	31.84
	6 hours	43.26
	12 hours	64.88
1200°C	2 hours	91.04

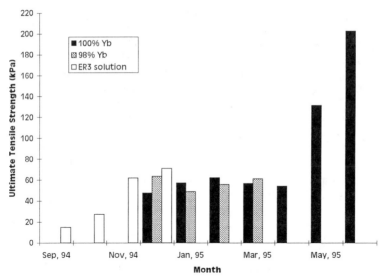

Figure 5. Fiber felt strength vs. time (average each month)

Burner Design

The burner design/configuration chosen was a cylindrical stirred reactor 12 cm in diameter and 24 cm long formed from a sheet of porous ytterbia felt. The end caps were stabilized zirconia, covered with ytterbia felt, that housed the burner nozzles for methane/oxygen injection into the reactor. An ideal stirred reactor is a fixed volume into which oxygen and methane are injected at a fixed total flow rate and temperature. There is instantaneous turbulent mixing and heating of reactants and reaction products to some reactor temperature T_R. The exhaust flow is a mixture of reaction products and some reactants at the temperature T_R. Most important, the entire reactor volume is heated to the same temperature. The inside surface of the ytterbia reactor walls is heated to T_R; therefore, the walls evenly emit radiation.

The maximum temperature of an ideal stirred reactor is limited to approximately 75% of the adiabatic flame temperature of a hydrocarbon fuel/oxidizer (ref. 2). The methane/oxygen adiabatic flame temperature is 2750°C and therefore the ideal maximum temperature of the stirred reactor is approximately 2,000°C. 2250°C is the melting point of the ytterbia ceramic matrix and if this 2000°C value is accurate, this limitation is compatible with the design. In the run that produced the 2,400 watts at 175.5 SCFH, failure of the burner occurred at 214 SCFH with melting at points on the ceramic reactor wall. It has not been determined if this was due to thin spots in the fiber matrix causing uneven heating or due to particulate impurities contaminating the ceramic and causing localized eutectic melting. The ytterbia did not appear to have melted on a large scale, only small pin holes.

The heating of the ytterbia porous matrix does cool the passing gas, with a resulting temperature gradient across the thickness of the ytterbia. The inside surface is at the reactor temperature, T_R, while the outer surface, the radiating surface, is at a lower temperature. Reducing the wall thickness of the ytterbia burner or reactor appeared to increase the temperature of the outer surface. This resulted in a 16% increase in the radiant emittance with respect to the emittance of a thicker material at the same methane flow rate.

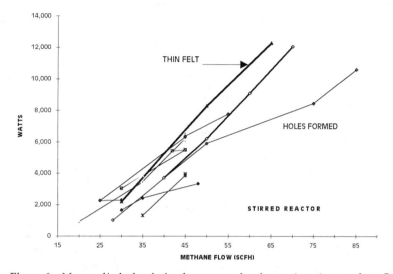

Figure 6 Measured/calculated stirred reactor total emittance (watts) vs. methane flow rate

Figure 6 indicates as much as 12,000 watts of radiant emittance from the stirred-reactor; however, it was found that the radiation was not solely selective emitter peak radiation. The radiant emittance was measured with a thermopile detector with a bandwidth from 0.4 to 5 micrometers and then calculated using the configuration factor identified in Appendix A. As a result, the measurement includes all contributions of longer wavelength emissions, that is, any in addition to the selective emitter radiant emittance. During initial testing, (when, for safety reasons, maximum fuel flow was not being used) it was found that at low methane flow rates a fraction of the emissions was due to combustion gas emission bands and/or carbon particle continuum radiation. These were indicated in spectra measured and shown in Figures 7 & 8. This long wavelength contribution decreases with increasing temperature, as will be apparent with the development of the temperature dependence of the selective emitter peaks. The strategy is to increase the temperature of the selective emitter until the other contributions are negligible in comparison. The ytterbia exponential temperature dependence and the ability to determine the quantitative advantage of increased temperatures is the point of the selective emitter temperature dependence development in this paper.

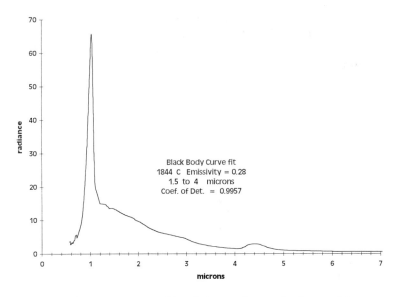

Figure 7. Stirred reactor spectrum with emission continuum of carbon particles

The spectrum of the stirred reactor (Figure 7) indicated a blackbody radiation continuum from 1.5 to 4 micrometers. The spectrum was similar to the description of the plumes of hydrocarbon burning rocket engines, which are identified as having a carbon particle continuum radiation (ref. 3). Carbon is not an equilibrium species in a high temperature hydrocarbon-oxygen system, but there is some question as to the possible non-stoichiometric fuel/oxygen mixture and the turbulence in the reactor is not conducive to equilibrium. A steady state concentration of carbon particles forming and combusting may be characteristic of this reactor using a fuel/oxygen ratio adjusted for maximum temperature operation.

The spectrum (Fig. 7) was curve fit to a blackbody indicating a temperature of 1844°C at an effective emissivity of 0.28. At the flow rate of 130 SCFH, the temperature of 1844°C is what one would expect in that particular experiment; the carbon and its greybody spectrum may represent a thermometer for measuring the reactor temperature. The carbon is burning exothermally but may not be significantly different in temperature than the exothermally burning combustion gases that fill the turbulently mixed reactor volume. The use of the carbon spectra as a thermometer needs further investigation. The temperature of an incandescent selective emitter (non-blackbody radiator) is very slippery to measure (optically). A platinum-rhodium thermocouple melted in this reactor in a vain attempt to measure the temperature directly; the high radiance reactor environment, whether blackbody or not, is also a difficult environment for temperature measurements.

The carbon continuum was absent in the spectrum of a burner different from a stirred reactor. Radiant emittance of an injector system combusting methane/oxygen injected onto the surface of ytterbia felt (Figure 8) showed the characteristic strong peak of the selective emitter and also illustrates the almost textbook-looking spectrum of hydrocarbon burner exhaust gases. The molecular emission bands of the exhaust gases did contribute significantly to the longer wavelength radiation. This can be an important consideration in narrow band system design. These results indicate that, regardless of the combustion conditions or burner type, the selective emitter spectrum will be accompanied by some degree of longer wavelength radiation. At the higher temperatures of methane-oxygen combustion the selective emitter peak amplitude should far exceed that of the additional contributions due to combustion. Lower temperature air breather systems must be analyzed carefully. The use of a filter, specifically a dielectric or plasma reflector/filter, would improve the system operation by reflecting the longer wavelengths back through the emitter heating the combustion gases to higher temperatures. From a system point of view the reflector is advantageous, the unreflected longer wavelength radiation would unnecessarily heat the PV cells thereby reducing their efficiency.

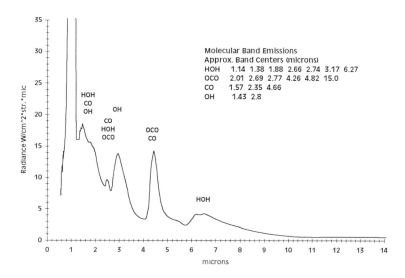

Figure 8. Full face injector spectrum showing combustion product emission bands

To a first approximation, the additional contributions in the long wavelength spectra are independent of the selective emitter peak emission. They are characteristic of the particular combustion design. The energy balance concerning the selective emitter addresses the emitter being heated and radiating energy.

Whereas, Figure 4 shows the total radiant output as a function of the input flow rate, Figure 9 is a power in-power out comparison. The methane input flow is converted to watts (normalized to the area of the radiating emitter). A silicon detector measurement and configuration factor yield the total radiant emittance (normalized to the area of the radiating emitter) output and this is what the PV cells can convert to power. Therefore, this data presents the radiant emittance of interest as a function of input energy. This measurement succeeds in eliminating the additional long wavelength contributions measured previously. The overall efficiency indicated (2 to 6%) includes the photon to watt efficiency of the silicon detector, not the silicon concentrator PV cells used in the generator. However, within the scatter of the data, the two efficiencies are the similar (4.5% and 2 to 6%). The normalization to the area of the emitter surface ideally allows extrapolation to any desired scale generator.

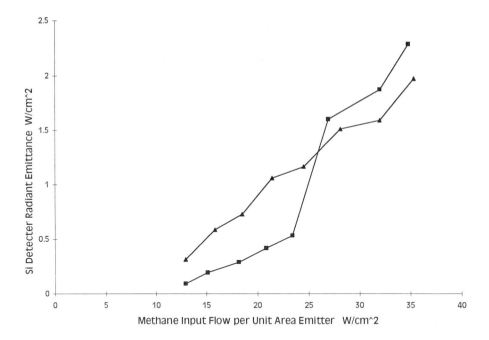

Figure 9 Stirred reactor power input vs. radiant output in silicon responsivity spectrum

The Apparent Temperature Exponent of Selective Emitters

Experimental evidence has suggested that selective emitters have a temperature dependence that exceeds that of a blackbody (ref. 4). The exact dependence is valuable for modeling calculations, thereby quantifying trends of spectral emittance with temperature. The temperature dependence of the emissive power or irradiance can be expressed as

$$\frac{e_2}{e_1} = \left(\frac{T_2}{T_1}\right)^n \qquad (9)$$

where n is the apparent temperature exponent. We say apparent temperature exponent because it will be seen below that n is not a constant, but rather a function of the wavelength, temperature, wavelength interval of the selective emitter, and temperature range.

From Eq. (13) the apparent temperature exponent can be expressed as

$$n = \frac{\ln(e_2/e_1)}{\ln(T_2/T_1)} \qquad (10)$$

The emissive power e is given by the integration of the spectral distribution of emissive power, to wit,

$$e = \int_{\lambda_1}^{\lambda_2} e_\lambda d\lambda = \int_{\lambda_1}^{\lambda_2} \varepsilon_\lambda e_{\lambda b} d\lambda = \bar{\varepsilon} \int_{\lambda_1}^{\lambda_2} e_{\lambda b} d\lambda \qquad (11)$$

where ε_λ is the spectral emissivity, $\bar{\varepsilon}$ is the average emissivity, and $e_{\lambda b}$ is the Planck spectral distribution of emissive power for a blackbody given by

$$e_{\lambda b}(\lambda, T) = \frac{C_1}{\lambda^5 \left(e^{C_2/\lambda T} - 1\right)} \qquad (12)$$

Here, C_1 and C_2 are the first and second radiation constants, respectively ($C_1 = 2\pi hc^2 = 3.7413 \times 10^{-12}$ and $C_2 = hc/k = 1.4338$).

Assuming that the average emissivity is independent of temperature, the apparent temperature exponent can be expressed in terms of the blackbody emissive power and the temperature only, i.e.,

$$n = \frac{\ln\left(\int_{\lambda_1}^{\lambda_2} e_{\lambda b}(\lambda, T_2) d\lambda \Big/ \int_{\lambda_1}^{\lambda_2} e_{\lambda b}(\lambda, T_1) d\lambda\right)}{\ln(T_2/T_1)} \quad (13)$$

or

$$n = n(\lambda, T; \Delta\lambda, \Delta T) \quad (14)$$

where λ is the (centerline) wavelength of the selective emitter and T is the (average) temperature. Thus, the apparent temperature exponent is seen to be a function of the wavelength interval and temperature range as well as the wavelength and temperature.

In general, Eq. (13) must be solved numerically. Solutions were carried out for a typical selective emitter (Ytterbia: $\lambda = 985$ nm, $\Delta\lambda = \pm 100$ nm) over a temperature range of 800-2800°K with $\Delta T = \pm 100$°K. The results are shown in Figure 10.

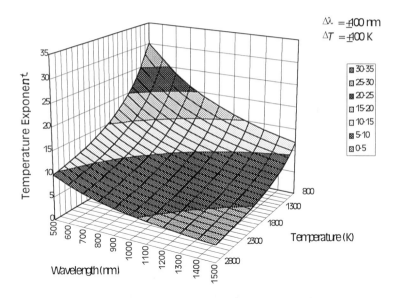

Figure 10. Calculation of the Apparent Temperature Exponent for Ytterbia.

Equation (13) can be solved analytically for a small wavelength interval and temperature range. Consider a wavelength interval of $d\lambda$ and a temperature range of dT such that

$$T_2 = T_1 + dT$$
$$e_2 = e_1 + de \qquad (15)$$

so that Eq. (10) can be expressed as follows (drop the subscript 1 for simplicity)

$$n = \frac{\ln(1 + de/e)}{\ln(1 + dT/T)} \approx T \frac{d \ln e}{dT} \qquad (16)$$

since $\ln(1+\varepsilon) = \varepsilon + O(\varepsilon^2)$. The blackbody emissive power in a small wavelength interval can be expressed as

$$e_b = \int_\lambda^{\lambda+d\lambda} e_{\lambda b} d\lambda = e_{\lambda b} d\lambda = \frac{C_1 d\lambda}{\lambda^5 \left(e^{C_2/\lambda T} - 1\right)} \approx \frac{C_1 d\lambda}{\lambda^5 e^{C_2/\lambda T}} \qquad (17)$$

where it has been assumed that $C_2/\lambda T \gg 1$ or $\lambda T \ll 1.44 \cdot 10^7$ nm°K. Substituting this in Eq. (10) there follows:

$$n = \frac{C_2}{\lambda T} = \frac{hc}{k\lambda T} \qquad (18)$$

Notice that this result depends only on the product of wavelength and temperature. We have found that this simple result gives very good approximations to the exact integration of Eq. (17) for a wide range of $\Delta\lambda$ and ΔT.

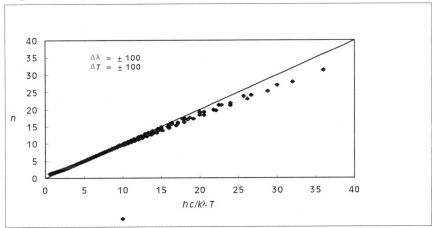

Figure 11. Apparent temperature exponent for ytterbia as a function of λt.

Figure 11 shows the results of Figure 10 replotted as a function of the single variable $hc/k\lambda T$. The approximation becomes progressively worse with decreasing hc/kλT or increasing λT because the assumption that $C_2/\lambda T >> 1$ is no longer valid. The approximation also breaks down with increasing hc/kλT or decreasing λT because the assumptions that $\Delta\lambda/\lambda$ and $\Delta T/T << 1$ are no longer valid. However, the approximate analysis is valid over wide range of λT values which are relevant to combustion systems with selective emitters.

CONCLUSIONS AND RECOMMENDATIONS

It is concluded that :

1. Quantum's TPV technology has achieved multi-kilowatt (2.4 Kw) power generation, proving the feasibility of the design concepts.

2. Quantum's technology for preparing selective emitter materials can provide emitter/burners capable of accommodating combined methane/oxygen flow rates of up to 525 SCFH during continuous operation at temperatures of 2000°C. (The present structure failed at flows above 630 SCFH.)

3. The use of SunPower (Sunnyvale CA) PV cells in the generator demonstrated the commercial availability of silicon concentrator cells capable of withstanding TPV environments.

4. The maximum power output of the laboratory generator can be improved by addressing sub-optimized generator features.
    - A. the emitter porosity/thickness optimization may result in as much as 10% improved power output,
    - B. recuperation of the exhaust gas heat can result in as much as 35% improvement,
    - C. the elimination of reactor carbon particle concentration may result in a few percent improvement,
    - D. the optimization of the fuel/oxygen injection and mixing in the reactor may result in a few percent improvement ,
    - E. the use of mirrors on the end caps would increase photon collection improving the output a few percent.

Optimizing all aspects of the design should result in a conservatively estimated 25-30% improvement in power output.

It is recommended that the following design improvements to the base system be undertaken:

1. the ytterbia felt porosity be optimized for the flow rates of the fuel/oxidizer in a particular burner so as to reduce the pressure drop and the temperature gradient across the reactor wall thickness

2. the long term strength and emittance of ytterbia fiber be evaluated under high temperature conditions.

3. the burner carbon particle phenomena be studied to gain better understanding of its cause and impact.

4. the fuel/oxidizer injection system be optimized for complete mixing.

5. work proceed on optimizing recuperation of energy from the burner exhaust gases.

ACKNOWLEDGMENTS

We are grateful for the funding provided by the U.S. Department of Defense, Advanced Research Projects Agency under contract # MDA972-93-C-0042 and for the guidance provided by the program manager, Dr. Bob Rosenfeld.

REFERENCES

1. D.C.White & H.C.Hottel, "Important Factors in Determining the Efficiency of TPV Systems, Part I" pg. 425, , and H.C.Hottel, "Part II, Radiative Transfer Efficiency of a Flat Plate TPV System: Analytical Model and Numerical Results" pg. 437, The First NREL Conference on Thermophotovoltaic Generation of Electricity, 1995, Am. Inst. of Phys., New York..

2. I. Glassman, Combustion, pg. 116-19, Academic Press, New York, 1977

3. G.A..Holmquist, "TPV Power Source Development for an Unmanned Undersea Vehicle," in The First NREL Conference on Thermophotovoltaic Generation of Electricity, pg. ###, 1995, AIP Conference Proceedings 321, Am. Inst. of Phys., New York.

4. Engineering Design Handbook, Infrared Military Systems, K.Seyrafi ed., AMC Pamphlet No. 706-127 May 1973, Department of the Army, Headquarters U.S.Army Materiel Command, Alexandria, VA

APPENDIX A

Configuration Factor

Configuration factors, also called view and angle factors are the tools by which radiation heat transfer is measured in an absolute sense. Consider, a finite surface radiating energy and some fraction of this intercepts another finite surface. The second surface is measuring the energy and with the accurate determination of the fraction intercepted, the total radiance from the first surface can be determined. The configuration factor is then, by definition, the ratio of the total radiant energy output by one surface to the energy received by another. To get configuration factors solutions to difficult integral solutions are necessary, however, today's PC's can accomplish the required numerical analysis.

For our case, assume uniform diffuse energy leaving the selective emitter surface where the emitter is a cylinder with the collector being a rectangle of equal length parallel to the cylinder or an element on the rectangular parallel plane. In ASME Papers A, Paper No. 56-A-144, H. Leuenberger and R. Person,"Compilation of Radiation Shape Factors For Cylindrical Assemblies", ASME Annual Meeting 1956, New York, NY there is reported the factor for a differential element b on the rectangular plane radiating on a cylinder c (F_{b-c}). Refer to the element b as "d" for the detector in the following discussion, such that, F_{b-c} is F_{d-c}. The computation of configuration factors involves integration over the solid angles by which surfaces can view each other. Because these integration's can be tedious and require numerical evaluation, it is desirable to use relations that already exist for configuration factors. The result is the ability to find some elementary shapes and their factors in the literature. The factor needed in this particular application is that for a cylinder radiating onto a detector. We have the opposite, a finite element of a plane (detector) radiating on a cylinder, F_{d-c}; however, reciprocity exists between the two, weighted with the areas such that:

$$F_{c-d} = \frac{A_d}{A_c} F_{d-c} = \frac{1}{A_c} \int_0^d F_{d-c} dA_d \qquad (A1)$$

Therefore, the configuration factor by L&P (shown below) need only be integrated over the area of the detector and multiplied by the appropriate constant. This was done to calculate the configuration factors used for the compilation of data.

$$F_{b-1} = \frac{SR}{S^2 + x^2} \left[2 - \frac{1}{\pi} \left[\cos^{-1} A + \cos^{-1} B - \frac{y}{R} \left(C \cos^{-1} D \right) - \left(\frac{L-y}{R} \right) \left(E \cos^{-1} F \right) + \frac{L}{R} \cos^{-1} \frac{R}{\sqrt{S^2 + x^2}} \right] \right] \qquad (A2)$$

$$A = \frac{y^2 - S^2 - x^2 + R^2}{y^2 + S^2 + x^2 - R^2} \qquad (A3)$$

$$B = \frac{(L-y)^2 - S^2 - x^2 + R^2}{(L-y)^2 + S^2 + x^2 - R^2} \qquad (A4)$$

$$C = \frac{y^2 + S^2 + x^2 + R^2}{\sqrt{(y^2 + S^2 + x^2 - R^2)^2 + 4y^2 R^2}} \qquad (A5)$$

$$D = \frac{R(y^2 - S^2 - x^2 + R^2)}{\sqrt{S^2 + x^2}\left(y^2 + S^2 + x^2 - R^2\right)} \qquad (A6)$$

$$E = \frac{(L-y)^2 + S^2 + x^2 + R^2}{\sqrt{\left[(L-y)^2 + S^2 + x^2 - R^2\right]^2 + 4(L-y)^2 R^2}} \qquad (A7)$$

$$F = \frac{R\left[(L-y)^2 - S^2 - x^2 + R^2\right]}{\sqrt{S^2 + x^2}\left[(L-y)^2 + S^2 + x^2 - R^2\right]} \qquad (A8)$$

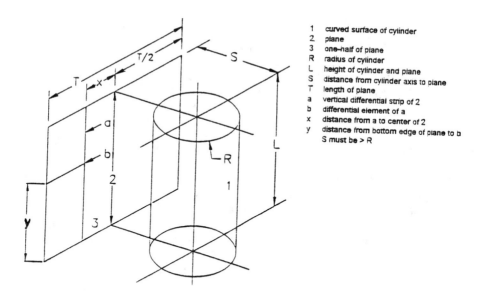

1   curved surface of cylinder
2   plane
3   one-half of plane
R   radius of cylinder
L   height of cylinder and plane
S   distance from cylinder axis to plane
T   length of plane
a   vertical differential strip of 2
b   differential element of a
x   distance from a to center of 2
y   distance from bottom edge of plane to b
S must be > R

Figure A1   Dimensional Layout for the Configuration Factor

APPENDIX B

Materials Processing, Fracture Mechanics & Grain Growth

Considering the fracture mechanics of the fibers, once microcracks are formed, the stress fields at their tips induce plastic deformation in the adjacent grain if the material is ductile. Crack growth will be slow, since the stress field can be relieved by plastic flow. For semibrittle materials, or in general when flow can take place only at high stress levels, stress builds up at the grain boundary until the strength of the solid is exceeded, and fracture takes place. Assuming the slip-band length is proportional to the grain size d, the fracture stress, σ_f, for this case may be expressed with the Petch relation (ref. B1) as:

$$\sigma_f = \sigma_o + k_1 d^{-1/2} \quad (B1)$$

In cases where the sizes of the initial flaws are limited by the grains and scale with the grain size in which brittle behavior is observed, the Orowan relation can be applied such that the strength should vary with grain size (ref. B1) as:

$$\sigma_f = k_2 d^{-1/2} \quad (B2)$$

In both cases, it is important to note that the fracture strength is inversely proportional to the square root of the grain size and therefore predicts that the strength should increase with decreasing grain size of the polycrystalline array.

When grains grow to such a size that they are nearly equal to the specimen size, grain growth is stopped. In a fiber, for example, when the grain size is equal to the fiber diameter, the grain boundaries tend to form flat surfaces normal to the axis so that the driving force for grain boundary migration is eliminated, and little subsequent grain growth occurs. Similarly, inclusions increase the energy necessary for the movement of a grain boundary and inhibit grain growth.

Whether or not primary recrystallization occurs, an aggregate of fine-grained crystals increases in average grain size when heated at elevated temperatures. The driving force for the process is the difference in energy between the fine grained material and the larger grain size product resulting from the decrease in grain boundary area and the total boundary energy. If all the grain boundaries are equal in energy, they meet to form angles of 120°. If we consider a two-dimensional example for illustrative purposes, angles of 120° between grains with straight sides can occur only for six-sided grains. Since grain boundaries migrate toward their center of curvature, grains with less than six sides tend to grow smaller, and grains with more than six sides tend to grow larger. For any one grain, the radius of curvature of a side is directly proportional to the grain diameter, so that the driving force, and therefore the rate of grain growth, is inversely proportional to grain size, d, (ref. B2):

$$d(d)/dt = k/d \tag{B3}$$

Inclusions increase the energy necessary for the movement of a grain boundary and inhibit grain growth. The boundary energy is decreased when it reaches an inclusion proportional to the cross-sectional area of the inclusion. The boundary energy must be increased again to pull it away from the inclusion. Consequently, when a number of inclusions are present on a grain boundary, its normal curvature becomes insufficient for continued grain growth after some limiting size, d_l, is reached (ref. B2):

$$d_l \cong d_i/f_{di} \tag{B4}$$

where d_i is the particle size of the inclusion and f_{di} is the volume fraction of inclusions. Although the relation is approximate, it indicates that the effectiveness of inclusions increases as their particle size is lowered and the volume fraction increases.

Although small grains are desirable for high strength, larger grains are desirable for creep resistance. The creep rate may be expressed with the Nabarro Herring relation where creep rate is proportional to d^{-2} or with the Coble relationship where creep rate is proportional to d^{-3}. In either case, it is clear that creep rate increases with decreasing grain size and therefore that there must be an optimal grain size for a dynamic balance between creep rate and strength. Although crystalline grains are desirable, it is clear that control of the grain size is crucial, particularly at the high operating temperature required for a TPV system. Control of grain growth is expected to provide greater durability in the use of emitter materials at high temperature over long periods of time by maintaining microstructural characteristics and therefore, mechanical properties.

B1   W.D.Kingery, H.K.Bowen, and D.R.Uhlmann, Introduction to Ceramics, John Wiley and Sons, New York, 1976, page 794

B2   W.D. Kingery, H.K. Bowen, and D.R. Uhlmann, Introduction to Ceramics, John Wiley & Sons, New York, 1976, pages 452-457

EXTENDED USE OF PHOTOVOLTAIC SOLAR PANELS

Guido E. Guazzoni* and M. Frank Rose **

* Physical Sciences Directorate
Army Research Laboratory, Ft. Monmouth, NJ
**Space Power Institute
Auburn University, AL 36849

ABSTRACT

The use of photovoltaic solar panels (and related generation of electric power) can be extended to a 24 hours per day under any environmental condition by equipping them with an artificial source of light, with emitting wavelengths matched to the photovoltaic solar panels, to be turned on in the absence of sunlight. This source of light can be obtained by heating a mantle to an incandescent temperature via the efficient, low polluting combustion of Natural Gas, Butane, Propane or other gaseous Hydrocarbon fuel.

INTRODUCTION

When sunlight is available, photovoltaic solar panels are a viable, environmentally clean, free source of electric power. Their use, to directly convert sunlight into electric energy, is increasing with improvements in solar cell conversion efficiency (around 14-15% for commercially available multicrystal Si cells) and with a reduction in cell manufacturing cost.

However, photovoltaic solar panels represent an investment with limited return because the panels can produce electric energy only for a few hours a day when atmospheric conditions permit. Solar panels and all their ancillary components have a limited life and, exposed to the environment, are subjected to both structural and performance degradation no matter if they are generating electric energy in sunlight or are idling in the dark.

To extend the use of photovoltaic solar panels to a complete, uninterrupted 24 hours per day under any environmental condition, the panels can be equipped with an artificial source of light to be turned on when sunlight is not available. This artificial source of light can be obtained by heating a mantle to an incandescent temperature via combustion of Natural Gas, Methane, Propane, Butane or any other gaseous or liquid hydrocarbon fuel. The radiating mantle considered for this study is made of ytterbia, a rare earth oxide selective emitter with a strong emission band centered around .95 µm.

EMITTER-BURNER TUBE SYSTEM

It is well known that the photovoltaic energy conversion obtainable using a selective emitter that radiates in the absorption band of a solar Si cell can be much greater than that obtained using a black/gray body radiator (1). Rare earth oxides, such as erbia or ytterbia, have been proven to be chemically stable, environmentally safe, and not polluting up to temperature close to their melting point (above 2000 °C).

Rare earth oxide selective emitters can be made through a specialized series of processes which begin with nitrates of the rare earth and end with rare earth oxide filament fibers. A special requirement for the mantle to be used in this application is the need for a robust, large area emitter that can be formed in suitable configurations (cylindrical in shape and possibly 15-30 cm long) capable of withstanding shocks and vibrations in both commercial and military applications.

The selective emitter mantle selected for this study is based on a manufacturing technology developed at the Space Power Institute, Auburn University (2). This manufacturing approach utilizes conventional paper making techniques to combine materials suitable as binders with the rare earth oxide fibers. As a result this technique allows for fabrication of large area, highly uniform, and robust sheets of emitter material of variable thickness and porosity, from which emitters can be formed in cylindrical configurations of various sizes. This application also envisions a tubular design air-gas burner, coaxial and an integral part of the cylindrical ytterbia emitter, suitable to be used with an array of commercially available (state-of-the-art) Si cell solar panels.

The solar cell array is comprised of six identical panels (each panel is 30 cm wide and 180 cm long), and engineered to be openly deployable for optimal operation in sunlight (Fig.1a). It is foldable to form a hexagonal enclosure around the burner-emitter tube for maximum light collection, when operation with the gas burner is needed (Fig. 1b).

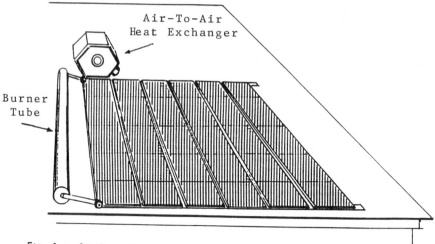

FIG. 1 A SIX-SOLAR PANEL ARRAY FOR OPERATION WITH SUNLIGHT

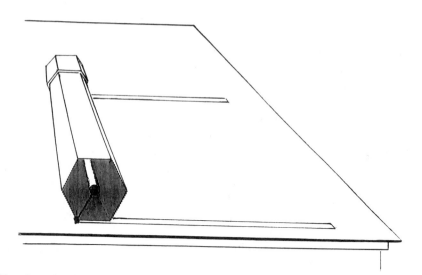

FIG. 1 B SIX-SOLAR PANEL ARRAY FOR OPERATION WITH YTTERBIA EMITTER

FIG. 1 CONCEPTUAL SYSTEM CONFIGURATION

The panel array characteristics and maximum electrical output in sunlight are listed in Table I.

TABLE I. Panel Array Characteristics.

Panel Array Total Surface	= 3.24 m²
Multicrystal Si Cell Efficiency	= 14%
Total Si Cell Active Area	= 2.75 m²
Maximum Array Output in Sunlight	= 404 W

The tubular burner is designed to provide both mechanical support and uniform heating of the ytterbia composite emitter. Fig. 2 depicts a possible configuration

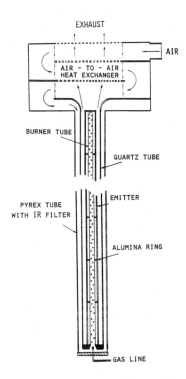

FIG. 2 YTTERBIA EMITTER-BURNER TUBE

for the emitter-burner tube which shall have approximately the same length as the solar panels and be engineered to be raised and operated at the central axial point of the hexagonal enclosure formed by the folding of the solar panels. The folding of the panels can be done manually or automatically operated via light or voltage sensors and electric motors. When positioned for operation, the emitter-burner tube latches on to an air-to-air heat exchanger (also depicted in Fig. 2). Fig. 3 is a conceptual design of the lower section of the emitter-burner tube that shows the gas line entrance, a venturi section for the primary air for combustion and the multiperforated burner tube. Combustion takes place in the space between the burner tube and the emitter inside surface. The combustion gases are forced through the emitter's porosity and leave from the emitter external surface. A concentric quartz tube of appropriate diameter surrounds the emitter to confine the combustion gases and channel them to one end of the emitter-burner tube. Here they enter the heat exchanger and preheat the incoming air for combustion. A larger concentric pyrex tube provides an annular passage for the preheated air which, traveling down counterflow with the combustion gases, reaches the other end of the emitter-burner tube to mix with the incoming fuel. An IR reflecting filter is deposited on the outside surface of the pyrex tube.

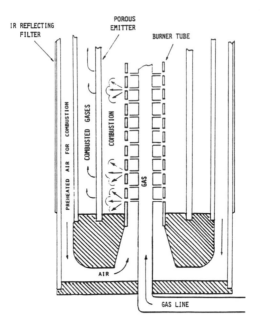

FIG. 3 EMITTER-BURNER TUBE LOWER SECTION

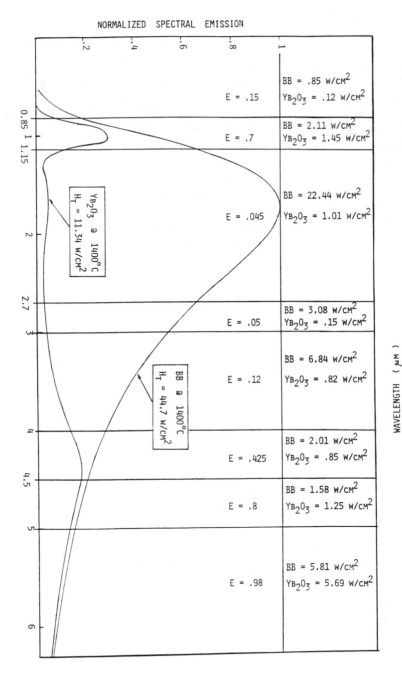

FIG. 4 SPECTRAL DISTRIBUTION OF THE RADIATION FROM BLACKBODY AND YTTERBIA (Yb_2O_3) COMPOSITE EMITTER

SYSTEM ANALYSIS

The special distribution of the radiation from a ytterbia composite emitter developed at the Space Power Institute and from a blackbody, both at 1400 °C, are presented in Fig. 4.

The two emission spectra are normalized to the peak value of the blackbody which occurs at approximately 1.73 µm, and are divided in eight wavelength ranges characterized by different values of the emittance of the ytterbia composite emitter.

Table 2 lists the value of the radiation density emitted by the blackbody and the ytterbia composite emitter in the eight wavelength ranges.

TABLE II. Density of the Radiation emitted by Blackbody and Ytterbia Composite Emitter at 1400 °C

Wavelength Range (µm)	Blackbody Radiating Density W/cm^2	Ytterbia Emitter Radiating Density W/cm^2
0 - 0.85	0.85	0.12
0.85 - 1.15	2.11	1.45
1.15 - 2.7	22.44	1.01
2.7 - 3.0	3.08	0.15
3.0 - 4.0	6.84	0.82
4.0 - 4.5	2.01	0.85
4.5 - 5.0	1.58	1.25
5.0 - ∞	5.81	5.69

The larger amount of radiation emitted above 4 µm by the ytterbia emitter is due to the high emittance value ($\cong$ 1) of the silicon dioxide fibers present in the emitter as structural material (Fig. 5).

The transmission of the quartz and pyrex tubes for radiation in the 0.3 - 3 µm wavelength range is slightly less than 100%.

Fig. 6 is a typical spectra transmission, including Fresnel reflection loss, of a commercial available quartz tube. In this wavelength band both pyrex and quartz absorb approximately 6% of the incident radiation. The remaining 94% is transmitted and reaches the IR reflecting filter.

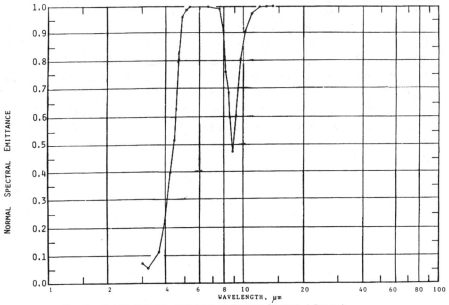

FIG. 5 NORMAL SPECTRAL EMITTANCE OF SILICON DIOXIDE (FUSED)

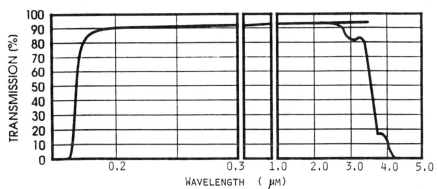

FIG. 6 TRANSMISSION OF QUARTZ

The filter selected for this application is based on a simplified combination of a transparent conductor (thin gold film), and two dielectric layers, one on each side, tailored to provide a high transmission below 1.2 µm and highly reflectivity (>99%) above 4 µm. The filter uses commercial available materials and a simple, inexpensive fabrication approach resulting in a low cost component essential for this specific application. The filter design relies on the thin gold film to provide reflectivity at the longer wavelengths. The cutoff is relatively slow but it falls in the 1.2 - 3 µm wavelength range where the emission of the ytterbia composite emitter is significantly low. Fig. 7 shows the transmission characteristic of this simple thin gold film-2 dielectric layer filter (3).

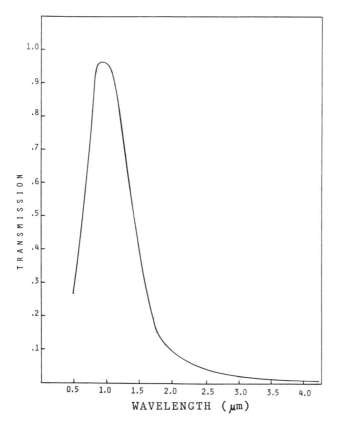

FIG. 7 TRANSMISSION OF THIN GOLD FILM-2 DIELECTRIC LAYER FILTER

The filter transmission characteristics applied to the 1400 °C spectral emission of the ytterbia emitter allow approximately 82% of the radiation in the 0 - 1.15 µm band (Si cell usable band) to reach the solar panel array. For a cylindrical emitter with a surface of 2260 cm^2 this radiation amounts to 2850 W.

Quartz shows "zero" transmission for radiation with wavelengths above 4 µm. A considerable fraction of the incident radiation above 4µm is reflected by the quartz inside surface back to the emitter and reabsorbed. The remainder of this radiation is absorbed by the quartz and reemitted in both directions, inward back to the emitter that, because of his high absorptivity in this wavelength region, reabsorbs it almost completely and forward to the pyrex tube which is in front of the IR. reflecting filter.

The same assumptions and rational can be used for the forward radiation which impinges on the inside surface of the pyrex tube. Therefore, both quartz and pyrex tubes act as reflectors for a large fraction of the energy emitted beyond 4µm, with a remaining small fraction reaching the IR reflecting filter.

Table III summarizes the processing of the total energy radiated by the ytterbia composite emitter over the entire spectrum. In Table III it is assumed that 75% of the radiated energy beyond 3µm is reflected back to the emitter surface by the combined action of the quartz, the pyrex, and the IR reflecting filter. The emitter reabsorbs this energy in accordance with the corresponding value of its spectral absorptivity ($\alpha = \varepsilon$).

From Table III we can see that only a fraction (11872 W) of the radiation reflected back to the emitter (16020 W) is reabsorbed. The fraction not reabsorbed (4148 W) is dispersed inside the system. Some of this energy is absorbed and carried away by the combustion gases and by the preheated air.

Because the emitter reabsorbs only 11872W, it needs a continuous supply of 13756 W from the combustion to remain in thermal equilibrium at 1400°C.

We define as emitter efficiency (EE) the ratio of the energy absorbed by the emitter from the combustion to the total energy generated by the combustion.

With an EE = 0.2 (20% of the thermal energy generated by the combustion is absorbed by the emitter, and reemitted as radiation), the supply of energy to be provided by the combustion to maintain the system in operation amounts to 68780 W and, with a fuel combustion efficiency of 98%, the needed fuel heat content amounts to 70183 W.

The energy carried away by the combusted gases is equal to the fraction of the combustion energy not used/absorbed by the emitter (68780W - 13756W = 55024W) plus the not-reabsorbed fraction of the radiation reflected back to the emitter (4148W). The energy in the exhaust amounts to 59172 W.

Table III. Evaluation of the Transfer of the Radiation Emitted by Ytterbia Composite Emitter

Spectral Distribution of the Energy Radiated by Ytterbia Composite Emitter 1400 °C
Cylindrical Emitter Surface = 2260 cm²

Wavelength Range (μm)	Radiated Energy W	Burner-Emitter End Losses W	Radiation Incident on Quartz Tube Inside Surface W	Radiation Absorbed and Reflected by Quartz Tube %	W	Radiation Absorbed and Reflected by Pyrex Tube %	W	Radiation Incident on IR Reflecting Filter W	Radiation Transmitted by IR Reflecting Filter %	W	Radiation Absorbed by IR Reflecting Filter %	W	Radiation Reflected by IR Reflecting Filter %	Radiation that Reaches the Ytterbia Emitter Surface W	Radiation Reabsorbed by the Ytterbia Composite Emitter α=ε	W	Radiation not Reabsorbed by Emitter W
(0.00 - 1.15)	3548	57	3491	6	209	6	197	3085	92.4	2850	4.5	139	3.1	95	0.7	66	29
(1.15 - 2.00)	1369	21	1348	6	81	6	76	1190	45	535	8	95	47	559	0.05	25	534
(2.00 - 2.70)	913	15	898	6	53	6	51	795	3	23	5	40	92	731	0.05	33	698
(2.70 - 3.00)	339	6	333	6	20	6	19	294	3	9	3	9	94	276	0.05	14	262
(3.00 - 4.00)	1853	30	1823					75% of this radiation is reflected back by the combined action						1367	0.12	164	1203
(4.00 - 4.50)	1921	31	1890											1417	0.43	602	815
(4.50 - 5.00)	2825	45	2780											2085	0.8	1668	417
(5.00 - ∞)	12859	205	12654											9490	0.98	9300	190
(0 - ∞)	25628	410	25218											16020		11872	4148

A fraction of the energy in the exhaust can be recovered via the preheating of the primary air for combustion. However, the recuperation of this energy is limited because of two major reasons: a) the quartz tube (Vycor, Suprasil, etc.) that confines the combusted gases and channels them to the heat exchanger has a maximum working (continuous) temperature of approximately 1200 °C. Therefore, the combustion gas that leaves the emitter surface at a temperature higher than 1400 °C must be expanded and cooled down to temperature below the 1200°C range, and b) the working temperature of the IR reflecting filter considered for this application is limited to values in the 350-400 °C range. The primary air for combustion can not be preheated to temperatures above these values.

With the design configuration, the low cost components, and the fabrication approaches selected the amount of waste energy recoverable via the preheating of the primary air is limited to 30-40 % of the total energy contained in the exhaust.

SYSTEM EFFICIENCY

Because of better matching of the Si cell spectral response with the spectral distribution of the ytterbia radiation in the 0.3-1.15 μm wavelength band than with sunlight, conversion efficiencies in the 20 -30 % range are expected even with commercial (multicrystal) solar cells. Table IV summarizes the radiation density and the electrical power output of the six-panel array folded in a hexagonal enclosure around the ytterbia composite emitter radiator.

Table IV. Six Panel Enclosure Power Output

Total Radiation on Panel Array Total Surface	= 3478 W
Panel Array Total Area	= 3.24 m^2
Radiation Density on Panel Array (compatible with maximum sunlight)	= 1073 W/m^2
Radiation Usable by Si Cell (0 - 1.15 μm)	= 2850 W
Si Cell Active Area (85% of Panel Area)	= 2.75m^2
Si Cell Usable Radiation on Cell Active Area	= 2422 W
Panel Array Electrical Output:	
Si Cell Conversion Efficiency 20 %	= 485 W
Si Cell Conversion Efficiency 25 %	= 605 W
Si Cell Conversion Efficiency 30 %	= 726 W

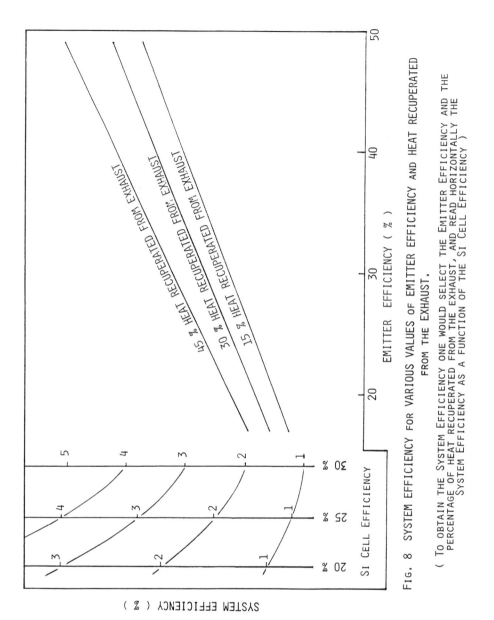

FIG. 8 SYSTEM EFFICIENCY FOR VARIOUS VALUES OF EMITTER EFFICIENCY AND HEAT RECUPERATED FROM THE EXHAUST.

(TO OBTAIN THE SYSTEM EFFICIENCY ONE WOULD SELECT THE EMITTER EFFICIENCY AND THE PERCENTAGE OF HEAT RECUPERATED FROM THE EXHAUST, AND READ HORIZONTALLY THE SYSTEM EFFICIENCY AS A FUNCTION OF THE SI CELL EFFICIENCY)

Fig.8 shows the overall system efficiency (electrical output divided by the fuel heat content) for three values of Si cell conversion efficiency. The system efficiency is plotted for various emitter efficiencies and amounts of heat recuperated from the exhaust.

CONCLUSION

The thermovoltaic energy conversion application analyzed is an example of a system based on established (commercially available) photocell technology, selective emitter composite experimental data, realistic structure design, and practically achievable configurations.

The assumption of a relatively low amount of heat recuperable that strongly impacts on the electrical overall system efficiency is based on the conservative assumption of operational limitations suggested by the use of state-of-the-art components such as commercial pyrex and quartz tubes, a low cost IR reflecting filter, etc. Their selection and related limited overall performances (working temperature) are imposed by realistic requirements of cost effectiveness and already demonstrated reliability.

The proposed extended use of photovoltaic solar panels also represents an examples of the thermophotovoltaic principle as a dual use technology with probably higher utilization potential and impact on the private sector than in the military. The amount (larger fraction) of "waste heat" not recoverable by the preheating of the primary air for combustion can be used in winter time to heat residential homes, industrial space, military and commercial shelters, etc., and to provide hot water all year. In this way, with the simple addition of a second section to the heat exchanger, the overall efficiency and cost effectiveness of the fuel used are drastically increased and determined by the ability to use the "waste heat" for other purposes than just for generation of electricity, with potential extension of the photovoltaic solar panel market to northern climates where their use is now strongly limited.

ACKNOWLEDGMENTS

The authors would like to thank Dr. William Biter, Sensortex Inc. for the helpful discussions and characteristic data on the thin gold film-2 dielectric layer IR filter.

REFERENCE

1. Guazzoni,G.E. "Rare-Earth Oxide Radiators for Thermovoltaic Energy Conversion," R&D Tech.Rep ECOM-3116 April 1969.

2. Adair,P.L. and Rose,M.F. "Composite Emitters for TPV Systems,"Proceedings The First NREL Conference on Thermophotovoltaic Generation of Electricity, 1994

3. Biter,W. Sensortex Inc. Private Communication

Electricity from Wood Powder
Report on a TPV Generator in Progress

Lars Broman*, Kenneth Jarefors*, Jorgen Marks**,
and Mark Wanlass***

*Solar Energy Research Center (SERC), University College of Falun Borlange (UCFB)
Box 10044, S-781 10 Borlange, Sweden

**Department of Operational Efficiency, Swedish University of Agricultural Sciences (SLU)
Herrgardsv 122, S-776 98 Garpenberg, Sweden

***National Renewable Energy Laboratory (NREL)
1617 Cole Blvd., Golden, CO 80401-3393, USA

Abstract: A joint project between NREL, SLU and UCFB aims at building a wood powder fueled TPV generator. The progress of the project is presented.

INTRODUCTION

Thermophotovoltaic conversion of IR radiation to electricity has been extensively studied since it was first suggested by P. Aigrin at MIT in 1960. During the period 1960 - 1993, some 150 papers and reports were published. The strongly renewed interest is obvious from the fact that during 1994 and 1995, so far about 90 more papers have been published (including those presented at this conference). A thermophotovoltaics bibliography with the entries categorized with respect to type and contents has recently been published [1].

A thermophotovoltaic generator of electricity requires efficient TPV cells with a bandgap of 0.55 - 0.6 eV and an emitter at high temperature. The temperature needed is dependent on several selectivity properties of the generator: emitter selectivity, filter selectivity, and backing selectivity of the TPV cell. Realistic assumptions regarding these properties require an emitter temperature around 1500 K.

BIOMASS FUEL AND BURNER

There are some different biomass fuels capable of producing a temperature of 1500 K in the flame, especially gas and liquid fuels of different kinds. When wood is the basis of the fuel, gasification or liquefying however consumes a higher percentage of the initial energy content than grinding and drying. This makes making wood powder, the only sufficiently uniform solid wood fuel, interesting for high temperature combustion [2].

Thus, much of the initial work in the present project has dealt with the problem of burning wood powder in such a way that a stable flame temperature of 1500 K is maintained. In a prototype pilot-scale 10 kW burner that includes feeder and combustion chamber, this goal has been achieved [3]. We have now initiated a project in co-operation with Borlange Energy, Ltd., aimed at installing and testing wood powder burners, rated in the order of 1-10 MW, in the company's existing plants, which produce hot water for the city's district heating systems.

CONVERTERS

Simultaneously, development work is being carried out on other parts of the TPV generator. Four converters for testing have been fabricated at NREL and sent to SERC. They are 0.6-eV $In_{0.65}Ga_{0.35}As$ epitaxially grown lattice mismatched on an InP substrate using metalorganic vapor-phase epitaxy (MOVPE). The front- and back-surface metallizations and sawing were done at Applied Solar Energy Corporation (ASEC) to produce converters with a total area of 1 cm^2. Two-layer antireflection coatings were then applied at NREL. The converters are mounted onto 50×50×2 mm^3 copper plates (electroplated with gold) by means of electrically conductive epoxy. Electrical leads are also affixed to the top grid metallization of the converters using conductive epoxy.

Modeling and characterization of the converters are in progress; internal quantum efficiency data have been measured at NREL and reflection losses have been measured at SERC, typically being about 10%. The diagram in Figure 1 shows absorptance, internal quantum efficiency and external quantum efficiency data for one of the NREL cells. It is worth noting that the reflectance does not increase at wavelengths over that corresponding to the bandgap (~2.0 µm). An efficient edge filter that reflects back the radiation with wavelengths greater than 2 µm to the emitter (or a selective emitter) is therefore necessary for high TPV system efficiency.

Efficiency measurements are also in progress, so far at radiation fluxes far below those of a working TPV generator. Incandescent wolfram spirals in bulbs are used as (greybody) emitters at appropriate temperatures. Very preliminary measurements indicate that the best converter's efficiency increases from 2.2% at a flux of 300 W/m² to 5% or better at a flux of 2 kW/m². This corresponds to an efficiency of approximately 15% for a spectrum where the flux above 2 µm is only 10 % of the blackbody spectrum.

Further testing of the converters is planned. The mounted converters have been configured to allow incorporation into a prototype TPV system based on a wood-powder-fired combustor.

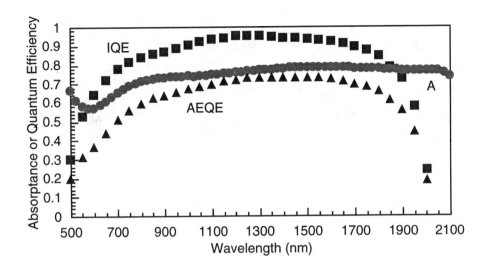

FIGURE 1. Absorptance (A), internal quantum efficiency (IQE), and absolute external quantum efficiency (AEQE) for one of the four NREL converters that are being studied.

OPTICS

Calculations on ideal filters [4] have shown that optimum bandgap E_g of the converter is strongly dependent not only on the (greybody) emitter temperature, but also on the efficiency of the edge filter between the emitter and the converter. As an example, E_g = 0.50 eV is optimum if the filter's reflectivity above the corresponding wavelength is 90 %, while E_g = 0.38 is optimum if the filter reflectivity is only 70 %.

A setup for testing the TPV system concept (IR emitter - selectively reflecting filter - TPV cell) is in the process of being designed. During the coming year, we plan to make and investigate multiple layer dielectric filters. Design of such filters will be facilitated by means of thin film design programs like TFCalc [5]. In order to limit the region of filter incident angles - which will make the edge filter act more efficiently - a special geometry of the internally reflecting tube that transmits the radiation is considered: a tube in the shape of two CPC cones joined at their large apertures and with the filter perpendicular to the CPCs' axes between the cones. As an example, C = 4 cones will make almost all rays reach the filter at incidence angles less than 30°.

CONCLUDING REMARKS

Work at SERC is so far not appropriately financed, since we have not yet been able to convince Swedish National Research Councils about the usefulness of TPV for power production. To be of interest, our R&D efforts have to be directed towards plants in the size of 1 MW$_e$ or larger. The probability of m^2-quantities of TPV cells, with high efficiency and long lifetime, being produced and marketed within the next 5-10 years must be shown to be large if we shall be able to receive any substantial funds.

ACKNOWLEDGEMENTS

Dr. T. J. Coutts is thanked for fruitful discussions. We also thank Jeff Carapella, Anna Duda, and Scott Ward for their contributions at NREL regarding converter fabrication.

REFERENCES

1. Broman, L., Thermophotovoltaics bibliography. *Progress in Photovoltaics* 3(1995)65-74.

2. Marks, J., Wood powder: an upgraded wood fuel. *Forest Prod. J.* 42(1992)52-56.

3. Broman, L. and Marks, J., Co-generation of electricity and heat from combustion of wood powder utilizing thermophotovoltaic conversion. *Proc. First NREL Conf. on Thermophoto-voltaic Generation of Electricity*, PP 133-138 (1995).

4. Broman, L. Calculations of optimum bandgaps of TPV devices for blackbody emitters and selective mirrors. (To be published.)

5. *TFCalc*. A thin film design program marketed by ARSoftware, 8201 Corporate Drive, Landover, Maryland 20785, USA.

Solar Thermophotovoltaic (STPV) System with Thermal Energy Storage

Donald L. Chubb*, Brian S. Good*, and Roland A. Lowe+

*NASA Lewis Research Center, Cleveland, Ohio 44135
+School of Technology, Kent State University, Kent Ohio 44242

Abstract. A solar thermophotovoltaic (STPV) system has both terrestrial and space applications because thermal energy storage can be utilized. Excellent properties (heat of fusion = 1800 j/gm and melting temperature = 1680K) make silicon the ideal thermal storage material for an STPV system. Using a one dimensional model with tapering of the silicon storage material, it was found that several hours of running time with modest lengths (~ 15cm) of silicon are possible. Calculated steady-state efficiencies for an STPV system using an Er-YAG selective emitter and ideal photovoltaic (PV) cell model are in the range of 15% - 17%. Increasing the taper of the storage material improves both efficiency and power output.

INTRODUCTION

Why would anyone be interested in a solar driven thermophotovoltaic (STPV) energy conversion system? The answer is thermal energy storage. With thermal energy storage the STPV system can operate when the solar flux is not available. As a result, STPV energy conversion is of interest for both terrestrial [1,2] and space power [3,4] applications. In the case of terrestrial application it is also possible to add combustion energy input to the system. Therefore, the energy conversion system can operate on energy input from the sun, stored thermal energy or combustion energy. Such a system would then be capable of 24 hour a day operation. Figure 1 is a schematic drawing of such an STPV energy conversion system.

In this paper, we consider the thermal energy storage portion of the system and also the overall performance of the STPV system. A one-dimensional model for the melting and solidifying of the thermal storage material is used to calculate the temperature change across the storage material as a function of time. Also, the overall efficiency and power output, as well as the various component efficiencies are calculated using a systems model described earlier at this conference [5].

THERMAL STORAGE ANALYSIS

As shown in Figure 1, thermal storage material is placed between the cavity receiver and the emitter of the thermophotovoltaic (TPV) system. During the time the solar flux is available the cavity receiver absorbs the solar energy which is then transmitted through the storage material to the emitter. When the storage material is in the solid state some of the input energy will be used in melting the storage material, as well as heating the emitter. There will be negligible radial thermal losses from the storage material if it is well insulated. Therefore, the storage material will begin melting at the receiver and a solid-liquid interface will progress from the receiver to the emitter until all the material is melted. As the solid-liquid interface moves toward the emitter, the emitter temperature T_E will

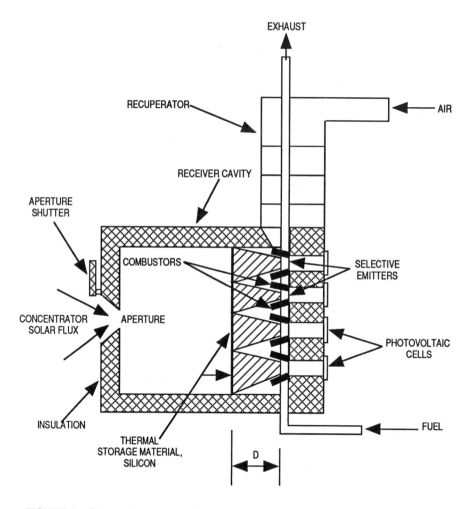

FIGURE 1. Schematic drawing of Solar Thermophotovoltaic (STPV) energy conversion system with thermal storage and auxiliary combustion heating

increase. When all the storage material has melted the emitter temperature will rise to some steady-state operating temperature, T_{ESS}, that will depend on the solar flux input.

When the solar flux is not available the shutter on the receiver is closed (see Fig. 1) and the storage material begins to solidify at the emitter. As the solid-liquid interface moves toward the receiver the emitter temperature slowly decreases until

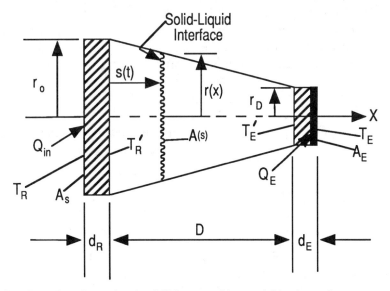

FIGURE 2. One dimensional solidifying - melting model for thermal storage material

all the storage material has solidified. At that time the emitter temperature will quickly drop if no solar flux or combustion energy is applied.

As shown in Figure 1 the storage material is tapered in going from the receiver the emitter. Tapering results in an increase in the thermal flux at the emitter. This has two benefits. First of all, it means the emitter can be maintained at a high temperature for a longer period of time during the solidification phase than for a geometry that has no taper. Secondly, during steady-state operation the temperature difference between the receiver and the emitter is reduced from what it would be for a geometry without taper.

The melting and solidifying of the storage material was modeled with the one dimensional model shown in Figure 2. We assume no radial thermal loss from the thermal storage material so that heat transfer occurs only in the $\pm x$ direction. We also neglect any convective heat transfer in the storage material. Therefore, the overall energy balance is the following:

$$A_s Q_{in} - A_E Q_E = A(S) \rho L_f \frac{dS}{dt} \qquad (1)$$

Where Q_{in} is the thermal input to the storage material at x = 0, Q_E is the energy flux at the emitter, ρ is the storage material density, L_f is the heat of fusion of the storage material, A_s is the thermal storage material area at x = 0, A_E is the emitter area, A(S) is the area of the solid-liquid interface, S is the interface location and t is time. Equation (1) together with the one dimensional heat conduction equation (we neglect convective heat transfer as already stated and assume steady-state conduction occurs on each side of the interface)

$$- k A(x) \frac{dT}{dx} = \text{constant} \qquad (2)$$

must be solved for T(x,t) and S(t). Appearing in Equation (2) is the thermal conductivity, k. The solution to Equation (2) is split into 3 regions.

$$T_R = T'_R + \frac{d_R}{k_R} Q_{in} \qquad -d_R \leq x \leq 0 \qquad (3a)$$

$$T_E = T'_E - \frac{d_E}{k_E} Q_E \qquad D \leq x \leq D + d_E \qquad (3b)$$

Where k_R is the thermal conductivity of the receiver back plate and k_E is the thermal conductivity of the material between the storage material and the emitter. In the tapered region ($0 \leq x \leq D$), the solution to Equation (2) depends on A(x). We considered two cases for A(x). First of all,

$$A(x) = 2 W r(x) = 2W \left| r_o - \frac{r_o - r_D}{D} x \right| \qquad \text{1D taper} \qquad (4a)$$

where W is the width of the storage material, D is the length of the storage material and r_o and r_D are defined in Figure 2. We call A(x) defined by Equation (4a) "one dimensional" taper, (1D), since A(x) = 2 Wr (x) is rectangular and only the dimension perpendicular to the x axis is changing. The second case for A(x) we call "two dimensional" taper, (2D), since A(x) = π [r(x)]² is circular and is changing in two dimensions.

$$A(x) = \pi[r(x)]^2 = \pi\left[r_o - \frac{r_o - r_D}{D}x\right]^2 \quad \text{2D taper} \quad (4b)$$

Notice that for both 1D and 2D taper r(x) is assumed to be a linear function of x.

Using the boundary condition $T = T_c$ at $x = S(t)$, where T_c is the melting temperature, Equation (2) can be integrated using either equation (4a) or (4b) for A(x). When that result is combined with Equations (1) and (3) the following results are obtained for the location of the solid-liquid interface, S(t), and the emitter temperature, T_E. for 1D taper the results are the following.

$$\frac{S(t)}{D} = \frac{1}{R-1}\left\{R - \exp\left[\frac{[R-1]}{D}g(T_E)\right]\right\}, \quad R = \frac{r_o}{r_D} = \frac{A_S}{A_E} \quad (5a)$$

$$\frac{dT_E}{dt} = \frac{Q_E^2}{\rho L_f k_s}\left(\frac{A_S}{A_E}\frac{Q_{in}}{Q_E} - 1\right)\left\{\frac{\exp\left[-\frac{2(R-1)}{D}g(T_E)\right]}{1 + \frac{T_c - T_E}{Q_E}\frac{dQ_E}{dT_E}}\right\} \quad (5b)$$

And for the case of 2D taper the results are the following.

$$\frac{S(t)}{D} = \frac{1}{R-1}\left\{R - \left[1 - \frac{R-1}{D}g(T_E)\right]^{-1}\right\}, \quad R = \frac{r_o}{r_D} = \sqrt{\frac{A_S}{A_E}} \quad (6a)$$

$$\frac{dT_E}{dt} = \frac{Q_E^2}{\rho L_f k_s}\left(\frac{A_S}{A_E}\frac{Q_{in}}{Q_E} - 1\right)\left\{\frac{\left[1 - \frac{R-1}{D}g(T_E)\right]^4}{1 + \frac{T_c - T_E}{Q_E}\frac{dQ_E}{dT_E}}\right\} \quad (6b)$$

Where,

$$g(T_E) = \frac{k_s}{Q_E}(T_c - T_E) - \frac{k_s}{k_E}d_E \quad (7)$$

and T_c is the melting temperature of the storage material, k_s is the thermal conductivity of the solid form of the storage material and k_E is the thermal conductivity of the material between the emitter and storage material. Notice the importance of the taper ratio, R, in determining S(t) and $T_E(t)$. Equations (5) and (6) apply for both melting and solidifying of the thermal storage material. However, during solidification, $Q_{in} = 0$.

Before Equations (5b) and (6b) can be integrated, Q_E and Q_{in} must be known. The energy flux at the emitter, Q_E, is the difference between the emitted plus reflected energy and the radiation incident on the emitter.

$$Q_E = \int_0^\infty \left[\varepsilon_{E\lambda} e_b + r_{E\lambda} q_{Ei} - q_{Ei} \right] d\lambda \tag{8}$$

Where $e_{E\lambda}$ and $r_{E\lambda}$ are the emitter spectral emittance and reflectance e_b is the black body emissive power [6] and q_{Ei} is the incident radiation on the emitter. Since $e_{E\lambda} = 1 - r_{E\lambda}$, equation (8) becomes the following.

$$Q_E = \int_0^\infty \left[\varepsilon_{E\lambda} \left(e_b - q_{Ei} \right) \right] d\lambda \tag{9}$$

Therefore, in order to calculate Q_E both $\varepsilon_{E\lambda}$ and q_{Ei} must be known. The spectral emittance, $\varepsilon_{E\lambda}$, depends on the emitter material and therefore is an input quantity. However, q_{Ei} will be determined by the optical properties of the components (window, filter, PV cells) downstream of the emitter, as well as the emitter optical properties. To simplify the analysis we assume the following.

$$Q_E = \varepsilon_E \sigma_{Sb} T_E^4 \tag{10}$$

Where σ_{sb} (5.67 x 10^{-12}w/cm² K⁴) is the Stefan-Boltzmann constant and ε_E is an "effective" total emitter emittance. Note that if $q_{Ei} = 0$ then ε_E is exactly the total emittance of the emitter. For a TPV system using the rare earth (RE) doped yttrium aluminum garnet (YAG) selective emitter, calculations based on the TPV model of references 3 and 5 indicate that $\varepsilon_E \approx .15$.

For the energy flux input, Q_{in}, the following applies.

$$A_s Q_{in} = \eta_c \eta_R \eta_{th} Q_{SOLAR} A_c \tag{11}$$

Where η_c is the concentrator efficiency, η_R is the receiver efficiency, η_{th} accounts for conductive and convective heat loss from the receiver-storage material combination, Q_{solar} is the solar input (.1w/cm^2) and A_c is the concentrator area. We assume all the efficiencies are constants. Since $\eta_R \sim 1 - T_R^4$ this is not necessarily a good approximation. However, for most cases of interest $T_F(t)$ changes are less than 200K so that η_R = constant will not result in large errors.

In a TPV system the most important parameter for determining performance is the emitter temperature, T_E. Therefore, the length of thermal storage material, D, that can be used will be set by the minimum emitter temperature, T_{EMIN}, that is chosen. This D value can be determined by setting $T_E = T_{EMIN}$ and S=D in Equations (5a) or (6a). Therefore, from Equations (5a) and (6a) the following are obtained.

$$D = \frac{R-1}{\ln R} g(T_{EMIN}) \quad \text{1D taper} \tag{12a}$$

$$D = R g(T_{EMIN}) \quad \text{2D taper} \tag{12b}$$

As Equations (12a) and (12b) show, for given R and T_{EMIN} the 2D taper case will have a larger D value than the 1D case.

SILICON THERMAL STORAGE RESULTS

If a dimensionless temperature, T_E/T_c, and Equation (10) are substituted in Equations (5a) and (6b) then the following parameter multiplies the right hand side of Equations (5b) and (6b).

$$\frac{1}{\beta} = \left(\frac{\varepsilon_E \sigma_{sb} T_c^4}{\rho L_f k_s T_c} \right)^2 \tag{13a}$$

The parameter, ß, has the dimensions of time and will determine the time that $T_E \geq T_{EMIN}$ during solidification of the storage material and the time $T_E \leq T_c$ during melting of the storage material. Rewrite Equation (13).

$$\beta = \frac{(pL_f D)\left(\frac{k_s T_c}{D}\right)}{\left(\overline{\varepsilon}_E \sigma_{sb} T_c^4\right)^2} \tag{13b}$$

The numerator of Equation (13b) is the product of the heat of fusion energy that can be stored and the thermal conduction rate for a thickness D. The denominator is the square of the energy flux leaving the emitter. As Equation (13) indicates, the melting time, t_{MELT}, and solidification time, t_{SOLID}, are very sensitive to the melting temperature and emitter effective emittance ($\overline{\varepsilon}_E$-2). To a lesser extent t_{SOLID} and t_{MELT} depends on the product $\rho L_f k_s$. Therefore, in order to obtain a long running time on the thermal storage material (large t_{SOLID}) a large value for this product is desirable.

The first requirement on the storage material is that T_c must be in the operating range of an efficient TPV system (1500K $\leq T_E \leq$ 2000K). The second requirement is that the product $\rho L_f k_s$ should be large. For the temperature range of interest there are several storage material candidates. However, because of its very large heat of fusion (L_f = 1800 j/gm) and favorable melting temperature (T_c = 1680K), silicon (Si), appears to be the ideal candidate. Si also has good thermal conductivity [7] (k_s = .2 w/cmK, k_l = .6 w/cm K). Therefore, for $\overline{\varepsilon}_E$ = .15, the value of ß = 8.6 hr for Si. As a result, it is expected that running times during solidification (t_{SOLID}) of this magnitude are expected.

Silicon results for D and t_{SOLID} for 1D taper are shown in Figure 3 for several taper ratios, R. The value of t_{SOLID} is obtained by the numerical integration of Equation (5b) with Q_{in} = 0 for Si (properties given above) and Q_E given by Equation (10). The initial condition for the integration is $T_E = T_c$ at t = 0. Also, silicon carbide (SiC) with k_E = .15w/cm-K and d_E = .5 cm was assumed for the material between the emitter and storage material. The integration is stopped when $T_E = T_{EMIN}$ and therefore t = t_{SOLID}. For an efficient TPV system it is desirable that $T_E \geq$ 1500K. Therefore, for T_{EMIN} = 1500 Figures (3a) and (3b) show that $10 \leq D \leq 17$ cm and $3 \leq t_{SOLID} \leq 11$ hr depending on the value of R. The largest values of R yield the largest D and t_{SOLID} values.

The time history of the solid-liquid interface location, S, and the emitter temperature, T_E, for both solidification and melting for several 1D taper ratios is

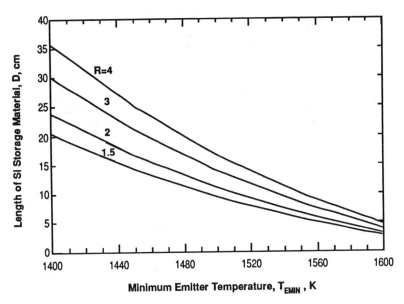

3a) Length of Si thermal storage material, D, resulting in $T_E \geq T_{EMIN}$

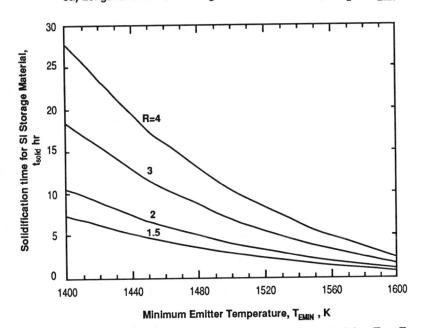

3b) Solidification time, t_{solid} of Si storage material while maintaining $T_E \geq T_{EMIN}$

Figure 3. Silicon thermal storage results for D and t_{solid} for several one dimensional taper ratios, $R = A_s/A_E$. Emitter effective emittance, $\bar{\varepsilon}_E$ = .15, SiC between emitter and Si (k_E .15 w/cm -K and d_E = .5 cm).

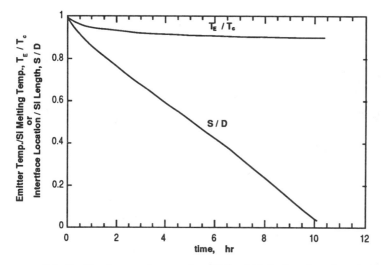

4a) Emitter temperature and solid-liquid interface location during solidification of silicon storage material.

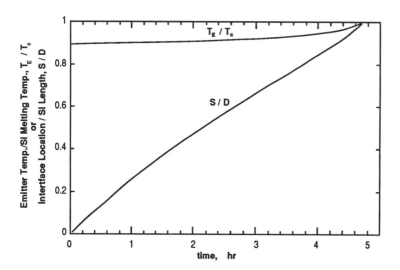

4b) Emitter temperature and solid-liquid interface location during melting of silicon storage material. Concentrator efficiency, $\eta_c = .87$, receiver efficiency, $\eta_R = .9$, thermal efficiency, $\eta_{th} = .95$, solar flux, $Q_{solar} = .1 w/cm^2$ and $A_c/A_E = 200$.

Figure 4. Silicon thermal storage results for T_E/T_c and S/D for one dimensional taper ratio, R = 4. Emitter effective emittance, $\overline{\varepsilon}_E = .15$, $T_{EMIN} = 1500K$, D=17 cm, $T_c = 1680K$, $t_{solid} = 10.4$ hr and $t_{melt} = 4.9$ hr. SiC between emitter and Si ($k_E = .15$ w/cm-K, $d_E = .5$cm).

shown in Figures (4a) and (4b). For the melting case, Equation (5b) was numerically integrated using Equation (11) for Q_{in} and the initial condition $T_E = T_{EMIN}$ at t = 0. The integration was stopped when $T_E = T_c$ at $t = t_{MELT}$. For both melting and solidification, S/D was calculated using Equation (5a). Note that $t_{MELT} \sim 1/2\ t_{SOLID}$. The solid-liquid interface moves with nearly constant speed except at the beginning and end of either the melting or solidifying phases. Also, T_E/T_c remains nearly constant except at the beginning of the solidifying phase or the end of the melting phase.

Similar results to those shown in Figure (3) and (4) are obtained for 2D taper. However, for given values of R and T_{EMIN} the values of D, t_{SOLID} and t_{MELT} are considerably larger for 2D taper than 1D taper. For $T_{EMIN} = 1500$ K and R = 2 the 2D taper results are $t_{SOLID} = 8.5$ hr, D = 15 cm, $t_{MELT} = 4.1$ hr while the 1D taper results are $t_{SOLID} = 4.1$ hr, D = 11 cm, $t_{MELT} = 2.3$ hr. For both 1D and 2D taper, considerable running time during solidification ($t_{SOLID} > 10$ hr) can be achieved with modest amounts (D $\leq$ 20 cm) of Si thermal storage material.

STEADY-STATE STPV PERFORMANCE WITH Si THERMAL STORAGE

The STPV computer model [3,5] can currently be applied to only steady-state operation. It is being modified to apply to the time dependent melting and solidifying phases of operation. However, we have used the model to calculate the steady-state operation when the Si storage material is in liquid form and $T_E = T_{ESS}$. The system used in the calculations is similar to the system being experimentally investigated at McDonnell-Douglas [3]. This system uses an erbium (Er) - YAG thin film selective emitter [8]. A single strong emission band centered about $\lambda_b = 1.55\mu m$ exists for the Er - YAG emitter with greatly reduced emission outside this band. However, as a result of the YAG host material [9] there is large emission for $\lambda > 5\mu m$. Therefore, in the computer model the Er-YAG emitter spectral emittance is approximated by a four band model.

1) $\varepsilon_{uE} = .2 \quad 0 \leq \lambda \leq \lambda_u$ (1.45μm) – below emission band (14a)

2) $\varepsilon_{bE} = .7 \quad \lambda_u \leq \lambda \leq \lambda_l$ (1.65μm) – emission band (14b)

3) $\varepsilon_{lE} = .2 \quad \lambda_l \leq \lambda \leq \lambda_c$ (5.0μm) – above emission band (14c)

4) $\varepsilon_{\infty E} = .9 \quad \lambda_c \leq \lambda < \infty$ – cutoff region (14d)

The values of $\bar{\varepsilon}_E$ have been experimentally determined for $\lambda \leq 3.0$ μm [8] and Equations (14a) - (14c) are approximations to that data.

For the PV cell model it is assumed that the quantum efficiency is one and that the cell bandgap energy, E_g, is matched to the Er-YAG emission band, that is $E_g = hc_o/\lambda_l$, where h is the Planck constant and c_o is the speed of light. As a result, the cell performance will be the maximum possible. We therefore call this the ideal cell model [3,5]. It may first appear that the ideal cell model is a poor approximation. However, the indium gallium arsenide ($In_x Ga_{1-x}$ As) PV cell can be varied in bandgap energy ($.36 \leq E_g \leq 1.4eV$) [10,11] to match the emitter. Also, the quantum efficiency in the region of interest ($\lambda_u \leq \lambda \leq \lambda_l$) is large (>.8) for the $In_{.53} Ga_{.47}As$ ($E_g = .75$ eV(1.65 μm)) cell, which is matched to the Er - YAG selective emitter [10,11]. Therefore, the ideal cell model is not a poor approximation for the Er - YAG selective emitter and $In_{.53}Ga_{.47}As$ PV cell TPV system.

In addition to the ideal PV cell model we also assume that a reflector such as gold is applied to the backside of the PV cell to reflect the long wavelength ($\lambda > \lambda_l$) radiation back to the emitter. Such a cell structure is possible as long as the cell is not strongly absorbing for $\lambda > \lambda_l$. Therefore, for the PV cell the reflectances in the four bands are the following.

1) $r_{uc} = .02$ $0 \leq \lambda \leq \lambda_u$ below emission band   (15a)

2) $r_{bc} = .02$ $\lambda_u \leq \lambda \leq \lambda_l = \lambda_g$ emission band   (15b)

3) $r_{lc} = .8$ $\lambda_l \leq \lambda \leq \lambda_c$ above emission band   (15c)

4) $r_{\infty c} = .8$ $\lambda_c \leq \lambda < \infty$ cutoff region   (15d)

With the emitter and PV cell properties stated above the efficiency, η_T, and output power density, P_{EL}/A_{PV}, where A_{PV} is PV cell area, can be calculated using the STPV system model [3,5]. The efficiency is the product of the component efficiencies.

$$\eta_T = \eta_c \eta_R \eta_{th} \eta_{Ef} \eta_{PV} \qquad (16)$$

The concentrator, η_c, receiver, η_R, and thermal, η_{th}, efficiencies were introduced earlier.

$$\eta_R = 1 - \frac{\sigma_{sb} T_R^4}{\eta_c Q_{SOLAR}} \left(\frac{A_R}{A_c}\right) \qquad (17)$$

Where A_c/A_R is the concentration ratio for the solar concentrator. This model for the receiver assumes the receiver behaves like a black body cavity, (receiver loss = $\sigma_{sb} T_R^4 A_R$). Since this is the maximum possible receiver loss, the model will predict a conservative value for η_R. The emitter-filter efficiency is defined as follows.

$$\eta_{Ef} = \frac{A_{PV} \int_0^{\lambda_g} q_{ic} \, d\lambda}{A_E Q_E} \qquad (18)$$

Where q_{ic} is the incident power on the PV cell, which is calculated by the computer model [5]. The emitter-filter efficiency is the ratio of the "useful" radiation ($\lambda < \lambda_g$) that reaches the PV cell to the thermal power input to the emitter. Finally, the photovoltaic efficiency, η_{PV}, is the following.

$$\eta_{PV} = \frac{P_{EL}}{A_{PV} \int_0^{\lambda_g} q_{ic} \, d\lambda} \qquad (19)$$

In the calculations to be presented we assume $\eta_c = .87$ and $\eta_{th} = .95$. The value for η_c is the expected result for the McDonnell-Douglas concentrator [1,2,3]. If the receiver and thermal storage material are well insulated then the convective and conductive heat transfer losses will be small so that $\eta_{th} = .95$ is a reasonable approximation. The expected concentration ratio, A_c/A_R, for the McDonnell-Douglas concentrator is 5000. This is the value used in the calculations. We also assume that all the radiation leaving the emitter reaches the PV cell and similarly all the reflected radiation leaving the PV cell reaches the emitter, (view factors = 1 and $A_E = A_{PV}$).

From the Si thermal storage results discussed earlier (Fig. 3) a Si length of D = 15 cm will yield reasonable values for running time, t_{SOLID}, during solidification.

Therefore, D = 15 cm was used in the calculations. Also, it was assumed that the material at the ends (see Fig. 2) of the Si is SiC with $k_R = k_E = .15$ w/cm-K and $d_R = d_E = .5$ cm.

As stated earlier, T_E is the most important parameter for determining the performance of a TPV system. In the case of an STPV system it is the ratio of the concentrator area to emitter area, A_c/A_E, that is most important in determining T_E. Therefore, the STPV results for the Er - YAG emitter plus ideal PV cell STPV system shown in Figures 5-8 are plotted as functions of A_c/A_E. Figure 5 shows the receiver and emitter temperatures for several values of the 1D taper ratio. As can be seen, increasing R has the desirable effect of reducing T_R while increasing T_{ESS}. Also the T_R reduction is greater than the T_{ESS} increase. Obviously, material temperature limitations will not permit large values of A_c/A_E.

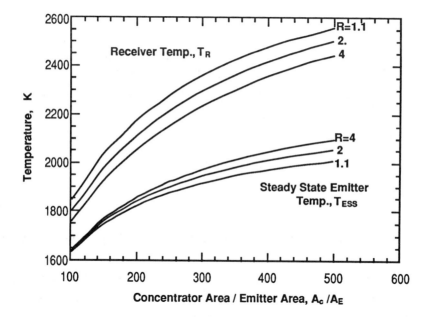

FIGURE 5. Receiver and emitter temperatures for Er-YAG emitter and ideal PV cell for several STPV system 1D taper ratio, $R = A_s/A_E$. Concentration ratio, $A_c/A_R = 5000$, concentrator efficiency, $\eta_c = .87$, thermal efficiency, $\eta_{th} = .95$, length of Si storage material, D=15 cm, emitter emittances, $\varepsilon_{EU} = .2$, $\varepsilon_{Eb} = .7$, $\varepsilon_{EI} = .2$, $\varepsilon_{EOO} = .9$, PV cell reflectances, $r_{cI} = .02$, $r_{cb} = .02$, $r_{cI} = .8$, $r_{coo} = .8$.

In Figure 6 the component efficiencies η_R, η_{Ef} and η_{PV} are shown for two values of R. Increasing R causes the largest improvement in η_R. This occurs because T_R decreases with R and $\eta_R \sim 1 - T_R^4$ (Eq. (17)). The emitter-filter efficiency, η_{Ef}, increases with increasing R because T_E increases with R. There is negligible increase in η_{PV} with increasing R because η_{PV} increases slowly with incident power, q_{ic}. The rapid decrease in η_R with increasing A_c/A_E while η_{Ef} and η_{PV} increase means there will be an optimum A_c/A_E for maximum η_T.

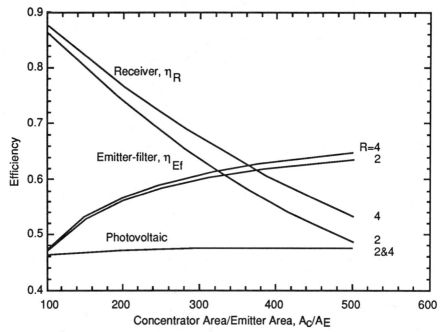

FIGURE 6. Component Efficiencies for several 1D taper ratios, R=A_S/A_E, same conditions as Figure 5

Figure 7 shows the total efficiency, η_T. As can be seen, optimum A_c/A_E is in the range 140 - 200 depending on the taper ratio, R. Also, η_T increases with R. In Figure 8 the output power density is presented. Again, P_{EL}/A_{PV} increases with increasing R. However, for A_c/A_E near the optimum (maximum η_T) the increase in P_{EL}/A_{PV} with increasing R is small.

Based on the results shown in Figures (3, 5, 7, 8) we conclude that using the largest taper ratio, R, was most desirable. It causes the receiver temperature to decrease and the solidification run time, emitter temperature, efficiency and power density all to increase; all desirable results. The limit on R will be determined by the maximum value of A_S that can be tolerated. With A_c/A_E and A_c fixed the only way to increase R is by increasing A_S. However, increasing A_S (the maximum area in the TPV system) will block more of the concentrator central region solar flux.

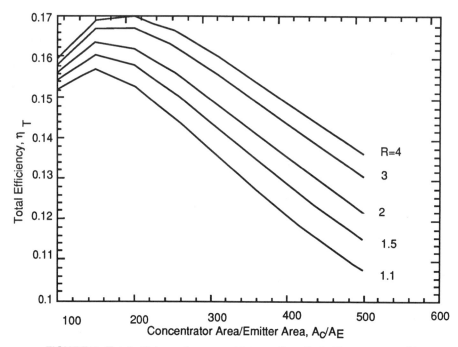

FIGURE 7. Total efficiency for several taper ratios, $R=A_S/A_r$, same conditions as Figure 5

This is an undesirable result since the concentrator is most efficient in the central region. One possible method for alleviating this problem is to change the geometry from linear to radial. In other words, rotate the storage material, TPV system modules 90° in Figure 1 so that the modules are perpendicular to the receiver axis and located around the circumference of the receiver. In this case the receiver diameter can be greatly reduced and the maximum diameter of the system will be determined by D.

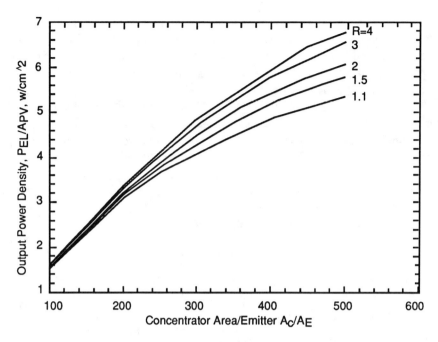

FIGURE 8. Output power density for several taper ratios, $R=A_S/A_E$, same conditions as Figure 5

CONCLUSION

An analysis of an STPV system using silicon as the thermal storage material was completed. The large heat of fusion (1800 j/gm) and desirable melting temperature (1680K) make silicon an ideal thermal storage material for an STPV system. Long running times (~10 hr) with modest lengths (~15 cm) of silicon were calculated using a one dimensional model for the melting and solidifying of the storage material. These results occur because the storage material is tapered from the receiver to the emitter. Taper increases the thermal flux at the emitter and results in several desirable results. Increasing taper ratio results in decreased receiver temperature and increasing solidification run time, emitter temperature, system efficiency and power density.

Steady state performance of the STPV system using silicon thermal storage and an Er-YAG selective emitter plus ideal PV cell model TPV system was calculated. These results show there is an optimum ratio of concentrator area to emitter area for maximum efficiency. The optimum values for this ratio range from 140 to 200 depending on the taper ratio. The maximum efficiency calculated was 17%. Increases in the efficiency can be obtained by further improvements in the Er-YAG emitter, as well as larger PV cell reflectance for the long wavelength radiation not convertible to electrical energy by the PV cell.

REFERENCES

1. Stone, K. W., Leingang, E. F., Kusek, S. M., Drubka, R. E. and Fay, T. D., "On-Sun Test Results of McDonnell-Douglas Prototype Solar Thermophotovoltaic Power System," "First World Conference on Photovoltaic Energy Conversion, Waikoloa, Hawaii, Dec. 1994.
2. Stone, K. W., Kusek, S. M., Drubka, R. E. and Fay, T. D., "Analysis of Solar Thermophotovoltaic Test Data from Experiments Performed at McDonnell-Douglas," AIP Conference Proceedings 321, The First NREL Conference on Thermophotovoltaic Generation of Electricity, Copper Mountain, CO, July 1994.
3. Stone, K. W., Leingang, E. F., Drubka, R. E., Chubb, D. L., Good, B. S. and Wilt, D. M., "System Performance of a Solar Thermophotovoltaic System for Space and Terrestrial Application," Proceedings of the 30th Intersociety Energy Conversion Engineering Conference, Orlando, FL, July 31 - August 4, 1995, to be published.
4. Chubb, D. L., Good, B. S., Flood, D. F. and Lowe, R. A., "Direct Solar Thermal-to-Electric Energy Conversion Using Thermophotovoltaics,"Solar'95, National Solar Energy Conversion Using Thermophotovoltaics," Solar '95, National Solar Energy Conference, Minneapolis, MN, July 1995.
5. Good, B. S., Chubb, D. L. and Lowe, R. A., "Comparison of Selective Emitter and Filter Thermophotovoltaic Systems," 2nd NREL Conference on Thermophotovoltaic Systems," 2nd NREL Conference on Thermophotovoltaic Generation of Electricity, Colorado Springs, CO, July 1995.
6. Siegel, R. and Howell, J. R., *Thermal Radiation Heat Transfer*, 2nd ed., Washington, DC, Hemisphere, 1981.
7. Wood, R. F. and Jellison, Jr., G. E., "Melting Model of Pulsed Laser Processing," Semiconductors and Semimetals, vol. 23, Academic Press, 1984.
8. Lowe, R. A., Chubb, D. L., Farmer, S. C. and Good, B. S., Appl. Phys. Lett., 64, 3551, 1994.
9. Thomas, M. E., Joseph, R. I. and Tropf, W. J., Appl. Optics 27,239 (1988).
10. Wilt, D. M., Fatemi, N. S., Hoffman, Jr., R. W., et al., Appl. Phys. Lett., 64, 2415 (1994).
11. Wilt, D. M., Fatemi, N. S., Hoffman, Jr., R. W., et al., "InGaAs PV Device Development for TPV Power Systems," AIP Conference Proceedings 321, The First NREL Conference on Thermophotovoltaic Generation of Electricity, Copper Mountain, CO, July 1994.

Testing and Modeling of a Solar Thermophotovoltaic Power System

Kenneth W. Stone
McDonnell Douglas
5301 Bolsa Ave
Huntington Bch., CA 92647

Donald L. Chubb
David M. Wilt
NASA Lewis Research Center
21000 Brookpark Rd.
Cleveland, OH 44135

Mark W. Wanlass
National Renewable Energy Lab
1617 Cole Boulevard
Golden, CO 80401

Abstract. A solar thermophotovoltaic (STPV) power system has attractive attributes for both space and terrestrial applications. This paper presents the results of testing by McDonnell Douglas Aerospace (MDA) over the last year with components furnished by the NASA Lewis Research Center (LeRC) and the National Renewable Energy Lab (NREL). The testing has included a large scale solar TPV testbed system and small scale laboratory STPV simulator using a small furnace. The testing apparatus, instrumentation, and operation are discussed, including a description of the emitters and photovoltaic devices that have been tested. Over 50 on-sun tests have been conducted with the testbed system. It has accumulated over 300 hours of on-sun time, and 1.5 MWh of thermal energy incident on the receiver material while temperatures and I-V measurements were taken. A summary of the resulting test data is presented that shows the measured performance at temperatures up to 1220° C. The receiver materials and PV cells have endured the high temperature operation with no major problems. The results of this investigation support MDA belief that STPV is a viable power system for both space and terrestrial power applications.

INTRODUCTION

McDonnell Douglas Aerospace (MDA) has been performing solar energy research and development for over 23 years for both space and terrestrial applications. During this time, MDA was the design integrator for the Department of Energy 10 MW_e Central Receiver program and jointly developed a 25 kW_e Stirling Dish system with USAB of Sweden. Five years ago, MDA became interested in solar thermophotovoltaics (STPV) for space applications. Models were developed, small scale laboratory tests were conducted to verify the models, and the models were used to investigate system performance. The results of this initial work showed encouraging performance characteristics for both space and terrestrial applications. In 1992-93, MDA fabricated a testbed, unit that was mounted on the MDA 90 kW_t concentrator shown in Figure 1, and a series of tests were conducted [1]. The resulting test data was used in the design of a second testbed unit. This unit was fabricated in late 1994 and testing started in December of that year. The objective of this testing was to obtain information on different STPV components, such as receiver losses, material

FIGURE 1. MDA 90 kW_t dish concentrator

© 1996 American Institute of Physics

performance, cell performance, etc., which could be used to verify analytical models. The analytical models would be used to design and estimate the performance of a STPV system. NASA LeRC has furnished both emitters and PV cells and, just recently, NREL has furnished three PV cells mounted on a single plate.

STPV OPERATION

The principle of operation of a TPV system has been described numerous times [2]. The principle of operation of a STPV system is the same except concentrated sun light is used as the heat source. The concentrated sun light is provided by a concentrator as illustrated in Figure 2.

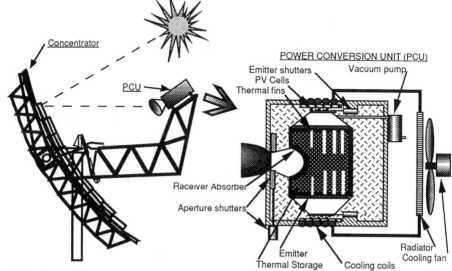

FIGURE 2. Configuration of a solar thermophotovoltaic power system.

A dish concentrator tracks the sun and the reflective surface of the dish reflects the sun's energy to the focal point. A power conversion unit (PCU) is located at the focal point which converts the sun's energy into electrical energy. One possible TPV PCU configuration is shown on the right in Figure 2. A non-imaging or compound parabolic concentrator is used at the entrance to the TPV cavity to further concentrate the solar energy. The use of this non-imaging concentrator allows a reduction in the cavity opening size and can reduce the accuracy requirements on the concentrator reflective surface and tracking system. The optical focal point or aperture entrance is located at the narrowest part (exit) of the non-imaging concentrator just before the receiver cavity. At the aperture entrance, the sun flux density reaches a level of 15,000 to 20,000 suns. Since this level of flux intensity is higher than what most materials will withstand, the TPV receiver is located 6 to 10 inches beyond the aperture entrance to reduce the flux level. The receiver has a half spherical shape so that the peak flux of the concentrated beam is reduced and the flux is distributed uniformly over a larger area. A pair of aperture shutters are located just behind the aperture entrance and before the receiver plate. These shutters are closed at times when the sun's irradiance is not available in order to reduce the radiative and convective losses from the cavity. It appears that thermal storage of the sun's energy would be a very attractive feature for several possible applications of a STPV system. Silicon has thermal characteristics that make it a very attractive storage medium. The melting temperature of 1608° K (1407° C) is a good performance operating temperature for STPV. Silicon's high heat of fusion of 1800 joules/gm (0.5 Wh/gm) will allow a large amount of energy to be stored in a small mass[3]. For example, 50 kW$_e$h of energy can be stored in 285 kg of silicon (assuming 35% TPV efficiency) which would have a volume of 0.12 m^3. Just behind the receiver plate is a chamber containing the storage material. At this point in the conceptual design, there are a number of TPV conversion configurations that are being considered. One of the possible configurations is illustrated in Figure 2 where an emitter material is placed around the outside of the thermal storage chamber. Thermal fins are placed in the storage media to conduct the stored energy in the center to the emitter surface. Since most emitters being considered have a very small power density, this is done in order to obtain a large surface area without requiring a large diameter of the TPV unit. A large diameter of the TPV unit would decrease the use of the concentrator surface near the vertex which is the most effective part per reflective area. Therefore, it is desirable, although not a hard requirement, to keep the TPV diameter as small as possible. Because the PV cells are expected to be very expensive but can operate at a much higher power density than will be radiated per unit area from the emitter surface, an IR reflective surface is used to concentrate the IR energy on the PV cells to reduce system cost. Emitter shutters are located just before the PV cells. These shutters are used to modulate the electric power output. When the electric demand load is low, some of the shutters are closed to reduce the electrical power generation. The shutters have an IR reflective surface to reflect the energy back towards the emitter so that this thermal energy will not be lost.

The PV cells are mounted on a plate that has cooling coils mounted on the back side. A cooling fluid is pumped through the coils and then through a radiator where a fan transfers the waste heat to the ambient atmosphere. A vacuum pump is used to reduce the atmospheric pressure in the emitter chamber area in order to reduce the convective losses.

DESCRIPTION OF TESTING CONFIGURATIONS

The test objectives were to:
- Demonstrate that material can be heated to the temperatures required for TPV operation with concentrated solar energy using present state-of-the-art solar concentrating technology that is economically viable.
- Demonstrate that representative materials that might be used in a commercial system will withstand the operation at high temperatures.
- To obtain test data that could be used to verify analytical models which would be used in the design of a commercial system.
- To provide a platform to test different emitters and cells in a representative environment.

After conducting a number of tests with the first testbed unit described in [4], it was decided to make several modifications to the design in order to increase the over all performance. The main problem with first testbed unit was the receiver plate was at the focal point. This results in more convective losses because the receiver plate was open to the ambient wind, and very high peak flux density on the center of the receiver plate. The peak flux was increased to a high level in order to achieve the desired operating temperature. Some erosion was observed in the center of the receiver plate which did bubble and smoke at times. Another problem was the thermal losses out the rear of the assembly were higher than desired. To rectify these problems in the second testbed unit, the receiver plate was moved 4 inches behind the focal point and increased in size as shown in Figure 3. The power in the reflected beam would now be distributed over a larger area reducing the peak flux level. The depth of the receiver/emitter plate was increased to provide more thermal mass. The back side of the plate was covered with several layers/types of high temperature insulation, a IR reflector, and another silicon carbide plate to reduce the thermal losses. The resources have not been available to conduct a detail design of the solar receiver system, define material requirements, or search for high temperature materials that would meet these requirements. A silicon carbide plate was used for the receiver/emitter plate. Two of these plates were bonded as shown in Figure

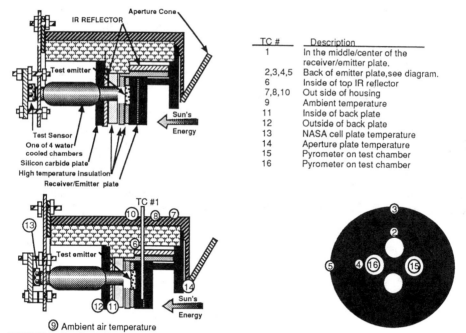

FIGURE 3. Configuration of Testbed II and location of thermocouples.

3 and used as the receiver/emitter plate. Ideally the receiver surface should be curved as shown in Figure 2 and discussed earlier.

There are 4 test chambers mounted on the emitter side of the receiver/emitter plate which can be used to test four different combination of emitters and PV cells. There are a total of 16 thermocouples, mounted at various locations as defined and shown in Figure 3, that are used to obtained temperature measurements. In addition there are 8 data lines to measure current-voltage of the mounted PV cells.

The testbed system was mounted on the MDA 90 kW_t concentrator shown in Figure 1. This concentrator consists of 82 mirrored facets as shown in Figure 4 where each facet is approximately 1 m^2. Each facet has a spherical curvature and can be adjusted such that the reflected beam can be aimed at a specific spot on the receiver. The mirrors were adjusted for a single aimpoint for these experiments. MDA has developed a very accurate and fast method of aligning the facets [5] but unfortunately the system was not available at this time. Therefore an alternate method, which is not as accurate, had to be used. An estimate of the resulting peak flux distribution with all of the mirrors aimed at a single focal point is shown in Figure 4 as a function of the distance from the focal point. The flux on the receiver plate was varied by covering the mirrors with a white wash paint. The first test was conducted with all the mirrors covered to ensure that the power reflected off the white washed mirrors was insignificant. The initial series of experiments were conducted with only mirror #1 uncovered. Mirror #2 was uncovered for the next series of experiments. After these series of tests, the mirrors were uncovered two at a time in the sequence shown in Figure 4 in order to obtain a symmetrical flux distribution on the receiver.

Different emitters and cells that have been or will be mounted in the testbed II unit are:

NASA emitter - A single crystal erbium doped yttrium aluminum garnet, Er-Yag. Erbium, Er, ions replace yttrium, Yt, ions in the doped garnet. For this emitter 40% of the Yt is replaced by Er.

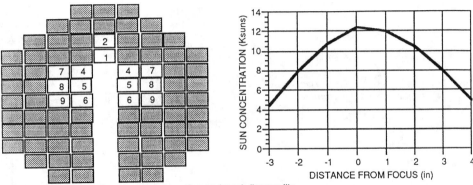

FIGURE 4. Concentrator mirror configuration and peak flux profile.

Emission from the Er-Yag emitter is dominated by the Er emission band centered at a wavelength of 1.5 micrometers.

Amonix PV cell - A high concentrating silicon cell provided by Amonix, Torrance, California. This cell has a 200 sun efficiency in the range of 23%.

NASA PV Cell - The NASA photovoltaic cell [4] is an iridium gallium arsenide ($In_{.53}Ga_{.47}As$), which is lattice matched to an iridium phosphide substrate. The cell has a bandgap energy of 0.75 eV with a cut-off wavelength of 1.65 µm. The external quantum yield is shown for this cell in Figure 6.

NREL PV Cell - The TPV converters provided by NREL are fabricated from epitaxially grown layers of GaInAs, which are deposited on, and lattice mismatched to, InP substrates (Serial No. W525A). The room-temperature bandgap for the devices is approximately 0.58 eV. Specially designed layers are included in the converter structure to enhance the collection of excess photogenerated charge carriers. The converters also incorporate dual-layer anti reflection coatings and electroplated gold contacts on the front surface. The chip is affixed to a copper mounting plate with electrically conductive epoxy. This makes the plate the common back contact for the converters. Each device has a nominal total area of 0.1024 sq cm and an active area of 0.0597 sq cm. The external quantum yield is shown for this cell in Figure 6.

In addition to on-sun testing, laboratory tests were conducted using a small furnace. The objectives of this testing were to verify performance of the testbed unit, further characterization of materials and PV cells, and investigate modifications to the testbed. The furnace was constructed with the same material and high temperature insulation as used in the testbed II unit. The laboratory configuration is shown in Figure 6. A PV cell was mounted in front of a 2.5 cm diameter hole in one side of the furnace on a water cooled plate. In some test configurations, tubes were mounted between the furnace opening and the PV cell to reduce the IR radiation loss.

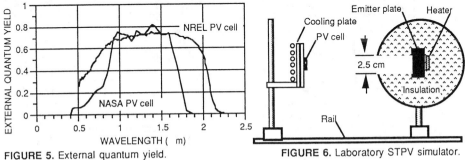

FIGURE 5. External quantum yield.

FIGURE 6. Laboratory STPV simulator.

TEST RESULTS

There were 10 tests conducted with the testbed I system with a total of 30 on-sun hours A maximum receiver temperature of 1350 C was obtained. A general summary of the testing conducted with the testbed system and with the laboratory STPV simulator is provided in Table 1.

TABLE 1. Summary of on-sun and laboratory testing.

On-sun testing with testbed II	
Number of tests	Over 50
Number of on-sun time	300 hours
Peak sun irradiance	1006 W/m^2
Peak temperature (center)	1220° C
Time above 800 ° C	122 h
Time above 900° C	100 h
Time above 1000 ° C	69 h
Time above 1100° C	37 h
Time above 1200° C	5 h
Maximum thermal energy/test	97 kWh
Total thermal energy on receiver	1.5 kWh
Emitters used	silicon carbide NASA
PV cells used	NASA Amonix

Laboratory testing	
Number of tests	Over 20
Hours of testing	80 h
Peak temperature	1225° C
Time above 800 ° C	60 h
Time above 900° C	55 h
Time above 1000 ° C	32 h
Time above 1100° C	10 h
Time above 1200° C	2 h
Emitters used	silicon carbide
PV cells used	NASA NREL Amonix

The objective of the initial testing with the testbed II was to obtain information on the thermal performance of the receiver. This was accomplished by measuring the temperature as a function of the input thermal power. Different levels of thermal inputs were achieved by covering the mirrors with a white wash paint as discussed earlier. The first test was done with all of the mirrors covered to determine how much energy was being reflected off the painted reflective surface. Less than a 10° C rise of the receiver plate temperature occurred, indicating that very little energy reflected from the painted surface was intercepted by the receiver. Examples of the temperature histories for different solar input levels are shown in Figure 7. The temperature shown is TC #1 which is located in the center of the receiver/emitter plate as shown in Figure 3. A temperature of 1220° C has been reached with only 14 of the 82 mirrors uncovered. The mirrors were uncovered in the sequence shown in Figure 4. All of these tests were started at low sun irradiance resulting

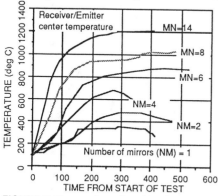

FIGURE 7. Temperatures time histories.

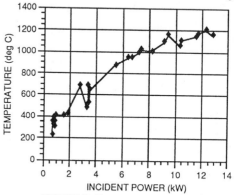

FIGURE 8. Temperature vs. incident power.

from either early morning start or initial low cloud cover. The number of mirrors is only a rough indication of the amount of thermal energy because of variation in the sun irradiance level. A plot of the peak temperature as a function of estimated average thermal power reflected to the receiver is shown in Figure 8. The temperature appears to be leveling off as would be expected because of 1) the IR losses are increasing, 2) the existing receiver aperture is larger than would be required, 3) more spillage because the mirrors were not aligned accurately, 4) no aperture reflector was used.

The receiver system has survived the over 300 hours of on-sun testing with no major problems. The surface of the receiver plate was examined and measured periodically and no surface erosion has been observed. Also no out-gassing from the system has been observed. During test 2.39, a hot spot was observed because of the high sun irradiance level and number of mirrors uncovered which resulted in some bubbling on the surface and left a small area rough. The assembly was moved 2 cm further from the focal point during the test and the bubbling stopped. A curved received plate, as discussed earlier and shown in Figure 2, would have prevented this hot spot.

A crack did form in the front receiver plate during the early tests with only four mirrors uncovered. It is believed that the crack resulted from 1) a water leak in an aperture coil at the focal point and 2) a uneven flux distribution with only four mirrors uncovered. The water leak cooled the bottom of the receiver plate which resulted in high thermal stress and caused the crack. In the 30 tests that have been conducted subsequently at 10 times the thermal energy level, there has been no change in the crack and it does not appear to have effected the performance.

Although it has not caused any problems, the receiver/emitter plate has settled about 2 cm in the vertical direction. The receiver/emitter plate was set on a block of silicon-carbide (1.5cmx1.5cmx7cm) on top of a 10 cm layer of insulation. Under the high heat and the weight of the receiver/emitter plate, the small block settled into the insulation. This has not caused any problems in the performance of the system.

After uncovering 10 or more of the mirrors there appears to be more spillage on the aperture cone (entrance to the cavity receiver) than expected. This is based upon visual observations and not actual measurements at this time. This additional spillage is the result of the temporary mirror alignment system that was used. Although the spillage has not caused any problems, it has increased the energy losses from the system.

The temperature of the emitter plate and variation of temperature over the emitter surface is important to the performance of the system. The slot in the dish reflective surface (lower center of dish has no mirrors as shown in Figure 4 results in an uneven flux distribution on the receiver plate that could produce an uneven temperature distribution on the emitter surface. Therefore, four thermocouples were placed on the emitter plate as shown in Figure 3 to measure the temperature distribution. An example of the emitter surface temperature is shown in Figure 9. The mean emitter temperature is about 100° C lower than the center plate temperature which also means the receiver surface is probably 100 to 200° C warmer, i.e.,. 1300° to 1400° C. The temperature variation over the emitter surface is less than 140° C with the top vertical TC being the lowest as a results of the slot in the mirror reflective surface. Outside of this area, the temperature variation is less than 50° C. The NASA emitter was mounted in the top test chamber. The surface temperature of the NASA emitter has not been measured yet but it is expected to be 150-200° C lower than TC2.

The NASA emitter has been mounted in the top chamber throughout the 300 hours of testing. The testbed unit has not been disassembled for close inspection of the NASA emitter but periodic inspection through the length of the cooling tube has been done. In an inspection after test 32, the surface of the emitter was observed to have been damaged. There is an apparent crack across the middle and some dark areas as illustrated in Figure 10. The majority of the surface appeared to be unchanged from when it was installed. This appearance has not changed since the observation made after test 32. It is believed that all or part of this damaged could have occurred as the results of a water leak that was discussed earlier.

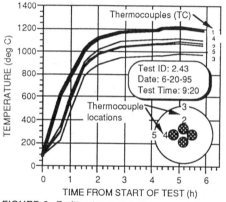
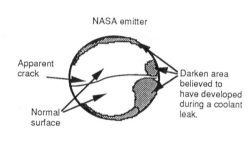

FIGURE 9. Emitter temperature.

FIGURE 10. Emitter condition after water leak.

The Amonix and NASA cells have been mounted on testbed II and I-V tests have been conducted on them at different temperature levels over the last six months. The NREL cells were just received and has not been mounted and tested on-sun. The I-V measurements for the Amonix Silicon PV cell are shown in Figure 11 with center plate temperatures ranging from 955° C to 1177° C. The emitter surface temperature is estimated to be a couple of hundred of degrees below these temperatures. The peak power per unit area for the Amonix cell is also shown in this figure as a function of the temperature. Also shown in this figure is a theoretical fit of T^4. The fill factor for the low temperature is 15% and for the higher temperature is 23%. The low current densities, low power level, and low fill factor are as expected for the following reasons. 1) As described earlier, the cell was mounted in the center of a copper tube over 35 cm from the emitter surface. As will be discussed later, the copper tube did not serve well as an IR reflector. Without a good IR reflector, the power level would be expected to be very low at this distance. 2) At these temperatures the power below the bandgap of a silicon cell would be expected to be very low.

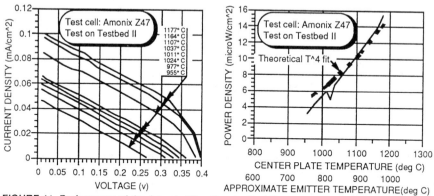

FIGURE 11. Performance of the Amonix PC cell on testbed II.

Over 20 I-V measurements have been taken on the NASA 261-3 PV cell while mounted on testbed II during the last 6 months. Examples of the I-V curves are shown in Figure 12 for a temperature range from 950° C to 1150° C. Also shown in this figure is the peak power as a function of temperature and the theoretical fit of T^4 to the data. The fill factor is approximately 45%. Again the low current densities, low power level, and low fill factor are as expected for the same reasons as discussed above. As was expected, the overall performance of the NASA cell is higher than the silicon cell because of the higher bandgap.

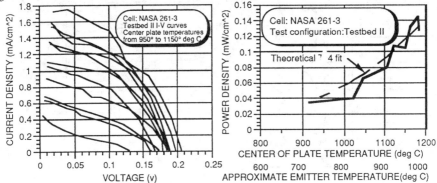

FIGURE 12. Performance of the NASA PC cell on testbed II.

LABORATORY TESTING

Over 30 laboratory bench tests with over 100 hours of time at high temperatures have been conducted on the Amonix, NASA, and NREL cells. One of the tests conducted with the laboratory STPV simulator was to verify the measured performance of the testbed unit. The PV cell was mounted at the same distance from the emitter as in the testbed system. The short circuit current was measured with the cooling tube in place and without the cooling tube. Approximately the same Isc was measured as with the testbed system. When the cooling tube was removed, the Isc dropped about 45%. If the cooling tube was polished, the Isc increased. After a series of tests it was concluded that the testbed II measurements were consistent with the current configuration, i.e., distance from emitter and oxidize cooling tube. To improve the testbed performance, the rear of testbed system would have to be modified.

The STPV simulator has been used to investigate how the testbed system should be modified and to create a data base to compare with the performance of the testbed system. The sensitivity of the Isc to the distance from the emitter is shown in Figure 13. At a distance of 2.5 cm from the emitter, the Isc exceeded 1 A/cm². In another experiment, insulation was placed around the NASA cell mounting plate with a small opening to the cell. The resulting I-V curves as a function of the emitter surface temperature are shown in Figure 14. At 1200° C the insulation around the cell degraded and started covering the surface of the cell. The output power for this case is shown in Figure 15.

The NREL cells have just been received and testing and characterization have just started. The cells were mounted on the cooling plate at a distance of 4.5 cm from the emitter in the open air. I-V tests were conducted on the cells from 700° C to 1150° C. Examples of these preliminary I-V curves and output powers are shown in Figure 16.

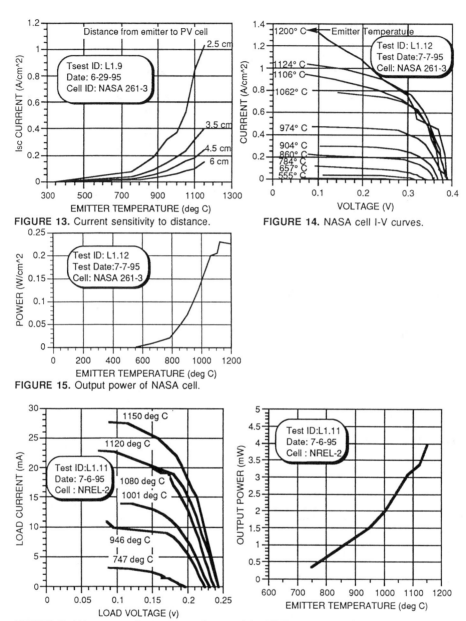

FIGURE 13. Current sensitivity to distance.

FIGURE 14. NASA cell I-V curves.

FIGURE 15. Output power of NASA cell.

FIGURE 16. I-V curves and output power for one of the NREL cells.

CONCLUSION

The STPV testbed II unit has reached TPV operating temperatures of 1000° C with less than 10% of the available solar energy. The representative receiver and insulating materials that have been used in the testbed II unit have endured the high temperature operation after 300 hours of on-sun testing and with over 1.5 MWh of solar energy incident on the receiver system. The receiver surface has no measurable surface erosion. A high intensity hot spot on the receiver did results in some bubbling of a small area but moving it further from the focal point eliminated the bubbling. This would not have occurred with a properly shaped receiver surface. The spillage on the aperture cone is higher than it should be because of the quick mirror alignment method that was used.

PV cells and a emitter have been mounted on the testbed unit and I-V tests were conducted. The results were as expected for the test configuration used. The Amonix and NASA PV cells have shown no performance deterioration after several hundred hours of operation. The NREL cell has not been mounted in the testbed unit at this time. Characterizing the performance of the NASA emitter has not been completed or has the testbed unit been disassembled for detail examination.

Initial characterization using the STPV laboratory simulator have been conducted on the Amonix, NASA, and NREL PV cells. In these tests, emitter surface temperatures as high as 1250° C have been reached and current densities of 1.3 A/cm^2 and power densities of 0.23 W/cm^2 have been measured.

Because of the experience with the STPV testbed and STPV laboratory simulator over the last year, MDA still believes that a STPV system can compete in both the space and the terrestrial market place. MDA plans to continue this investigation into the performance of a STPV power system. The testbed unit will be modified to increase its performance, analysis of test data and development of analytical models will continue.

REFERENCES

1. Stone, K. W., Kusek, S. M., Drubka, R. E., Fay, T. D. "Analysis of Solar Thermophotovoltaic Test Data From Experiments Performed At McDonnell Douglas," The First NREL Conference on Theromphotovoltaic Generation of Electricity, AIP Conference Proceedings 321, Copper Mountain, CO, July 1994.
2. Iles, P. A., "Photovoltaic Principles Used in Thermophotovoltaic Generators," The First NREL Conference on Theromphotovoltaic Generation of Electricity, AIP Conference Proceedings 321, Copper Mountain, CO, July 1994.
3. Chubb, D. L., Good, B. S., Lowe, R. A., "Solar Thermophotovoltaic (STPV) with Thermal Energy Storage," The Second NREL Conference on Theromphotovoltaic Generation of Electricity, AIP Conference Proceedings, Colorado Springs, CO, July 17-19, 1995.
4. Jain, R. K., Wilt, D. M., Landis, G. A., Jain, R., Weinberg, I., Flood, D. J., "Modeling of Low-Bandgap Solar Cells for Thermophotovoltaic Applications," The First NREL Conference on Theromphotovoltaic Generation of Electricity, AIP Conference Proceedings 321, Copper Mountain, CO, July 1994.
5. Stone, K. W., Blackmon, J. B., "Application of the Digital Image Radiometer to Optical Measurement and Alignment of Space and Terrestrial Solar Power Systems," The 28th Intersociety Energy Conversion Engineering Conference, Atlanta, Georgia, Aug 8-13, 1993.

SESSION II:
MARKETS AND APPLICATIONS

ns for Thermophotovoltaics

M. Frank Rose

Space Power Institute
Auburn University, Alabama

Abstract: Thermophotovoltaics (TPV) is a promising method of producing small portable electric power units for a variety of applications. Numerous studies indicate that overall efficiencies greater than 15% are possible. There are a number of competing technologies, in various stages of development, which are capable of similar efficiencies. In this paper, thermoelectrics (TE), alkali metal thermal to electric converter (AMTEC), small motor-generators, and fuel cells will be discussed in terms of the state of the art, cost, complexity, and as competition to units based on TPV.

INTRODUCTION

There is considerable emphasis on the development of lightweight, efficient, reliable "portable" electric power units for a variety of applications. This emphasis is motivated, in part, by military requirements for more power for the sophisticated weaponry it intends to introduce into the battlefield in addition to extending the operating time of devices already in the inventory. There is also a growing civil market for compact portable power units. In this case, reliability, non-polluting technology, and cost are major drivers. Table 1 lists typical power requirements for a representative sampling of applications.

TABLE 1. Typical Power Requirements for a Representative Sampling of Applications

>Watts	>Kilowatt	>10 Kilowatts	>100 Kilowatt
Telephones	Tools	Yachts	Advanced radar
Home electronics	Rec. vehicles	RPV	Spacecraft
Computers	Wheelchairs	Golf carts	Electric bus
Navigation buoy	Actuation	Electric Cars	Weapons
Soldier system			

Note that the term "portable" is relative and ranges from "hand held" to "man portable" to something which would fit on a "flat-bed truck." The sampling of potential applications for small portable power units indicates that there should be a market for any technique which is mass and cost effective with respect to the conventional technologies such as batteries. Many of the applications listed in Table 1 are currently powered by batteries. In many cases, there would be decided advantages if the battery could be replaced by a "fueled system." In terms of stored energy, fuels have energy densities on the order of 10,000 Watt-hrs/kg. The conversion system determines the ultimate energy density available. Fueled systems have the advantage over batteries in that they can be replenished from a fuel store, have long shelf life, and should have a life independent of the

© 1996 American Institute of Physics

number of "charge-discharge cycles." Once the mass penalty has been paid for the converter, the system mass is fuel dominated and size determined by how much fuel the application demands. Further they can be repaired.

There are numerous technologies which are potentially capable of application in the areas listed in Table 1. Over the years, the Space Power Institute has conducted a series of workshops, the Prospector Series (1), to assess the state-of-the-art in portable power technology with a focus on Army applications. Much of the material discussed in this paper comes from the following Prospector proceedings which are available through the Army Research Office or from the Space Power Institute:

> Mobile Battlefield Power Workshop, 1990
> High Energy Density-High Power Density Power Sources R&D, 1992
> Small Engines and their Applicability to the Soldier System, 1992
> Small Fuel Cells for Portable Power, 1994

PORTABLE POWER SOURCES

There are numerous potential power systems which would be competitive with thermophotovoltaics for the applications listed in Table 1. There are an enormous range of options within any of the power technologies. To keep this paper within bounds, only representative technologies within a class will be discussed.

Battery Technology

The low end of the scale is almost the exclusive domain of battery technology. Of course, battery technology has applications all across the spectrum. In the proceedings from Prospector III, High Energy Density-High Power Sources R&D, numerous battery technologies were discussed. Almost all of the Government funding agencies have major battery programs which are coordinated through an interagency power group. A current battery technology capable of high energy density is based upon Lithium chemistry. Indeed, commercial versions are beginning to appear regularly. The hypothetical energy density is about 6200 Whr/kg, somewhat more than half the energy density available from fuels. Laboratory experimental primary (non-rechargeable) devices have reached 1000 Whr/kg. The current laboratory state-of-the-art in secondary (rechargeable) batteries is in the range of 200-400 Whr/kg. Note that these energy densities are comparable to that of conventional explosives and if abused, batteries with this energy density pose a significant safety hazard.

The battery kinetics dictate the rate at which energy can be withdrawn. Increased power density is associated with increased safety hazards due to internal heating etc. It is impossible to maximize both energy density and power density simultaneously. The range of power densities for the Lithium technology extends above 100 W/kg. The upper end of that range is bought at the expense of energy density and safety. In general, it may be possible to reach about 50% of the theoretical energy density in practical primary devices which would have to be employed with care. Current technology is on the order of 25%. The most promising long term Lithium technologies are the Lithium-solid polymer electrolyte, Li/SO2, and the nLi/(CF)n which appear capable of energy densities on the order of 1000 Wh/kg with unknown power density and safety.

The advanced battery technologies have the following problems:

- Safety
- Limited shelf life
- Limited charge-discharge cycling
- Minimal repair capability
- Low power density
- Cost
- Disposal

As an example of current cost of military batteries, the cost for one kWh of energy from the army BA5590 is approximately $375.00. The comparable cost for a Ni-Cd(single charge) is on the order of $2,105.00.

Fuel Cells

There are an enormous number of fuel cell concepts under active consideration within the R&D community. Fuel cells are particularly attractive in that they do not employ thermodynamic cycles in the conversion mechanism and hence are not limited to Carnot efficiencies. Laboratory prototypes have been operated at efficiencies as high as 75%. In the proceedings of Prospector VI, numerous options, and the state-of-the-art for fuel cells were discussed and this description is taken from the deliberations of that workshop. For the purposes of this paper only two fuel cell types, Hydrogen PEM and the Methanol PEM, will be described. They are particularly suited to a wide range of applications as listed in table 1 and the state-of-the-art is reasonably well defined in small sizes.

Hydrogen PEM

There are a number of firms which will custom manufacture small units. Analytic Power Corporation has a "century series of units" which range in power level from 150 -10 kW. These units are greater than 50% efficient and the manufacturer claims that with chemical hydride power packs, the units have energy densities in the range of 1.7-3.0 kWhr/kg. The three methods of fuel storage are chemical hydrides, high pressure tanks and cryogenic tanks. The current state-of-the-art for storage in high pressure tanks is kevlar wrapped cylinders storing gas at 10,000 psig with a safety factor of 1.5. At that pressure, the storage density is on the order of about 1.3 kWhr/kg which is comparable to chemical hydrides. Cryogenic storage for small portable units is less attractive due to the need for superinsulation to provide some long term storage. The current state-of-the-art for a hydrogen PEM cell using air as the oxidizer is approximately:

Power level   150 W
H2-ambient air with filtration
Forced ambient air convective cooling
Efficiency   Approximately 65% at 80 ASF
Weight   3.6-4.5 kg
Volume   Approximately 2.5 ltr
Lifetime   > 500 hrs
Power Density   Approximately 50 W/kg

The cost for a unit of this type is on the order of $8,000.00. Costs would be much lower if mass produced. It was estimated that the cost could be lowered by more than an order of magnitude with improvements in design, new materials, and volume production.

The major advantages of Hydrogen PEM fuel cells are:

Non polluting
Quiet
Low temperature operation
Highly efficient
Well developed technology
Minimal moving parts
High Energy Density

The major disadvantages of the Hydrogen PEM fuel cells are:

Requires a new fuel to be introduced in large quantities for mass market
Fuel storage and handling technology immature
Technology, while well developed, is not mature
Cost

There are numerous opportunities to improve this technology to the point where it could be successfully inserted into a consumer market. Manufacture and storage on a local scale of hydrogen is a pacing item as well as minimizing the need for expensive catalysts in the manufacture of electrodes. If sufficiently small "crackers" could be integrated into the individual fuel cells, it would lessen the fuel problem by allowing the use of diesel fuel at a significant mass penalty.

Methanol PEM

Methanol fuel cells, while less well developed than Hydrogen fuel cells, is an attractive system primarily due to the fact that Methanol is a readily available fuel which is easy to store and transport. It requires only a suitable container which can be operated at STP. There are two types of Methanol fuel cells under active investigation. For this comparison, only the direct oxidation method will be discussed. When compared to Hydrogen-Oxygen fuel cells, the theoretical energy density is approximately 15% that of the Hydrogen-Oxygen system. Nevertheless, it is theoretically possible to obtain about 6.3 kWhr/kg from the Methanol-air system. The Army and ARPA have major programs in this technology. The state-of-the-art in laboratory cells is as follows:

Single Cell	0.33 V @ 800 mA/cm2
	0.44 V @ 640 mA/cm2
5-Cell Stack	0.45 V @ 300 mA/cm2
Life Tests	> 200 hrs continuous Tests (single cell)
	> 500 hrs intermittent (single cell)
	>400 hrs @ 48 W (5 cell stack)
Efficiency	25-30 % @ < 300 mA/cm2
Fuel Efficiency	>70% @ 300 mA/cm2

Based upon the above data, it was estimated that a 250 watt, 1 kWhr unit would weigh approximately 4.7 kg. It is impossible to estimate cost due to the immature state of the technology.

There are a number of issues pacing the development of Methanol fuel cells. The most pressing are:

Poor anode kinetics
Methanol crossover
Fuel loss
Cathode Polarization

There are well defined approaches to solving each of these problems in a manner which will make methanol fuel cell technology competitive with other power systems. The advantages for Methanol fuel cells are the same as those for Hydrogen with the added benefit of a safe readily available, easily handled fuel.

Alkali Metal Thermal To Electric Converter (AMTEC)

The AMTEC converter, often referred to as the Sodium heat engine, is a device which can convert heat directly into electricity with the only moving part being a stream of liquid metal. In practice, a closed container is filled with liquid sodium and is divided into two regions by an electromagnetic pump and a solid electrolyte which is Sodium ion permeable. A highly porous highly conducting electrode is attached to the solid electrolyte. If the two sections of the container are kept at differing temperatures, there will be a pressure gradient across the membrane which will cause the Sodium ions on the high temperature side to migrate through the membrane to a lower pressure part of the container. In that way, a potential difference is established across the membrane which can be used to do work. Since this is essentially a heat engine, the efficiency is determined by the temperature difference between the two regions of the system. In general, the thermal input from some source (nuclear, fossil fuel, solar) is at a temperature in the range of 900-1300 K. The low temperature heat sink is kept in the range of 400 - 800 K. Working with this temperature differential, it should be possible to build units which are about 25-35% efficient. To a first approximation, the efficiency is independent of size. This technology is immature and roughly in the same state of development as TPV. The current state-of-the-art is approximately:

Single Cell efficiency	Approximately 20%
Life testing	Approximately 14000 hours
Cost Studies	$300-$500 per kwe for fully developed and mature technology
Power Density	>0.5 kW/kg

A mature AMTEC technology would be competitive with the other technologies described in this paper. The principle advantages of AMTEC are:

- Long life potential
- High power density
- Few or no moving parts
- Readily adaptable to fuel type
- Involves only one "exotic" material (beta alumina)
- Low cost

The current programs focus on lifetime of the components, thermal isolation, and efficient thermal coupling to the high temperature side of the devices. Practical demonstrations would involve arrays of cells in series/parallel arrangements.

Motor-Generators

This technology is old and established and forms the basis for most of the non-battery portable power technology available today. In the proceedings from Prospector IV, Small Engines and their Applicability to the Soldier System, the technology and potential advancements, in small motor-generator technology is described. Existing engines such as those used in the conventional "Honda" auxiliary power plants are notoriously inefficient, noisy, polluting and of limited life. Within the workshop proceedings, numerous small engines and advanced concepts were discussed. The current state-of-the-art in small engines in terms of specific power and specific fuel consumption are on the order of 0.7 W/cm^2 and .7 g/Whr. Near term state-of-the-art is on the order of 150 W/cm^2 and .37 g/Whr. It was estimated that a concerted research effort could produce a micro-engine with the following characteristics:

4 Stroke, spark ignition, JP fuels

Specific weight	1.5 W/gm
Fuel consumption	0.3 g/whr
Noise level	less than 65 db @ 1 m
life	>1000 hrs between maintenance cycles

Generators for use with engines of this type are typically greater than 95% efficient. Noting that the fuel consumption is on the order of 0.3 g/Whr, the efficiency of a unit of this type would be on the order of 10% overall. The units can run on a variety of fuels with minor adjustments. Currently a Honda EM350 can be purchased for less than $500.00 per unit. This unit operates at 150 watts and has a weight on the order of 12 kilograms dry.

Thermoelectrics

The use of thermoelectric elements for power generation is a well established technology and has been the basis for numerous spacecraft power systems utilizing radioisotope heaters. Reliability is outstanding with continuous operating lifetimes measured in years.

There have been several attempts by the Army to produce a fieldable power system based on thermoelectrics. These units in general were less than 5% efficient but offered the attractive features of quiet reliable operation. The most recent design study for a thermoelectric power system was funded by the U. S. Marine Corp (2). The study finished a preliminary design of a 500 W system using segmented Lead Telluride modules. The design study indicated that the following were achievable:

Power	500 W
Voltage	24 V DC and 110 V AC
Reliability	10,000 hrs MTBF
Maintenance	1hr for 5,000 hrs operation
Noise	< 50 db at 1 m
TEG	Lead Telluride
Hot Junction	832 K
Cold Junction	377 K
Mass	<20.4 kg
Volume	<0.4 m3
TEG Eff.	Theoretical 9.6% Actual 8.9%

No estimate of total system efficiency was given. In practice, it would be hard to recuperate more than about 60% of the energy left in the combustion gasses which would again produce a total system efficiency no greater than about 5%. Advanced thermoelectric materials might eventually produce elements with conversion efficiencies greater than 12%.

CONCLUSIONS

There are a number of promising power systems emerging from the laboratories which, at first glance, offer an attractive alternative to the conventional motor-generator sets. Cost and the need for exotic fuels in some cases have been major drivers, limiting applications to niche markets. Market penetration and mass production should lower the prices, of what are now custom devices, to a level where they can compete successfully. These concepts, along with TPV, are not mature enough for major market penetration, are too costly, or require a new fuel infrastructure for wide spread use.

ACKNOWLEDGEMENTS

The support of the Army Research Office through sponsorship of the Prospector Series of Workshops and through Contract #DAALO39260205-1.

REFERENCES

1.
    a. Prospector I: "Thermal Management of Space-Based High Power RF Sources," sponsored by USAF, LANL, RADC, SPI, March 20-22, 1990. Edited by M.F. Rose.

    b. Prospector II: "RTG Power Applications Workshop," sponsored by ARO, DOE, and JPL, March 22-25, 1992. Edited by M.F. Rose.

    c. Prospector III: "High Energy Density, High Power Density Power Sources R&D Workshop," sponsored by ARO, May 26-28, 1992. Edited by M.F. Rose.

    d. Prospector IV: "Small Engines and Their Applicability to the Soldier System," sponsored by ARO, November 10-12, 1992. Edited by M.F. Rose.

    e. Prospector V: "Microelectromechanical Systems -- Their Applicability to the Soldier System," sponsored by the ARO, under Contract #DAAL03-91-C-0039, July 1, 1993. Edited by M.F. Rose.

    f. Prospector VI: "Electric Actuation," sponsored by Auburn University's CCDS, NASA Marshall, the USAF Wright Laboratories, and ARO, March 21-23, 1994. Edited by M.F. Rose.

    g. Prospector VII: "Small Fuel Cells for Portable Power Workshop," sponsored by SPI and ARO, October 31-November 3, 1994. Edited by M.F. Rose.

2. Bass, J.C., Elsner, N.B., and Leavitt, F.A., "The Preliminary Design of a 500 Watt Thermoelectric Generator," Proc. IECEC, P 586-591, AIAA-94-4197-CP, August 1994.

Grid-Independent Residential Power Systems

Robert E. Nelson

ThermoLyte Corporation
45 First Avenue
P.O. Box 8995
Waltham, MA 02254-8995

Abstract. A self-powered, gas-fired, warm air furnace is evaluated as a candidate for the autonomous generation of electrical power. A popular, commercial residential furnace is analyzed for electrical power requirements. Available energy conversion concepts are considered for this application, and the thermophotovoltaic (TPV) option is selected due to reliability and cost. The design and the internal components peculiar to the TPV converter will be covered. Operating results, including NO_x emission, will be summarized. This work was sponsored by the Basic Research Group, Gas Research Institute, Chicago, IL.

INTRODUCTION

Reliable electrical power is a valuable commodity. Remote electrical power generation is required at locations far from the electrical utility's grid, and in some cases, independent generation of electricity at residences on the grid may be justified for reasons of reliability or cost. We shall evaluate the processes available for electrical power generation and evaluate one application where thermophotovoltaic power generation in a gas-fired appliance appears to be especially attractive.

AVAILABLE TECHNOLOGIES

Many energy conversion options should be considered for this application. The traditional, and in many cases, most cost effective alternatives include machines such as internal combustion engines (Otto and Diesel cycles), external combustion engines (Stirling cycle), steam engines (Rankine cycle), and gas turbines (Brayton cycle). All of these systems tend to generate noise and vibration, to experience wear from friction and failing lubrication, and to require periodic maintenance. A substantial industrial base exists for the manufacture of many sizes of internal combustion engines, but small size Brayton machines are not commercially available.

There are many direct energy conversion concepts that are characterized as containing no essential moving parts. These concepts, which may be powered by a wide choice of fuels, offer the prospect of silent and reliable service over long periods of time. The first that we will cite is thermoelectric energy conversion, where an electric field is generated in a metal or semiconductor in the presence of a thermal gradient. This is a well established technology, insofar as this conversion option has been available for hundreds of years. This technology yields power supplies that are typically heavy, inefficient, and costly. Thermoelectric systems remain a power solution for niche applications, but have failed to identify a large scale application.

Thermionic devices, where electrons are ejected from a hot emitter and collected on a cool electrode, require a complex system design with specialized materials. The major application for this costly conversion technique appears to be in space power applications with radioactive isotope or nuclear reactor types of thermal sources.

Another energy conversion process utilizes ion conducting membranes in an alkali-metal-thermal-to-electrical-converter (AMTEC). This device works on high pressure sodium vapor and may contain heat pipe components. This concept is presently under development.

The last concept that we shall consider is thermophotovoltaic (TPV) energy conversion. TPV is the newest of the major direct energy conversion techniques, insofar as it was invented in 1960 at the Massachusetts Institute of Technology.[1] The energy source, which may be a chemical flame, concentrated solar energy, or a nuclear reactor, is used to elevate an emitter to incandescent temperatures (Figure 1). The radiant energy from the emitter is captured on an array of photovoltaic cells, such as solar cells, which convert the incoming radiant energy directly into d.c. electrical power, which is available at terminals on the photovoltaic array. This process, in general, is lightweight, silent, reliable, and is expected to have a high conversion efficiency, insofar as these systems typically exhibit a high Carnot (theoretical) efficiency. These devices are conceptually simple and easy to build, in that mechanical tolerances among the major components are typically not critical.

---

[1] D.C. White and R.J. Schwartz, "P-I-N Structures for Controlled Spectrum Photovoltaic Converters," Proceedings of NATO AGARD Conference, Cannes, France. March 16-20, 1964. Published by Gordon & Breach Science Publishers, London. 1967. pp. 897-922.

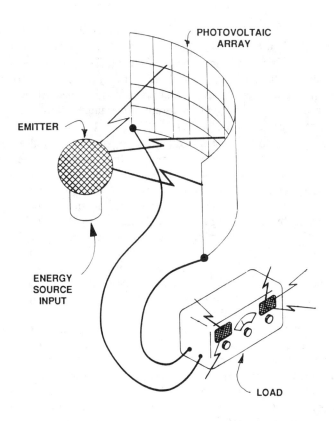

FIGURE 1. A basic thermophotovoltaic power generation system

Early research on TPV systems[2] employed emitters fabricated out of materials capable of surviving high temperature operation. Refractory materials such as alumina or silicon carbide were typically employed, and these materials, when heated to approximately 1500°C, exhibit gray body emission properties. That is, they are very broad band emission sources. Unfortunately, all conventional photovoltaic devices are relatively narrow band converters, in that they efficiently convert radiation into electrical power only when the wavelength of incoming radiation is slightly less than the equivalent wavelength of the photovoltaic device's absorption edge. Obviously, all the radiant energy from these broad band sources at wavelengths beyond the photovoltaic's absorption edge is incapable of conversion, and this long wavelength energy is potentially wasted. In addition, short wavelength emission with wavelengths substantially less than the absorption edge wavelength cannot be converted efficiently.

---

[2]J.J. Werth, "Thermo-Photovoltaic Converter with Radiant Energy Reflective Means," U.S. Patent No. 3,331,707. Filed July 31, 1963. Issued July 18, 1967.

Later TPV development at Fort Monmouth, New Jersey[3] recognized that TPV efficiency could be improved if a selective emitter were used rather than a broad band emitter. Guazzoni[4] and coworkers identified the rare earth elements as potential narrow band emitter candidates, but they were unable to make practical rare earth oxide emitter structures that could survive thermal stress when cycled between room temperature and operating temperatures high enough for efficient TPV performance (1700 to 2000K).

More recent work[5] at The Gillette Company employed rare earth oxide emitters in fibrous form to yield thermally stress-tolerant emitter structures that exhibited exceptional selectivity. An essential feature of this work was emitter configurations composed of small diameter (10 μm) filaments. A filament this size is too small to develop any significant thermal stress across a fiber diameter. Thermal stress built up along the axis of a fiber is easily relieved by a flexing of the fiber. The small diameter filaments also couple very well to flames, inasmuch as there is a very high thermal transfer from the flame's exhaust products to the fine filamentary array. In fact, the fibers are in approximate thermal equilibrium with the convective stream. The small filament diameter also limits the gray body emission that is common with oxide ceramics at elevated temperatures. Finally, the combination of a high thermal transfer and a small filament volume results in a very rapid thermal response. The thermal time constant of our fibrous emitters is of the order of 20 milliseconds, which permits rapid turn-on and load following capabilities for these kind of emitters.

Small diameter filaments have a disadvantage in that they are susceptible to fracture when mechanically stressed. Ceramic fibers fracture because flaws, voids, or microcracks, which arise typically during the fabrication process, propagate when the ceramic filament is stressed. We have developed some low cost synthesis techniques[6] that minimize the density and severity of these flaws, and, consequently, enhance substantially the impact resistance of our fibrous structures.

---

[3]G.E. Guazzoni and S.J. Shapiro, "Spectral Emittance of Neodymium, Samarium, Erbium, and Ytterbium Oxides at High Temperature," Research and Development Technical Report ECOM-3281, United States Army Electronics Command, Fort Monmouth, NJ. May 1970.

[4]G. Guazzoni and E. Kittl, "Cylindrical Erbium Oxide Radiator Structures for Thermophotovoltaic Generators," Research and Development Technical Report ECOM-4249, United States Army Electronics Command, Fort Monmouth, NJ. August 1974.

[5]R.E. Nelson, "Thermophotovoltaic Technology," U.S. Patent No. 4,584,426. Filed Septemberb2, 1983. Issued April 22, 1986.

[6]W.J. Diederich and R.E. Nelson, "Refractory Metal Oxide Processes," U.S. Patent No. 4,883,619. Filed August 16, 1982. Issued November 29, 1989.

Typically, our fibrous emitters withstand g-forces that are about two orders of magnitude greater than conventional commercial counterparts.

The rare earth elements are of particular interest in selective emitters. All of the rare earth elements, with the exception of promethium, form stable oxides that are refractory. The rare earth elements, furthermore, have a unique atomic structure in that there are electron vacancies in the inner electron shells. These vacancies permit electronic transitions that occur, energetically, in the visible and near-IR portions of the electromagnetic spectrum. We have identified a number of rare earth oxides in fibrous form that exhibit selective emissions that are compatible with available photovoltaic converters. For example, when heated, erbia (erbium oxide) exhibits a very narrow band emission centered at 1.55 μm that is a suitable emitter for germanium or indium-gallium-arsenide photovoltaic devices. Erbia has a less significant emission around 1 μm, as well as a monochromatic emission in the visible that gives this emitter its characteristic green glow. The 1.55 μm emission is especially selective. The full width at half maximum (FWHM) is only 65 nm, which is extraordinarily narrow for an emitter of thermal origin. Holmia (holmium oxide) and neodymia (neodymium oxide) emit selectively at 2 μm and 2.5 μm, respectively. Another emitter of special significance in TPV applications is ytterbia (ytterbium oxide). Ytterbia has only one selective emission, and that occurs at 0.98 μm with a FWHM of 150 nm. This emission is well situated to illuminate silicon photovoltaic cells since silicon has an absorption edge at about 1.15 μm. The superposition of an ytterbia emission spectrum onto the spectral responsivity of a crystalline, silicon (solar) cell is demonstrated in Figure 2. A 1600°C, gray (or black) body spectral exitance is shown also in Figure 2 to demonstrate the large amount of nonconvertible emission from this broad band emission source.

It is especially significant that we have uncovered a selective emitter for a silicon photovoltaic cell for our TPV power supply. It is our experience that the cost of the photovoltaic array dominates the overall cost of typical TPV systems. Not only is silicon a very efficient photoconverter, but it is also the most cost effective photovoltaic cell material. Silicon is a well understood semiconductor and is available from multiple and competitive sources all over the world. Silicon is used in large quantities in telecommunications and computer products and, to a lesser extent, in solar cells. Silicon devices have established a noteworthy trend of offering more and more performance at less and less cost as silicon processing techniques improve. While many of the processing improvements may not be directly applicable to photovoltaic structures, a silicon photovoltaic cell does benefit from the huge silicon infrastructure currently in place. Our cost projections for silicon photovoltaic cells allow us to conclude that TPV will be the lowest cost option among the direct energy conversion candidates.

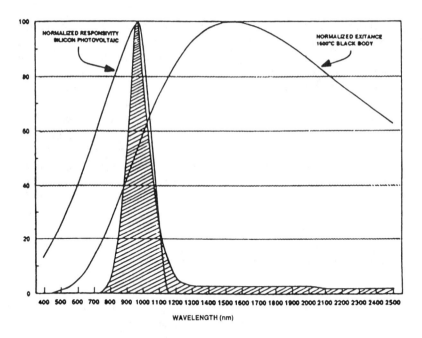

FIGURE 2. Normalized spectral exitance of fibrous ytterbia emitter shown as cross-hatched area

APPLICATION

TPV is well suited to provide the electrical power required in a gas-fired, warm-air furnace. There have been instances when consumers have been inconvenienced or even endangered when home heating systems have shut down during cold weather because of a loss of electrical power due to weather conditions or accidental damage to power lines. Under these circumstances, the gas utility is particularly frustrated because it cannot provide all the benefits of its service when electrical power is not available. Self-powered gas furnaces provide an obvious solution to this problem. Efficiency requirements may be relaxed in this situation, because waste heat that is sometimes lost is used in this application for space heating. In fact, there will be an increase in gas load and a small drop in electrical load as consumers switch to this kind of equipment. The market for residential heating systems is very competitive, and consumers are very cost conscious. A low cost, which was noted earlier, is one of the advantages of TPV energy converters. Silent operation and long term reliability with a minimum requirement for servicing are additional advantages.

An existing residential, gas-fired, warm air furnace was chosen for conversion to self-powering. A 100,000 BTU/hr Lennox G23 Series is the model selected and is illustrated in Figure 3.

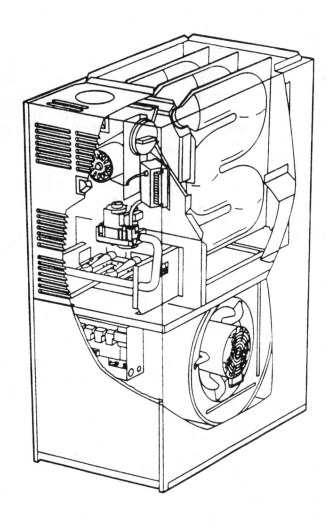

FIGURE 3. A typical residential gas-fired warm air furnace

TPV POWER SUPPLY

A planar emitter/photovoltaic array was chosen for this application since this geometry is well adapted to the existing furnace design. We replaced the conventional 4 inshot burner assembly with our planar emitter array (Figure 4). The combustion products from the emitter pass through the main heat exchanger on to the exhaust (flue collector) in a fashion similar to the existing design. The selective ytterbia emitter in filament form is supported by a porous ceramic substrate. This emitter configuration (Figure 5), which is called a supported continuous fiber emitter (SCFE), delivers a high level of uniform silicon-convertible exitance in planar form. We employ a cellulose support process that is based on textile precursors to fabricate the fibrous emitter structures. We use continuous filament rayon yarn (150 to 1200 denier/50 to 300 filaments) that is impregnated with aqueous metal salt solutions (ytterbium nitrate in this case). The treated yarn is then tufted into the porous ceramic support much like a rug-making process that yields an uncut looped pile. After tufting, the treated rayon is converted into a ceramic by means of a controlled heat treatment process. At low temperatures and with a controlled gas ambient, the nitrate salt in the rayon filament converts into the oxide. Continued heat treatment at higher temperatures pyrolyses away the rayon, which leaves in a gaseous form (water vapor and carbon dioxide), until an oxide skeleton remains which has the morphology of the textile precursor, but with greatly reduced dimensions. A final high temperature sintering step further densifies and strengthens the ceramic filaments. For these SCFE structures, we have chosen a molded cordierite ported tile (Hamilton Porcelain #242) as the substrate. The very uniform hole pattern in these tiles dictated this choice. Although these prototype emitters were tufted manually, we anticipate that future emitters will be machine tufted, and a very uniform hole pattern is required for that process. After sintering, the ytterbia fibers were bonded to the plenum side of the substrate with alumina cement. The same alumina cement may be used to plug some of the remaining ports in some regular fashion to adjust emitter porosity for combustion stability. Four tiles constitute the emitter with a radiating area of 514 cm^2 (13.5 cm x 38 cm).

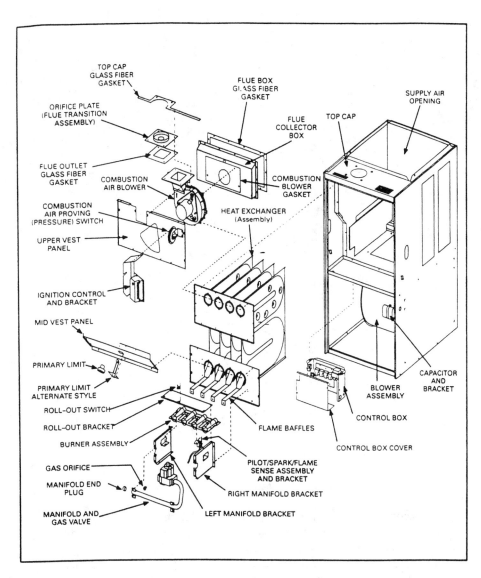

FIGURE 4. Internal components of a typical gas-fired warm air furnace

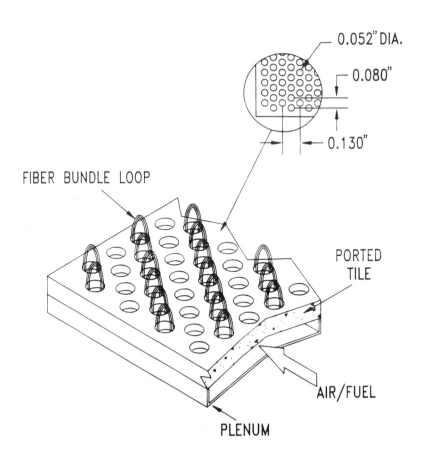

FIGURE 5. Supported continuous fiber system

The silicon cells are mounted on a plane parallel to the emitter (Figure 6). The spacing between these two planes is kept small to enhance the view factor. The cells are thermally bonded to a forced-convection cooled heat sink. An extruded aluminum heat sink assembly (Wakefield Engineering 510 Series) has a substantial finned section cooled by the flow of return air to the furnace. The silicon cell array consists of 10 sections that each contain 20 series-connected cells. Each section generates about 12v d.c., and we have the option of placing all 10 sections in series for 120v d.c. output, or all 10 sections in parallel for an output of 12v d.c. Other combinations of series/parallel connections are clearly possible. The preferred method of electrical isolation in cell mounting, as developed by the cell supplier, Applied Solar Energy Corporation, is to utilize a thin layer of intrinsic silicon that is metallurgically bonded to an aluminum subplate. The silicon cells are bonded to metallized lands on the intrinsic silicon layer. The aluminum subplates, which contain the 20 silicon cells in series, are, in turn, bolted to the heat sink base with a layer of thermal conducting grease at the interface.

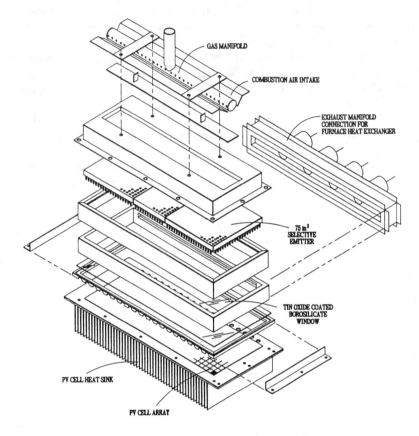

FIGURE 6. Exploded view of TPV power source for residential warm air furnace

Two layers of glass thermally protect the silicon cell array. The first layer, parallel to the emitter array, prevents the escape of the combustion products which must be routed to the main heat exchanger in the furnace. The emitter side of this first layer of borosilicate glass may be coated with an IR-reflecting layer such as tin oxide. All oxide ceramics, including the rare earths, are emissive at wavelengths beyond about 10 µm because of a lattice vibration phenomenon. A plasma filter, such as a high conductivity tin oxide layer, reflects long wavelength radiation back to the emitter source where this radiation is readily absorbed. A second layer of soda lime glass located between the first glass layer and the photovoltaic cell array cuts down convective transfer to the silicon cells.

POWER CONTROL SYSTEM DESIGN

The TPV emitter power supply generates 450 watts of electrical power. For efficiency and to avoid power-conditioning apparatus such as inverters, a brushless d.c. motor was chosen for the major power load, the blower motor. The vent motor was also a brushless d.c. version. For convenience, 24 volt motors were chosen, but 120 volt versions may be more cost effective because of their use in current high-efficiency designs. A small amount of additional electrical power is required for ignition, controls, and safety monitoring.

There is more complexity in the control system for this self-powered furnace as compared with a conventional line-powered system. For this reason, we are proposing a microprocessor-based control unit (Figure 7) to take into account the maintenance of a battery that is required for the ignition sequence, as well as for the operation of the main blower motor for a short period of time after the heating cycle is complete. In addition, after each ignition, the furnace must operate long enough to recharge the battery independently of the space heating requirement of the dwelling. System monitoring for overheating motors, heat sinks, and the like is easily accomplished in this versatile control system. Desirable safety monitoring for carbon monoxide generation or oxygen depletion may also be included.

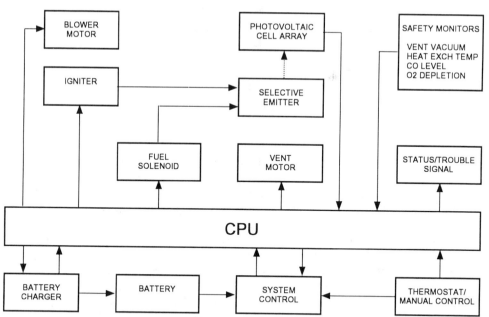

FIGURE 7. Self-powered warm air furnace control system

PROJECT STATUS

The TPV power supply containing the emitters, fuel/air delivery system, and glass shields has been installed in the warm air furnace and operated. The silicon convertible exitance has been measured. A subsection of silicon photovoltaic cells has been mounted on the forced convection heat sink, and the cell array's open circuit voltage, short circuit current, and fill factor have been determined. The measurements will be completed when the full complement of silicon photovoltaic cells is available.

CONCLUSIONS

The combination of cost effective silicon cells and selective emitters will provide a low cost basis for TPV, which will satisfy the cost constraints of the gas appliance market. High efficiency silicon photovoltaic cells and selective emitters, in turn, will become cost effective once a manufacturing base and a demand are established for these two components.

The TPV self-powered warm air furnace is a technically and economically viable product. There are, however, developmental and evaluation issues that must be completed before manufacturing status can be achieved.

APPENDIX

NO$_x$ EMISSION

INTRODUCTION

The operation of selective ytterbia emitters at high temperatures has several benefits in a thermophotovoltaic (TPV) power generator utilizing silicon concentrator cells. The high exitance increases silicon conversion efficiency and reduces silicon area per unit watt of electrical power generated. The ratio of silicon-convertible exitance to nonphotoconvertible exitance also increases with emitter temperature, which leads to higher conversion efficiencies. There is a significant concern, however, with elevated emitter temperatures, and that is the emission of NO$_x$.

STRUCTURE EVALUATED

A representative supported continuous fiber emitter (SCFE) composed of pure ytterbia fibers was selected for a measurement of NO$_x$ emission. The fibers of this emitter were supported by a 1.5 cm thick slice of Celcor (extruded cordierite from Corning) with a pore density of 47 pores per cm^2 (300 pores per square inch). Half of the pores, in a uniform pattern (alternate diagonals, actually) in the substrate were filled with a ceramic fiber bundle loop that extended approximately 8 mm above the emissive side of the substrate surface. A pictorial of the configuration is shown in Figure A1. The fiber bundle consisted of 300 filaments of ytterbia with a filament diameter of about 10 µm. On the plenum side of the substrate, the fiber bundles were dressed close to the substrate and anchored in place with alumina cement. The Celcor substrate, which was square in cross-section and measured 15.2 cm (6") along each side, had fiber bundle loops installed in a 13 cm (5") x 13 cm (5") active central area. That is, there was a 1.3 cm (0.5") margin around the periphery of the emitter with no fiber bundle loops and with those pores plugged by alumina cement on the plenum side of the substrate. The active area of this emitter was 161.3 cm^2 (25 in^2). The emitter structure was clamped to a flat sheet of stainless steel (1.5 mm thick) which had a 13 cm x 13 cm cutout to accommodate the active area of the emitter. Simple spring clips mounted on posts around the outer border were utilized to hold down the supported continuous fiber emitter firmly against the stainless steel mounting plate. A square annular gasket fabricated from 1.5 mm thick Fiberfrax (15 cm x 15 cm outside dimension; 13 cm x 13 cm inside dimension) served to seal the emitter to the stainless steel mounting plate. The mounting plate, in turn, was attached at its outer periphery to a plenum into which was introduced the air/fuel premix. The air/fuel premix was supplied via a 2.5 cm stainless steel tube (thin wall) attached to the back side of the plenum, as shown in Figure A2. A 7 cm diameter disc of stainless steel was positioned over the fuel/air mixture entry point inside the plenum to diffuse the incoming gas stream. Also, a chromel/alumel thermocouple was attached to the center of the Celcor substrate on the plenum side to monitor substrate temperature during operation. The thermocouple wire was routed through a small hole in the plenum wall and was sealed with RTV.

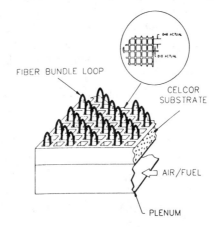

FIGURE A1. Supported continuous fiber system

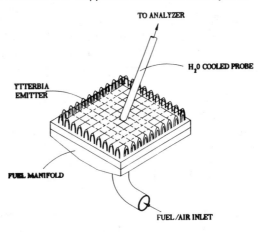

FIGURE A2. Water Cooled Quartz Probe Collects Exhaust Products Sample From Horizontally Oriented SCFE

TEST PROCEDURE

In order to avoid custom fixturing and to minimize the effects of ambient air dilution on the exhaust products, the plane of the supported continuous fiber emitter was oriented horizontally with the emission direction pointed up (vertically). The plenum was connected to the natural gas fuel supply, and the combustion air came from a laboratory compressor after passing through a water trap, an oil trap, and a pressure regulator. The air and fuel were metered with the aid of calibrated rotameters and supplied to the plenum as a mixture. A hot wire igniter (silicon carbide) was the ignition source. The tip of a water-cooled quartz probe was located in the center of the emitter about 1 cm above the ceramic fiber loops. The body of the probe was angled about $45°$ off vertical so that the combustible Tygon tubing attached to the end of the probe was out of the hot exhaust gas stream. The Tygon tubing from the probe was routed to an ice-cooled trap to condense the water vapor in the exhaust gas sample stream. The gas sample was analyzed quantitatively for several gas species. CO and CO_2 were measured with LIRA 3000 Infrared Gas Analyzers that sense the infrared absorption of these gases. NO was measured by a chemiluminescent reaction with ozone in an analyzer made by Thermo Electron's Environmental Instruments Division, Model 10. NO_x is measured by converting catalytically to NO at a temperature of $650°C$ and subsequently measuring the aforementioned chemiluminescent reaction. Oxygen was measured with a Thermo Instruments Model No. WDG-P Oxygen Analyzer that senses the voltage across a zirconia element that is generated by a difference in oxygen partial pressure on the two sides of the sensor. The element is operated at a high temperature ($700°C$) and ambient air is in contact with one side of the sensor. Prior to all testing, all the equipment was energized and allowed to equilibrate for one hour. The infrared instruments were calibrated with a test gas containing known concentrations of CO and CO_2 in nitrogen. The NO_x analyzer was calibrated with a test gas containing a known concentration of NO in nitrogen.

The emitter was powered at a number of fuel rates, and the primary combustion air was varied. Substrate temperatures (plenum side) were measured with the aforementioned chromel/alumel thermocouple.

EXPERIMENTAL RESULTS

The densities of the measured exhaust gas constituents, as well as the substrate temperatures for a range of fuel rates and excess air, are listed in Table A1. The flame temperature (adiabatic) has been inferred from the measured CO_2 percentage, and the calculated estimates are listed. Blue flame NO_x generation over the aforementioned temperature range is about 5 times higher than we have measured, Apparently, the ytterbia fibers serve to lower the combustion temperature from the blue flame condition to yield the lower emission of NO_x.

TABLE A1. Emissions Testing: Ytterbia Supported Continuous Fiber Emitter

Fuel Rate	Fuel Rate	Excess Air Ratio	CO_2	Adiabatic Flame Temp. (Calc.)	Adiabatic Flame Temp. (Calc.)	CO	NO	NO_x	O_2	Substrate Temp.
(Btu/ft^2-hr)	(watts/cm^2)		(%)	(°F)	(K)	(ppm)	(ppm)	(ppm)	(%)	(°C)
115,000	36.3	1.08	11.25	3457	2176	300	12	17	3.1	283
115,000	36.3	1.18	10.35	3300	2089	140	8.0	11	4.2	256
115,000	36.3	1.27	9.90	3221	2045	100	6.8	9.8	5.0	230
115,000	36.3	1.36	9.40	3126	1992	70	5.4	7.6	5.6	204
173,000	54.5	1.13	11.40	3483	2190	630	13	20	3.0	233
173,000	54.5	1.22	10.50	3330	2105	280	9.0	14	3.9	209
173,000	54.5	1.32	9.90	3221	2045	190	7.0	11	5.0	177
173,000	54.5	1.41	9.10	3070	1961	130	5.8	8.4	6.0	143
230,000	72.6	1.15	11.25	3457	2176	900	11	21	3.2	187
230,000	72.6	1.25	10.35	3300	2089	450	8.6	15	4.3	171
230,000	72.6	1.32	9.90	3221	2045	330	7.6	13	5.2	147
230,000	72.6	1.41	9.40	3126	1992	240	6.5	10	6.5	123

Utility Market and Requirements for a Solar Thermophotovoltaic System

Ken Stone
McDonnell Douglas Aerospace
5301 Bolsa Avenue
Huntington Beach, California 92647

Scott McLellan
Arizona Public Service
P.O. Box 53999
Phoenix, Arizona

Abstract. There is a growing need for clean affordable electric power generation in both the U.S. and internationally and solar thermophotovoltaic (STPV) can meet the needs of this market. This paper investigates the utility grid market applicable to a solar thermophotovoltaic power generating system. It finds that a large international electrical market and a smaller U.S. electrical market exist today but the U.S. market will grow by the year 2005 to a level that would easily support the high production level required for solar systems to be cost effective. Factors which could influence this market and the system characteristics considered by utilities in selecting future power systems such as levelized energy cost, dispatchability, environmental, etc. for both the grid and remote market are discussed. The main competition for this market and the operating performance of this competition are described. A conceptual design of a STPV power system is presented, the operation is described, and how the performance meets the utility requirements is discussed. The relationship between the cost of the TPV conversion unit and the system efficiency of the STPV system is given for both the grid and remote markets that it must meet in order to be competitive.

INTRODUCTION

McDonnell Douglas Aerospace (MDA), located in Huntington Beach, California, and has been involved in solar energy research and development for over 23 years. MDA was the design integrator for the Department of Energy (DOE) 10 MW$_e$ Solar One Central Receiver program and teamed with United Stirling AB of Sweden on a company funded Stirling Dish program (both shown in Figure 1). MDA has been investigating solar thermophotovoltaics (STPV) for both space and terrestrial application for the last five years (Ref 1).

Arizona Public Service (APS), located in Phoenix, Arizona is the largest public utility in Arizona and has a long-term goal to use solar technologies for power generation. The APS strategy is to begin with small scale tests, and gradually install larger and larger systems as the technologies become more cost effective. APS has been involved in the

FIGURE 1. MDA Dish at Solar One.

© 1996 American Institute of Physics

Solar One Central Receiver Program and is a participant in the follow on Solar Two Central receiver program. APS is installing various photovoltaic in on and off grid applications. Aps is also involved in the Dish Stirling development program.

APS and MDA are currently working together to define the future electric power market, type of markets, requirements for the different markets, the competition for the different markets, and evaluate some of the promising solar conversion systems for these markets.

MARKET OVERVIEW

Is there a market for a STPV system? Surveys indicate that a major international market exists at the present time and that a major domestic solar region market will emerge in the early 2000s. The U.S. market timing coincides with the availability of a commercial STPV system. This market is composed of both grid connected and non-grid connected or remote power systems. Also, between now and the year 2000 there will be a domestic market that can be used to support a STPV commercialization development program. For instance, Arizona Public Service has a goal of 12 MW of cost effective renewable energy systems between now and the year 2000.

The DOE SOLAR 2000 (Ref. 2) and Utility Data Institute (UDI) (Ref. 3) both report that nearly 550 to 600 GW of new generating capacity may be required worldwide from 1990-93 to 2000-03 which is a global rate of 1.9% per year. UDI states about half of all the 1993-2002 capacity block worldwide is not yet under construction. The U.S. will need over 100 GW of new generating capacity in the 1990s and another 90 GW in the first decade of the next century. Developing countries will require about 350 GW of capacity. Annual Energy Outlook for 1994 (Ref. 4) predicts that the U.S.'s electrical demand is expected to grow 1 to 1.5 percent per year to the year 2010.

The Southwest U.S. has the greatest resource potential for solar generation, however currently this region has a reserve generating capacity of about 25%. The utility industry is anticipating a competitive environment due to potential restructuring that may increase retail access to generation resources. These factors may cause the actual additions to vary greatly from current forecasts.

In a paper by DOE (Ref. 5), the increased capacity that will be required in the three Southwest states of California, Arizona, and Nevada is shown in Table 1. This increased capacity consists of peaking, intermediate, and base load capacity which are defined in Figure 2. Also shown in this table is the number of 30 kW STPV units required to meet this market. The percent of this market that STPV could capture would be related to how well it meets the market requirements which will be discussed later. An example of the planned yearly addition for Arizona's five major public utilities is shown in Figure 3 (Ref. 6) which is based upon 1992 data. APS's 1994 long range forecast for new generation is shown in Figure 4 and the California Energy Commission's forecast (Ref. 7) for California is shown in Figure 5.

Seventy-eight percent of the southern solar region market, for the years 2000-2002, discussed above is committed but UDI (Ref. 3) states that half of the worldwide required capacity is not yet under construction.

TABLE 2-1. Domestic Market in Solar Region

State	1996-2000	2001-2005	2006-2010
Arizona	60 MW	750 MW	1,000 MW
California		2,800 MW	7,617 MW
Nevada		560 MW	1,200 MW
Total Req.	60 MW	4110 MW	9,817 MW
No 30KW Units	2000	137,000	327,233
Units per year	500	34,250	81,808

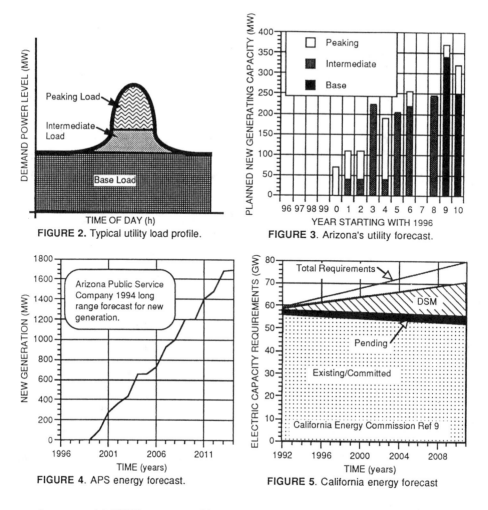

FIGURE 2. Typical utility load profile.

FIGURE 3. Arizona's utility forecast.

FIGURE 4. APS energy forecast.

FIGURE 5. California energy forecast

A commercial STPV system would most likely not be available until after the turn of the century at which time a small percentage of the available market would be a very sizable market. Between now and then, there is a market that can be used to defray some of the cost of development. There is a proposed 1000 MW Solar Enterprise Zone (SEZ) in Nevada which can be used in the development of the STPV system. This project is managed by Corporation for Solar Technology and Alternative Resources (CSTAR) which is a public corporation charged with developing solar and renewable energy technologies. They currently have a request for proposal (RFP) for 100 MW as a first step toward the 1000 MW. The results of a CSTAR survey indicates that DOE and DOD facilities will require in excess of 255 MW of power in the coming years.

The California Energy Commission (Ref 6) has been encouraging the development of renewable energy and states that 672 MW of renewable resources will be added by California's major utilities SMUD, LADWP, SDG&E, SCE, and PG&E. SMUD in its 1993 Integrated

Resource Plan (IRP) Update, which has long been a strong supporter of renewable energy, continued its commitment to renewable energy. SMUD currently plans to add an additional 300 MW (potentially 10,000 STPV units) renewable resources in the 2000 to 2005 time frame.

For the international market, the World Bank reports that an average of $60 billion has been invested annually in the developing countries but over 2 billion people still lack adequate energy for economic growth and basic needs. The electricity demand is predicted to grow at a rate of 7% per year for the next 20 years. The Asian countries alone will require 285,000 MW of electric power by the year 2000 (Ref. 7). Some of the countries that are expected to add significant new capacity are Mexico, India, Pakistan, China, South America, and Mediterranean Africa.

MARKET TYPE

The market discussed above consists of both grid and off-grid remote systems. The grid connected systems will consists of base, intermediate, and peaking power. Therefore the power systems must be capable of meeting the requirement for these type of capacities economically.

In the past the US. utilities have built large central generating plants but now many of them are investigating smaller distributed systems. Many utilities have aging distribution systems. Small distributed generating systems could delay installing expensive distribution systems. Also as cities at the end of the distribution system grow, their needs may outgrow the capability of the distribution system. The size of this market has not been quantified but could be a substantial niche market for distributed type system such as a STPV power generating system.

Generally, when remote applications are discussed, one thinks of the international market. U.S. utilities also have a need for remote or off-grid generating systems. APS has customers that have or are planning a community development in remote areas where there is presently no distribution system. There are not enough customers at a particular location to support the high cost of a power distribution grid Therefore, APS is currently investigating remote type power systems. For the most part, these systems must be capable of generating power upon demand, 24 hours a day. APS has just started investigating this market and has not quantified the size of this market. Other utilities in the solar region have a similar situation.

The international market also consists of both grid and off-grid power generating system improvements. The industry growth in many of these developing countries has far exceeded the electric generating capacity. Cities with millions of inhabitants go without power each day which is hurting their economic growth. As is well known, there is a large international market for off-grid power generating systems. This market is vast and consists of systems of a few kilowatts to megawatt size systems that generate power upon demand to systems that generate power only a few hours a day, e.g. such as to pump water. Although many of these markets are economically poor and can hardly afford the first few kW, experience has shown that their need grows later and these markets can now find the resources to increase their capacity.

MARKET REQUIREMENTS

There are many factors that are considered by a utility in the selection of a new power plant and when they will add new generating capacity. Some of these factors are discussed below. Just what importance these factors play in the decision process will be dependent upon the particular condition of the utility at that time, regulatory constraints, population and industrial growth, competition, customer desires, etc. Some of the factors are:

Levelized energy cost (LEC) is a measure of the life cycle cost of the plant divided by the electrical energy generated over a time period of 20 to 30 years. LEC is defined as:

$$LEC = \frac{\text{annualized measure of total system resultant costs}}{\text{annual energy output expected from system}}$$

where system cost includes manufacturing cost, balance of plant cost, installation cost, operating and maintenance (O&M) cost, fuel cost, taxes, cost of money, etc. The actual LEC for a given system will vary with each utility because of the different economic base, i.e., tax rate, cost of money, labor cost.

Capacity factor is another important consideration and is defined as:

$$\text{Capacity factor} = \frac{\text{Annual generated energy}}{(\text{Rated power}) \times (\text{hours in a year})}$$

If power can only be generated during a part of the day such as when the sun is available, then the system would need some other appealing characteristic to be of interest to the utilities. This is a very limiting consideration for most renewable systems unless they have a storage system or have a hybrid mode of operation.

Dispatchability is the ability to generate power upon demand. As discussed above, the utility demand load will vary over the day. Typical load profiles for Sacramento Municipal Utility District (SMUD) and APS are shown in Figure 6 for the summer and the winter period. For most utilities the peak demand load does not correspond with the peak solar noon irradiance. Therefore if the power output of the plant is proportional to the solar profile, the system will not match the summer demand profile and the utility must rely on some other power source. As shown in this figure the peak summer load for both utilities occurs in the late afternoon and at a different time for each utility. The utility will also experience a large fluctuation during the day as illustrated in Figure 7 by a typical load profile for a residential home in the SMUD area. The fluctuations result mainly from weather changes.

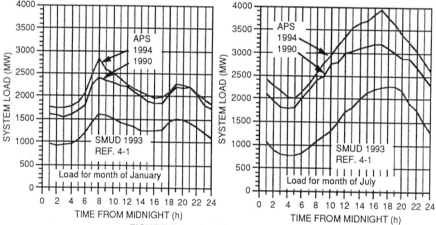

FIGURE 6. Load profiles for winter and summer.

Power modularity is a characteristic that may become much more important to utilities in the future. It is the ability to add power in small increments. It takes a long time (environmental, studies, permits, construction) between the utilities decision to build a new power plant and theproduction of its first power (3.5 or more years), utilities must forecast their needs years in advance. Because of the uncertainty in the ability of utilities to forecast their future needs, when the plant comes on-line, the utility may not need all of the new capacity or it may need more. Power plants that have a short construction time such as modular solar plants, short environmental permitting time, and the ability to be added in small increments in the future have a very desirable feature. The ability to economically add power in small increments allows the utility to better follow its load growth demand profile without having to buy power from another utility or having excess power generating capacity.

Distributed power system is the ability to install small power generating plants at various sites. In the past, utilities have installed large central power generating plants which were normally located at some distance from their customers. This approach results in several cost disadvantages for the utilities, 1) the power distribution system is a very costly thing to install, 2) there is a transmission line cost which must be passed onto the rate payer, 3) a growth rate that outgrows the capability of the distribution

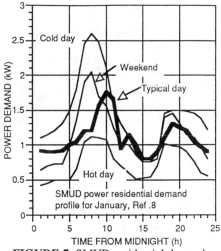

FIGURE 7. SMUD residential demand.

system which means a new line must be installed, 4) many utilities have aging power distribution systems that will not be able to handle the future demand load, 5) reduced site flexibility. Many utilities are investigating distributed power system that could be located strategically along their distribution systems and thereby avoid or delay installing or upgrading expensive power distribution systems and reducing line losses.

Installation time is the time from when the utility makes a decision to install a plant until the plant is on line generating power. This is the time to conduct environmental studies, obtain the required permits, install the plant, and checkout the plant. During this time the utility has invested resources, time, and money; therefore, the utility would like to keep this time a short as possible.

Environmental pollution is presently a consideration because of the high cost to meet current standards. Because of pollution credits this could become even more important in the future. In California Public Resources Code, Section 2500.1(c) directs the California Energy Commission (CEC) to include values for costs and benefits in calculating the cost effectiveness of energy resource options. These air emissions considered include nitrous oxides, sulfur dioxide, carbon monoxide, reactive organism and particulates. For every ton of air pollution that is expected to be produced by a plant in California, four tons of air pollution will need to be permanently removed from the load area before a new plant can be built and operated. In California a composite emissions values for out-of-state power emission are shown in Table 2. These costs are added onto the levelized energy cost in the planning of new plants. In this way, omissions costs influence the selection of new power plants. Nevada has a similar emission cost which utilities must consider.

Fuel price and dependence is a major concern to utilities. The major cost of operating their current power plants is the cost of the fuel. This cost is something that is not within their control and subject to inflation and the many factors which affect the fuel supply and demand. Those influences include foreign conflict, strikes, cold or hot streaks, hurricanes and tornadoes.

TABLE 2. Emissions costs

$/Ton/Year	Northwest	Southwest
Nitrogen Oxide	$730	$760
Sulfur Oxides	$1,500	$1,500
PM10	$1,280	$1300
ROG	0	$5
Carbon	$28	$28

Remote systems - The remote or off-grid market has a wide variety of requirements which can be divided into three categories. The first category of this market is for systems that only need to supply power when the sun is available. Applications might be for such things as water pumping, operating a small machine like a sewing machine, etc. It has been found that with time the needs of the customers in this category grow and before long they need more power and want it at times when the sun is not available. Much of this first category market then transitions into the next market category which wants power from early morning until late evening. The majority of this market requires electric power for lights, TVs, refrigerators, etc. which require the generating system to supply power at times when the sun is not available. The needs of this market category grow until the third market category is reached. This category requires power 24 hours a day in order to sustain the economic growth of their community and the luxuries they wish to enjoy. The power system must not only supply power 24 hours a day but respond to demand fluctuations in the load.

System maturity - Utilities will consider the risk involved with the product. Is it well proven? Will the manufacturer be around to stand behind the warranty? Utilities are very concerned about the O&M costs of power systems particularly ones that are not in widespread use in the utility market. Before a system can get wide commercial acceptance with utilities, it must have proven performance record. Even with proven systems, utilities are concerned that the manufacturer will be able to stand behind his product and supply spare components for many years.

MARKET COMPETITION

Natural gas powered systems in the continental U.S., in the form of combine-cycle (CC) and combustion turbines (CT), because of their high efficiency, adequate supplies, and low fuel prices, will be the dominant grid electric-generating power system installed in the Southwestern U.S. over the next 10 to 20 years. The LEC of such plants are very dependent upon the price of natural gas. Estimates of the natural gas price as predicted by different organization are shown in Figure 8 and in Table 3. The California Energy Commission forecasts that gas prices will remain low or increase only moderately over the next two decades, (1.1 percent per year expected average with a possible high rate of a 2.3 % and a low rate of a 0.6 % depending upon economical conditions. The natural gas price estimated by SMUD based upon Cambridge Energy Research Associates is shown in Figure 8. This price includes intrastate costs added in. An estimate of the levelized energy cost based upon APS experience is shown in Figure 9 as a function of the capacity

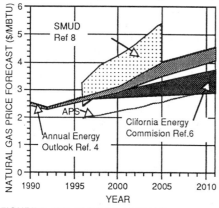

FIGURE 8. Natural gas price forecast.

TABLE 3. Natural gas forecast for 2010.

Forecast	Price
Gas Research Institute	$ 3.22 / MBTU
Annual Energy Outlook	$ 3.47 /MBTU
National Petroleum Council	$ 3.77 /MBTU
American Gas Association	$ 2.89 /MBTU
Data Resources	$ 3.88 /MBTU

factor and for a low and high gas price. The low gas price used was the APS estimate shown in Figure 8 and the high gas price used was the expected SMUD value which is very close to the DOE high economic growth estimate. A 3 % inflation rate was used. The O&M models and efficiency of the CC and CT systems were based upon APS experience. The California Energy Commission (CEC) in it's 1992 Energy Technology Status Report (Ref. 9) compares the estimated levelized energy cost for commercially available and not yet commercially available electricity generation technologies. An estimate by the CEC for available natural gas technologies are given in Table 4. This uncertainty in levelized energy cost presents problems to the U.S. utilities in controlling the price of their product and is one of the reasons they are interested in alternative energy sources.

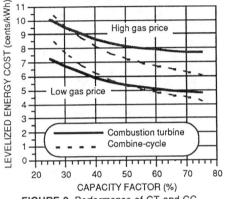

FIGURE 9. Performance of CT and CC.

TABLE 4. CEC LEC Estimate for gas systems.

	1989 Ref. yr (cents/kWh)	
	Constant $	Nominal $
Baseload(60-75% CF)		
1. Conventional Rankine	5.0-7.2	6.8-9.8
2. Supercritical Rankine	5.1-8.3	7.0-11.2
3. Conventional combine cycle	4.0-7.4	5.6-10.2
4. Steam Recuperated gas turbine	5.1-8.1	7.2-11.2
Intermediate(20-35% CF)		
1. Conventional combine cycle	5.3-11.9	7.2-15.8
2. Steam Recuperated gas turbine	7.1-14.1	9.7-19.1
Reference 9.		

The major competition for the remote US market is diesel powered generators and/or photovoltaic (PV) systems. PV systems are the main competition for the first category market discussed above and diesel generating systems are the main competition for the next two category markets. The major cost for such a system is the cost of the fuel, transportation of the fuel, and life overhaul cost of the diesel engine. Depending upon the location of the remote site, the transportation cost can be the most significant cost. The LEC for these markets in the US is not as easily defined but estimates range from $0.25/kWh to $1.00/kWh. In an effort to decrease the cost, PV panels are being integrated into the diesel generating systems. This not only decreases the cost of the fuel but also decreases the annual overall cost of the diesel generating system. A configuration of a hybrid PV diesel generating system is illustrated in Figure 10. Batteries and a converter are incorporated in order to respond to load fluctuations and provide some storage when the PV panels are generating power and there is no demand.

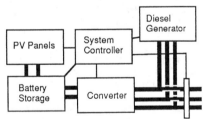

FIGURE 10. PV-Diesel system.

The competition for the international market is very dependent upon the region but oil based systems are probably the most prevalent competition for solar based systems in the developing countries. The LEC is difficult to quantify because many of the power companies are subsidized by the government.

DESCRIPTION OF A STPV SYSTEM

The operating principle of an STPV system is illustrated in Figure 11. A concentrator tracks the sun and focuses the energy on the receiver of the power conversion unit The sun's energy

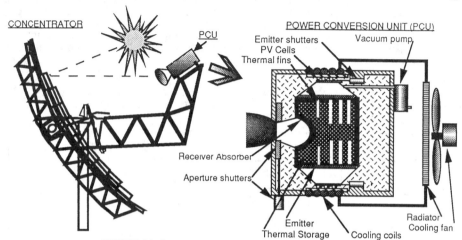

FIGURE 11. Components and operation of a STPV system.

is transferred from the receiver to the thermal storage unit. On the back side of the thermal storage unit is the emitter surface. Opposite the emitter is an array of PV cells which convert the IR radiation from the emitter to d.c. electrical energy. Cooling coils are located on the back side of the PV array. Fluid flows through the cooling coils removing the waste heat to keep the PV cells at an optimum operating temperature. The fluid flows though a radiator where a fan blows air through the radiator removing the waste heat to the atmosphere. An aperture reflector is used at the focus point to capture fringe energy and to reduce thermal losses by reducing the aperture opening. Just behind the aperture opening is a shutter system. When the suns energy is not available, the shutters are closed to reduce the energy loss from the cavity receiver. There is a second shutter system that controls the IR radiation to the PV cells. This allows the system to store and shift power production.

The TPV conversion system can also be configured to operate as a hybrid system. A combustion burner would be integrated into the design of the TPV unit that could burn natural gas, biomass gas, or hydrogen. The gas could be burned when the sun's energy is not available such as at night or on cloudy days. It could also operate in a mode that supplement the sun's energy when the incident solar irradiance is not sufficient to produce maximum power. Gas could be burned at a rate to make up for the low sun energy during early morning, late afternoon, cloudy days, or at night. The unit would provide a constant maximum power output 24 hours a day. A hybrid system would now be considered to have base load capability but with the advantage of decreasing the dependence upon the fuel source.

PERFORMANCE OF A STPV SYSTEM

The estimate of the peak power performance of a STPV system is shown in Figure 12. This performance estimate is based upon the testing and analysis performed by MDA and others. A typical sun irradiance profile for Barstow, CA and the generated power profile for a STPV only system is shown in Figure 13 for a summer day. As the sun rises in the morning the sun irradiance increases as the sun gets higher in the sky because of less atmosphere. A few hours after solar noon, the irradiance level starts decreasing. The power output from a solar only STPV system will follow the sun irradiance profile very closely as shown in the figure. There is a slight delay after sunrise before the emitter reaches operating temperature and power production begins. From this point on the STPV power output follows the sun irradiance profile very closely.

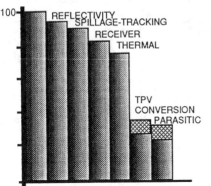

FIGURE 12. Peak power performance.

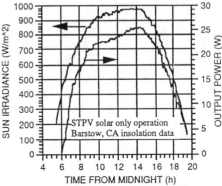

FIGURE 13. STPV power profile.

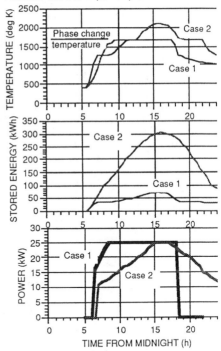

FIGURE 14. STPV with storage.

If thermal storage is added to the system, then the generated power profile can be controlled to provide a desired shape as illustrated in Figure 14 for two different cases. The object for case 1 was to generate maximum power over the day and for case 2 was to generate a varying profile over the day. The thermal storage material considered for this analysis was silicon which melts at 1682° K and has a heat of fusion of 396 cal/gm. The size of the concentrator was increased from the size used in Figure 13 in order to provide additional energy for the storage system. As the sun rises in the morning, the temperature and energy level in the thermal storage unit rises. When the temperature reaches the operating temperature, the emitter shutters are open for case 1 and partially open for case 2 and power production begins. As the sun irradiance increases, the power output increases directly for case 1 but the emitter shutters are used to control the power output level for case 2. The excess energy during this time for case 2 accumulates in the thermal storage system as shown. As the sun irradiance starts to decrease after solar noon, the energy from the thermal storage unit was used to maintain the desired output power level. When the sun sets, the aperture shutters are closed to stop IR losses out of the cavity receiver. As the power production continues, the temperature of the emitter continues to drop. When the emitter temperature reaches the lower operating level, the emitter shutters are closed to retain this energy over night as shown. For case 1 the maximum output power level is maintained until just after sundown and power generation is continued until after midnight for case 2.

The use of thermal storage with the STPV system can be used to reduce the utility peaking power problem discussed previously. The APS July demand profile is shown in Figure 15. If the output power from a solar only STPV power plant is subtracted from this demand profile, the peak is decreased and occurs later in the day as shown in this figure but a solar only system can not satisfy the peak demand entirely. In addition, the base load required has been decreased in the morning hours. The utility would have to decrease the base load generating capacity and use its peaking and intermediate generating capacity to generate evening and night power. With thermal storage and using the control logic of case 2 in Figure 14 the peaking load can be reduced to the base load level as shown in Figure 15.

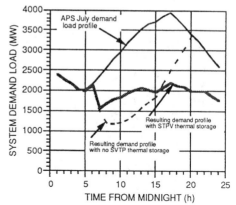

FIGURE 15. APS load with STPV.

COMPETITIVENESS OF A STPV SYSTEM

Levelized energy cost (LEC) - The STPV system described above consists of two basic elements, a concentrator and a TPV conversion unit. The cost of a dish concentrator that would meet the requirements of a TPV unit has been estimated by various manufactures because of the past years of development, detail design, fabrication, and testing. Due to lack of maturity of the TPV design, the many possible design variations, and components that are not currently manufactured, obtaining a reasonable cost estimate of a TPV conversion unit is difficult at this time. Requests for TPV cost information to date have only indicated that it should be in a viable range. In addition to the cost uncertainty, there is also an uncertainty about the power performance of the system. Therefore rather than trying to estimate the LEC with these uncertainties, it was felt it would be more useful to define the performance/cost that a TPV conversion unit must meet in order to be cost competitive with CT or CC systems. The best estimate at this time for the TPV conversion unit performance/cost is shown in Figure 16 for both utility grid application and a remote market application. A major effect upon the LEC of the system is the production level. The system installed cost decreases as the production level increases. The production level used in this analysis is for approximately 500 units per year or 1.5 MW per year is a very small part of the expected market.

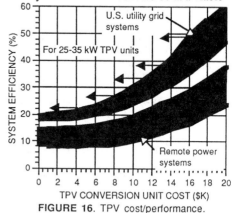

FIGURE 16. TPV cost/performance.

Capacity factor for a solar only system would be about 26%. If thermal storage is added and the size of the concentrator is increased to charge up the storage, then the capacity factor can be increased from 26% to 50% or higher and still remain a cost effective system. The capacity factor for a hybrid STPV plant would be approach 100% which is higher than a CT or CC plant. A STPV plant can actually achieve a higher capacity factor than a CT or CC plant because these large plants are normally shutdown for periodic maintenance annually. Also these plants do have single point failures which can take the entire plant off-line. Since a STPV power plant is made up of many small units, single or multiple units can be taken off-line without significantly changing the output power of the STPV plant. The only single point failure that can take the STPV plant off-line is with the utility substation.

Dispatchability - A solar only STPV power plant would have very little dispatchability as discussed above. A STPV system with thermal storage or gas hybrid would provide the dispatachability desired by utilities.

Power modularity - The power of a STPV plant can be increased in small increments (30 kW, 100 kW, 500 kW, etc.) If growth is designed into the field power distribution system, then the size of the plant can be increased at any time. In the matter of weeks, megawatts of power can be added to the plant. During the summer of 1994, APS peak demand exceeded its generating capacity, therefore it had to buy peak power from other companies. This generally results in additional cost to APS because of higher rates for the power. Since the peak demand is rarely exceeded at this time and CC and CT plants can only be added in large increments it is not cost effective to add a CC or CT plant at this time.

Distributed power system - Each STPV unit operates independent of the next unit. Therefore, units can be installed in small groups at various sites. This results in additional cost advantages to the utility from the site flexibility and the modularity of a STPV plant. The utility can locate small distributed power plants throughout their distribution system. This would not only allow the utility to add small increments of power to meet the peaking power requirements but would reduce transmission losses as well.

Installation Time - A STPV plant would have a very short installation time when compared to a CT or CC power plant. As illustrated in Figure 17, the time to complete the environmental studies is less, the time to obtain permits is less, and the time from the start of construction to first power is less. In Arizona the time to perform the studies and obtain the necessary permits ranges from 3.5 to 4.5 years for a CT or CC plant and longer for other types of plants. This time is generally much less for renewables. The California

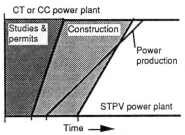

FIGURE 17. Installation time.

a Energy Commission gives special consideration to renewables in the research or early development phase and these requirements can be met in under a year. This will reduce the cost and effort of this phase for a renewable energy plant such as a STPV system. Although the actual construction time of a completed STPV plant may be similar to a CT or CC plant, the time to first power generation is much shorter. The time to install a single STPV unit is a matter of hours once the field site has been prepared (months) and foundations installed (days). Since each STPV unit operates independent of the next, the STPV unit can start generating power and income revenue long before the plant is completed and much before a CT and CC power plant is on line. For a 50 MW plant this could mean an additional $13 to $15 million for the shorter study/permit phase (1.75 years) and $3-$4 million of revenue for the shorter construction phase in a solar only operation.

Environmental pollution - The solar only STPV plant would not produce any environmental pollution. This would result in pollution credits available for non polluting power generation. In California and Nevada this would be a consideration in the future planning of utilities.

Fuel price and dependence - The addition of STPV power plants would make the utility less dependent upon both price fluctuation of the fuel and the availability of the fuel since the solar only STPV system does not require fuel.

Remote systems - The operating characteristics and the flexibility of the STPV system make it a very desirable candidate for remote applications. A solar only STPV system could replace the solar panels used in the hybrid PV-diesel system depicted in Figure 10. Estimates of most concentrating systems are generally less costly than flat plat PV systems, therefore, the total system cost would be less. If thermal storage were added to the STPV system, then the majority of the batteries could be eliminated. Battery cost, maintenance, and replacement are a major cost of the remote system defined in Figure 10. The incorporation of hybrid operation would eliminate the need for the costly diesel system and batteries altogether.

System maturity - The maturity of the system at this time is an issue. Except for a few isolated single TPV systems, there are no large scale systems to generate the data base required for performance and O&MV to be credible with the utilities.

CONCLUSION

There is a market for a solar thermophotovoltaics power generating system if it can meet the market requirements in terms of cost and performance. Although the US utility grid connected market does not emerge until the turn of the century, the international grid market exists today and is growing every year. The remote market, both in the U.S. and internationally exists today and is also growing each year. A small percentage of this estimated market could support a very high production level of STPV systems. The general operating characteristics of a STPV system can meet the wide range of requirements of this market. Although there is a lot of uncertainty in the LEC of a STPV system at this time, it appears the system can be competitive. The STPV system can meet the requirements of capacity factor and dispatchability. A STPV single unit of 25 to 50 kW operates completely independent of each other, therefore it is ideally suited for adding small power increment (power modularity) and adding small groups at various locations in the utility distribution area (distributed power system). The non-hybrid STPV unit is environmentally benign and makes utility planning easier with its shorter installation time. A STPV plant would also reduce the utility dependence upon fuel price fluctuations. A STPV system with thermal storage and/or with hybrid operation meets the requirements for the remote market. With a funded development program a STPV unit should be able to satisfy the utility risk factors.

REFERENCES

1. Stone, K. W., Kusek, S. M., Drubka, R. E., Fay, T. D., "Analysis of Solar Thermophotovoltaic Test Data From Experiments Performed At McDonnell Douglas," The First NREL Conference on Thermophotovoltaic Generation of Electricity, Copper Mountain, CO, 1994.
2. Annan, R. H., Solar 2000, Vol. I, Office of Solar Energy Conversion, United States Department of Energy, Feb. 1992.
3. Bergesen, C.A.E., World Directory of New Electric Power Plants, Utility Data Institute, UDI-2460-94, March 1994.
4. U.S. Dept. of Energy, Annual Energy Outlook 1994 With Projections to 2010, DOE94-005938, January 1994.
5. Gronich, S., Bradley, J, DeGroat, K, tanner, S., "Solar Power Commercialization in the Southwest: A Market and Technical Assessment," 4th ASME/JSME Thermal Engineering Joint Conference, Maui, Hawaii, March 19-24, 1995.
6. Imbrecht, C. R., 1992 Electricity Report, California Energy Commission, P104-92-001, January 1993.
7. Worldbank
8. 1995 Integrated Resource Plan, Volume II, Sacramento Municipal Utility District, March 1995.
9. Blevins, B. B., 1992 Energy Technology Status Report, California Energy Commission, November 1992.

Thermophotovoltaic Energy Conversion: Technology and Market Potential

Leon J. Ostrowski and Udo C. Pernisz
Dow Corning Corporation, Midland, MI 48686

and

Lewis M. Fraas
JX Crystals, Incorporated, Issaquah, WA 98027

Abstract. This report contains material displayed on poster panels during the Conference. The purpose of the contribution was to present a summary of the business overview of thermophotovoltaic generation of electricity and its market potential. The market analysis has shown that the TPV market, while currently still in an early nucleation phase, is evolving into a range of small niche markets out of which larger-size opportunities can emerge.
Early commercial applications on yachts and recreational vehicles which require a quiet and emission-free compact electrical generator fit the current TPV technology and economics. Follow-on residential applications are attractive since they can combine generation of electricity with space and hot water heating in a co-generation system. Development of future markets in transportation, both private and communal or industrial, will be driven by legislation requiring emission-free vehicles, and by a reduction in TPV systems cost. As a result of "moving down the learning curve," growing power and consumer markets are predicted to come into reach of TPV systems, a development favored by high overall energy conversion efficiency due to high radiation energy density and to high electric conversion efficiency available with photovoltaic cells.

INTRODUCTION

Improvements in the technology of alternative energy sources have met with increased demand for user-friendly and environmentally compatible methods for generating energy. For this technology to be commercially successful it will be required that those systems are economically competitive with existing alternatives. After reviewing the components of TPV systems, we focus at the external trends which are driving the commercial development of this technology and its maturation through its diffusion into an expanding applications base.

This paper is a summary of the material presented as a poster at the 2nd NREL Thermophotovoltaic Generation of Electricity Conference.

© 1996 American Institute of Physics

TECHNOLOGY OVERVIEW

Thermophotovoltaics (TPV) requires several stages of energy conversion: from the primary energy source to heat at the emitter temperature, then to radiative energy in a wavelength range compatible with the photovoltaic receivers which finally convert the primary energy to electricity. The photocell technology, evolved over many years of solar energy conversion applications, is highly developed but has to be modified to match the wavelength range of the radiation available at terrestrial temperatures. For our purposes, the primary source of energy is a gas or liquid which converts its chemical energy content into heat by combustion. Table I shows some of the alternatives available for each of the stages.

A schematic of a typical gas-fired TPV generator is shown in Figure 1. The emitter is fabricated from a high emissivity material and assembled for efficient heat transfer from the combustion zone to its surface; an example of one assembled from SiC ceramic parts is shown in Figure 2. The emitter is enclosed by dichroic reflectors to contain radiation not usable by the photovoltaic cells. Table II lists the key components and gives a rough estimate of the relative systems cost distributed to its key components.

An analysis of the overall systems efficiency is depicted in Table III; it shows estimates of typical performances and losses of components without attempting to indicate areas for optimization. Table IV compares solar energy conversion with thermophotovoltaic conversion emphasizing the latter method's high power density which allows such systems to be very effective. As a consequence of the high power volume density, comparatively small systems yield large amounts of electric power. Other advantages of TPV energy conversion are also listed such as operation free of CO and hydrocarbon emission with very low NO_x content as a result of controlled complete fuel combustion at relatively low temperature. Additionally, the system can be desigend for low heat loss through waste heat recuperation at the fuel intake, and by utilizing the sytem in a co-generation mode if desired.

MARKET POTENTIAL OVERVIEW

The external trends driving the commercial development of this technology are shown in Table V. The four major categories for these forces, namely economical, social, political, and technological, are primarily positive trends. Commercial viability will evolve when technical trends from two or more areas converge and when there are no major negative trends. It appears that the trends discernible in the TPV energy field are mainly driven by forces in the social and political arena.

Current economic trends are favorable but still weak. The economic trend will be strengthened by further technical development and a better understanding of the competing alternatives. In addition, the economy of scale will improve by market acceptance and through developments of the market infrastructure. The application areas for TPV technology are listed in table VI according to the major power user segments. To help mature the technology, one would first focus on high value applications, and next focus on high end user need. This is where the applications are likely to achieve success, allowing the infrastructure to begin its development. such markets are seen for military applications and high value recreational uses. The market assessments given in the table (under need, value, and size) reflect a qualitative ranking based on the authors' judgements; they could be perceived differently from different vantage points of the market.

An attempt at projections of market size and future systems cost was based on interviews with select people in the industry. The market analysis has shown that the TPV market, while currently still in an early nucleation phase, is evolving into a range of small niche markets out of which larger-size opportunities can evolve. Early commercial applications on yachts and for RV's which require a quiet and emission-free compact electrical generator fit the current TPV technology and economics. Specific end users needs can be identified and brought into the design of new systems packaged in line with such market segments. As the market becomes larger, its value decreases, an effect which will make the consideration of second-tier markets attractive: commercial emergency back-up power, residential co-generation, and remote power needs. Follow-on residential applications are especially attractive since they can combine generation of electricity with space and hot water heating in a co-generation system. Another application as a back-up system for extended power failures would not only replace gasoline-powered systems but could also provide uninterruptible power for electronic equipment. Large "day-after-tomorrow" markets are seen in power grid peak load shaving and in transportation. The latter comprises both communal and private as well as industrial applications and will be driven by legislative developments of emission-free vehicles and by a reduction in TPV systems cost. The technology is particularly suitable here since the comparatively low temperature maintained in a controlled chemical oxidation of the fuel prevents the formation of nitrous oxides and the emission of unburnt hydrocarbons.

As a result of "moving down the learning curve," growing power and consumer markets are predicted to come into reach of TPV systems, a development favored by high overall energy conversion efficiency due to the high radiation energy density and to the high electric conversion efficiency available with the photovoltaic cells. The available PV technology is at a cost of $6 - $10 per watt today. To break into the other markets will require that TPV be capable of equal or lower costs — ultimately requiring a cost of 1 $/W for power grid applications. As shown in Table VII, the market size is a function of the cost per watt of the TPV system. Figures for this quantity, and for the dollar value of the market, were obtained from knowledgeable experts in the TPV industry. The market size in terms of number of TPV units was calculated assuming a constant unit size of 10 kW; this then dictates the dollar value of the unit. The real market is, of course, more adequately described by a distribution of unit sizes around a mean.

ACKNOWLEDGEMENTS

The following have contributed to the material presented here: Larry G. McCoy, Timothy E. Easler, and Kenneth G. Gruszynski from Dow Corning Corporation, and Michael R. Seal, Vehicle Research Institute, Western Washington University.

TABLES

Table I. TPV Systems Assembly

Key Component	Alternative	Example
1. source of energy	chemical:	combustion of gas or fluid
	radiative:	capture sunlight with large concave mirror, focus
	nuclear:	heat from fission reactor
2. emitter	black body	
	heat reflecting filter	
	selective emitter	
3. photovoltaic cell	small band gap	III-V: GaSb, In GaAs
	standard PV cell	IV: x-Si, x-Ge

TABLE II. TPV System Components

Key Component	Function/Feature	% Cost	Implementation
Radiation Source	**fuel ⇒ heat**		
Burner	gas combustion	10	SiC burner/emitter
Emitter	SiC bulk, $\varepsilon = 0.95$	10	combination
Radiation Shifter	**heat ⇒ radiation**		
Dichroic Filter	reflects unusable heat back to emitter	10	quartz tube, ITO coat or other technology
Collector	**radiation ⇒ electr.**		GaSb, $\eta = 39\%$
PV cell		35	
Recuperator	gas pre-heat cell cooling	20	honey-comb SiC silicone fluids
Packaging	fuel regulator	10	
Housing			silicone sealants
Electronic Contr.	power conditioning	5	microprocessor, ROM
Sensors	operating conditions		power electronics

TABLE III.   TPV Efficiency

overall chemical-to-electrical conversion   17%

	power density [kW/m²]	efficiency	losses [kW/m²]
electricity generated	37		
		PV cell	39%
received at PV cell	94		
		optics	80%
radiated to PV cell	117		
convection filter			(70%) 9
reflected back	350		
radiated off	532		
blackbody equiv.	560	emissiv	95%
delivered to SiC	182		
		heat exchanger	85%
generated by combustion	**214**		

94

117

56

350   532

SiC   ε = 0.95
@ T = 1500 °C

fuel: nat.gas ==>

Table IV. Advantages of TPV Energy Conversion

	unit	natural sun	man-made source	
temperature	°C	5627	1400	1700
	K	5900	1673	1973
radiant power	W/cm²	0.1	44	86

very high energy density:
no radiative losses: all radiation captured by PV cells
low temperature combustion: no production of NO_x
— environment–friendly
complete combustion: no hydrocarbon or CO emission
— "Smog-Free"
low excess heat loss: recuperation/cogeneration

Table V. Driving Forces for Alternative Energy Source

economical	short term: niche markets	minimal impact
	long term: peak load power shaving	adv. over coal
	cogeneration	home/appliances
	more disposable income	
social	clean power	low NO_x & hydrocarbon
	high recreational value	quiet, light, mobile
	residential emergency power	
political	international trade	oil import/export
	energy conservation	resource management
	legislated alternate fuel mandates	govt. funding
	military applications	
technological	efficiency increases	system cost reduction
	burner/emitter unit improvements	

Table VI. TPV Markets: Applications Analysis

Application Use Trend	User Need	Market Value	Market Size	Ext. Funds	Competing Alternatives
Recreational • yachts • RV units	high	high	sml	—	gasoline engine generator, battery (rechargeable)
Commercial • emergency back up	high	med	sml		diesel generator
Residential • cogeneration • electr./heatg (hot water heater + electric power)	med	med	lrg	yes	electric power grid
Military	high	high	med	yes	intern. combustion (IC) engine
Remote • remote transmitter • cathodic protection • water pumping (3rd world dev.)	med	med	med		solar photovoltaics diesel generator
Transportation • low emission fleets • hybrid automobile & busses (publ.transp.)	med *⁾	low	lrg	yes	IC eng. & add-ons **fuel cell**
Electric Power • peak loading • grid extension	low	low	lrg		electric power grid
Space Missions • satellite	low	high	sml		photovoltaics, fuel cell, nuclear

*⁾ legislative action in several states will set regulatory minimal requirements

Table VII. Cost/Size Effect and Market Potential

	Energy Cost $/W	Typ. Size kW	Unit Cost $
Solar PV at 6... 10 $/W today ⇒	0.5	100	50 000
	1.5	10	15 000
	5	1	5 000
	15	0.1	1 500

Cost		Estimated Size US Market		Likely Markets to evolve
Energy $/W	System k$	# unit	G$ (10^9)	
5	50	6×10^3	0.3	military, recreational
2	20	8×10^4	1	commercial, remote
1	10	10^6	10	electric power, transportation, residential

assume constant system unit output of 10 kW

FIGURES

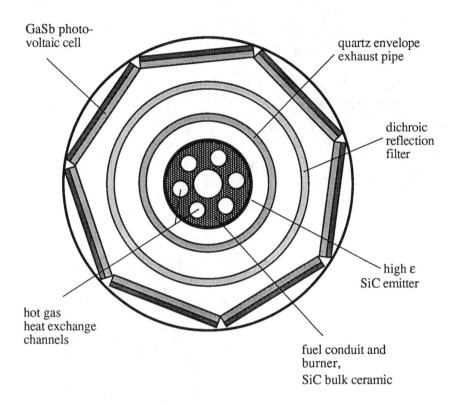

Figure 1

Schematic of thermophotovoltaic generator using integral burner/emitter of SiC and dichroic filters, e.g. quartz envelope and second cylinder with ITO (indium-tin oxide) filter. The GaSb photovoltaic cells and their positions are not drawn to size; the cooling features are omitted.

Figure 2

SiC emitter parts fired from the isopressed and machined green body.
Top left: Ledge on top (right) and lip on distance ring (left) allow the elements to be stacked at a pitch to obtain optimal mixing and contact to the SiC internal surface of the hot combustion gases. Diameter of the parts is 38 mm.
Botton right: TPV emitter stack (incomplete) assembled from the SiC parts.

SESSION III:
OPTICAL SYSTEM DEVELOPMENT

A Small Particle Selective Emitter for Thermophotovoltaic Energy Conversion

D. L. Chubb* and R. A. Lowe+

*NASA Lewis Research Center, Cleveland, Ohio, 44135
+Kent State University, Kent, Ohio 44242

Abstract. This paper presents an analysis of the performance of a selective emitter using small particles of rare earth compounds stable at high temperatures in a low emittance inert gas stream. An expression for the spectral emissive power excluding radiation scattering was derived to include the radiation in the emission band, produced by electronic transitions of the rare earth ion, and the continuum radiation outside the emission band produced by the rare earth host. Preliminary results suggest that a selective emitter based on suspended rare earth oxide particles will have high efficiency and merits further experimental investigation.

INTRODUCTION

Two basic types of TPV energy conversion concepts have been developed. The first type uses a grey body emitter with an optical filter that redistributes the spectral distribution of the energy radiated by the emitter to closely match the spectral response of the PV cell. Typically a short pass or bandpass filter is used whose cut-on wavelength coincides with the bandgap of the PV cell. To maintain the structural integrity of the filter and achieve high system efficiency the filter absorptance must be very low[1,2]. The second type uses a selective emitter which has an emission spectrum that closely matches the spectral response of the PV cell. In both systems an IR reflective coating on the filter or PV cells may be used to reflect unused long wavelength radiation back to the emitter to increase conversion efficiency. Recent advances in the performance of selective emitters[3-5] and low band-gap PV cells[6] offer hope for efficient TPV energy conversion using a variety of thermal sources such as combustion[7], radioisotope energy sources[8,9], and concentrated solar energy with thermal storage[10].

The early work of White and Schwartz[11] recognized the benefits of selective emitters for efficient thermophotovoltaic energy conversion however, finding an efficient selective emitter has been a difficult task. The most promising solid state selective emitters are compounds containing elements in the Lanthanide Series[12] (rare earths). The atomic structure of the doubly and triply charged rare earth ions such as Yb, Er, Ho, Th, Dy, and Nd accounts for their unique spectral emission characteristics. The orbits of the valence 4f electrons, whose electronic transitions determine the spectral emission properties, lie inside the 5s and 5p electron orbits. The 5s and 5p electrons "shield" the 4f valence electrons from the surrounding ions in the crystal. As a result, the rare earth ions in the solid state have radiative characteristics much like an isolated atom in a gas and emit in relatively narrow bands rather than in a continuum as do most solids. In a crystalline host each rare earth has a relatively strong emission band corresponding to a ground state transition of the 4f electron in the near IR. Emission outside this band is governed by the radiative properties of the host material.

Early spectral emittance[12] work on rare earth oxides showed relatively strong emission bands in the near IR ($0.9 < \lambda < 3.0$ μm). However, the emittance for photons below the bandgap of the PV cell was also significant and the efficiency of these emitters was low. Recently, Nelson and Parent[3,13] have reported significant improvements in efficiency with the rare earth oxide Welsbach mantle emitter. The small characteristic dimension (5-10 μm) of the rare earth oxide fibers used to construct the mantle is responsible for the low emittance for photon energies below the cell bandgap and high emitter efficiency. Rare earth doped yttrium aluminum garnet has also shown promising emission characteristics based on the thin film selective emitter principle[4,14]. As opposed to the Welsbach mantle which must be used in a combustion driven system, thin film selective emitters can be coupled with virtually any thermal source. In addition, it has been demonstrated that YAG and other crystalline hosts can be doped with multiple rare earth species to produce a selective emitter with multiple emission bands. As a result, greater power density over a wider wavelength interval can be achieved than with a single rare earth dopant.

As previously mentioned major improvements in the performance resulted from the small characteristic dimension of the fibers used to construct the Welsbach mantle emitter. This principle has also been employed in the design and experimentally verified in the performance of the thin film selective emitter.[21,22] A new type of selective emitter utilizing this principle is to use small particles suspended in a carrier gas. This article is an analysis of the potential performance of a selective emitter using small particles of rare earth compounds stable at high temperatures in a low emittance inert gas stream. An expression for the spectral emissive power excluding radiation scattering was derived to include the radiation in the emission band, produced by electronic transitions of the rare earth ion, and the continuum radiation outside the emission band dependent on the optical properties of the rare earth host. Using the derived expression for the spectral emissive power, the emitter efficiency and power density are calculated.

In order to determine the efficiency of a small particle selective emitter (SPSE) a derivation based on the continuum radiation transfer equations was performed. A comprehensive theoretical analysis of a SPSE must include radiation scattering effects. However, scattering is neglected in this analysis for several reasons. First, neglecting the size and material dependent scattering in the source term of the radiative transfer equation simplifies the theoretical analysis. Second, the derived efficiency is weakly dependent on scattering as previously shown in the analysis of the thin film selective emitter.[14] Although the power density and spatial characteristics of the SPSE are directly affected by radiation scattering these characteristics are more important in the design of an actual device. Since the purpose of this paper is to support the feasibility of a new concept rather than provide a complete theoretical analysis necessary for the successful design of a SPSE in a TPV convertor, radiation scattering effects will be left to future efforts.

ANALYSIS

Figure 1 shows the one dimensional model of the SPSE. The radiation is assumed to vary only in the z direction. This is an excellent approximation for an

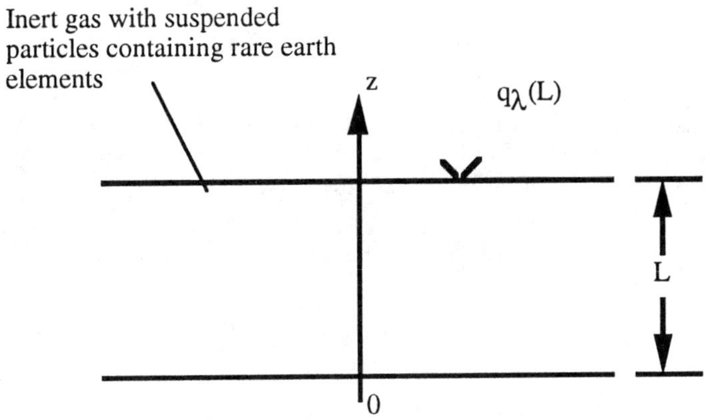

Fig. 1 Schematic representation of a small particle selective emitter (SPSE)

SPSE whose dimension perpendicular to the z direction is much greater than the thickness L. In this case the solution to the radiation transfer equation for the spectral emissive power $q_\lambda(L)$ is

$$q_\lambda(L) = \pi i_{\lambda_b} \varepsilon_\lambda = e_{\lambda_b} \varepsilon_\lambda \qquad (1)$$

where ε_λ is the spectral emittance and $i_{\lambda b}$ and $e_{\lambda b}$ are the black body intensity and emissive power. The spectral distribution of the emissive power, $e_{\lambda b}$, and the radiant intensity, $i_{\lambda b}$, in a vacuum are given as a function of absolute temperature and wavelength by

$$e_{\lambda b}(\lambda) = \frac{2\pi C_1}{\lambda^5 \left(\exp\left[\frac{C_2}{\lambda T}\right] - 1 \right)} \qquad (2)$$

where $C_1 = .595 \times 10^4 \ \mu m^4/cm^2$ and $C_2 = 14388 \ \mu m \ K$. The spectral emittance for an infinite slab[15] is the following

$$\varepsilon_\lambda = 1 - 2E_3[K_\lambda(L)] \qquad (3)$$

where $E_3[K_\lambda(L)]$ is the third order exponential integral of the optical depth, K_λ of the material. The general form of the exponential integral $E_n(x)$ is given by

$$E_n(x) = \int_0^1 u^{n-z} \exp\left[-\frac{x}{u}\right] du \qquad (4)$$

For the case n=3 the following approximation is used[15]

$$E_3(x) = \frac{1}{2}\exp[-\alpha x] \qquad (5)$$

where $\alpha = 1.8$ and Eq. (3) becomes

$$\varepsilon_\lambda = 1 - \exp[-\alpha K_\lambda(L)] \qquad (6)$$

The optical depth of the material is

$$K_\lambda(L) = \int_0^L \alpha_\lambda(z) dz \qquad (7)$$

where $\alpha_\lambda(z)$ is the spatially dependent spectral extinction coefficient. Additional assumptions used in this model are the following:
(1) the inert carrier gas and the small particles are in thermodynamic equilibrium
(2) the temperature is uniform and therefore α_λ is a constant
(3) the dimension perpendicular to L is infinite
(4) there is no radiation scattering

Since the spectral extinction coefficient, $\alpha_\lambda(z)$, is constant throughout the material the optical depth of the material is simply

$$K_\lambda(L) = \alpha_\lambda L \qquad (8)$$

The model used to determine α_λ is similar to the three band model used in the analysis of the thin film selective emitter.[14] In this analysis the spectrum is split into 4 regions with α_λ being a constant within each region. For a host material (particles) doped with a rare earth α_λ in each of the wavelength regions is defined as

1) $0 \leq \lambda \leq \lambda_b - \Delta_b, \qquad \alpha_\lambda = \alpha_c$

Where λ_b is the center of the rare earth emission band and Δ_b is the half width of the emission band.

2) $\lambda_b - \Delta_b \leq \lambda \leq \lambda_b + \Delta_b, \qquad \alpha_\lambda = \alpha_b$
3) $\lambda_b + \Delta_b \leq \lambda \leq \lambda_p, \qquad \alpha_\lambda = \alpha_c$
4) $\lambda_p \leq \lambda \leq \infty, \qquad \alpha_\lambda = \alpha_p$

The carrier gas is assumed to be transparent; a good approximation for inert gases such as helium at temperatures of interest (< 2000 K). The extinction coefficients α_b and α_c are properties of the host material and α_c is used to define the extinction coefficient above and below the emission band. In addition to rare earth doped oxide ceramics it is also feasible to use pure rare earth oxide particles in the SPSE. The rare earth oxides also exhibit the multiphonon absorption region similar to the rare earth doped oxide ceramics beyond a wavelength λ_p. As a result this analysis should apply to an SPSE using rare earth oxide or rare earth doped oxide ceramic particles.

For a volume of small particles the extinction coefficient is the following[18]

$$\alpha_\lambda = n_D \sigma_\lambda \qquad (9)$$

where n_D is the particle density and σ_λ is the spectral extinction cross section. For the case of no scattering and assuming spherical particles

$$\sigma_\lambda = \frac{\pi D^2}{4} a_\lambda = \frac{\pi D^2}{4} \varepsilon_{\lambda m} \qquad (10)$$

where D is the particle diameter, a_λ is the particle absorptance and ε_λ is the particle emittance ($a_\lambda = \varepsilon_\lambda$). For spherical particles the emittance is the following[15]

$$\varepsilon_{\lambda m} = 1 - \frac{2}{(\alpha_{\lambda m} D)^2}\left[1 - (\alpha_{\lambda m} D + 1)\exp(-\alpha_{\lambda m} D)\right] \qquad (11)$$

where $\alpha_{\lambda m}$ is the extinction coefficient for the particle material. By combining equations (9) - (11) the extinction coefficients are defined as

$$\alpha_b = n_D \frac{\pi D^2}{4}\left\{1 - \frac{2}{K_{bm}^2}\left[1 - (K_{bm} + 1)\exp(-K_{bm})\right]\right\} \qquad (12)$$

$$\alpha_c = n_D \frac{\pi D^2}{4}\left\{1 - \frac{2}{K_{cm}^2}\left[1 - (K_{cm} + 1)\exp(-K_{cm})\right]\right\} \qquad (13)$$

$$\alpha_p = n_D \frac{\pi D^2}{4}\left\{1 - \frac{2}{K_{pm}^2}\left[1 - (K_{pm} + 1)\exp(-K_{pm})\right]\right\} \qquad (14)$$

Here $K_{bm} = \alpha_{bm} D$ is the particle optical depth for the emission band, $K_{cm} = \alpha_{cm} D$ is the particle optical depth for the wavelength regions below and above the emission band, and $K_{pm} = \alpha_{pm}D$ is the particle optical depth for the wavelength range including the multiphonon absorption region.

As mentioned previously, the extinction coefficient α_p of the host material or pure rare earth oxide in the multiphonon absorption region ($\lambda > \lambda_p$) is large. Using the extinction coefficients α_b, α_c and α_p the emissive power, q_λ, in each of the wavelength regions can be calculated. For the emission band using eq. 1 the emissive power is defined as

$$\varepsilon_b = 1 - \exp(-1.8\alpha_b L) \tag{15a}$$

$$q_b = \int_{\lambda_b - \Delta_b}^{\lambda_b + \Delta_b} \varepsilon_\lambda e_{\lambda b} d\lambda = \varepsilon_b \int_{\lambda_b - \Delta_b}^{\lambda_b + \Delta_b} e_{\lambda b} d\lambda \tag{15b}$$

where eq. 6 has been used for ε_λ. Since $\Delta_b \ll \lambda_b$ we may assume that $e_{\lambda b}$ is constant in the integral. Therefore, a close approximation for q_b is

$$q_b \approx 2 e_{\lambda b}(\lambda_b, T) \varepsilon_b \Delta_b \qquad \lambda_b - \Delta_b \leq \lambda \leq \lambda_b + \Delta_b \tag{16}$$

In similar fashion the emissive powers for the other wavelength regions are the following.

$$\varepsilon_c = 1 - \exp(-1.8\alpha_c L) \tag{17a}$$

$$q_c = \varepsilon_c \left[\int_0^{\lambda_b - \Delta_b} e_{\lambda b} d\lambda + \int_{\lambda_b + \Delta_b}^{\lambda_p} e_{\lambda b} d\lambda \right] \quad \begin{array}{c} 0 \leq \lambda \leq \lambda_b - \Delta_b \\ \text{and} \\ \lambda_b + \Delta_b \leq \lambda \leq \lambda_p \end{array} \tag{17b}$$

$$\varepsilon_p = 1 - \exp(-1.8\alpha_p L) \tag{18a}$$

$$q_p = \varepsilon_p \int_0^\infty e_{\lambda b} d\lambda \qquad \lambda_p \leq \lambda \leq \infty \tag{18b}$$

Equations (16) - (18) indicate that the emissive powers are exponentially dependent on the emitter thickness L. Finally, having defined the emissive powers in the four wavelength regions the emitter efficiency, η_E, can be calculated. The efficiency is defined as the ratio of the emitted power in the emission band, q_b, to the total emitted power.

$$\eta_E = \frac{q_b}{q_b + q_c + q_p} \qquad (19)$$

RESULTS

Using this model we provide results for η_E and q_b as a function of temperature for an SPSE using rare earth doped (Yb, Er, Th, Ho, and Nd) oxide ceramic particles. For the oxide ceramics such as YAG, spinel, sapphire, and yttria, α_λ is typically low[16,17] (< 0.1 cm^{-1}) in the region of transparency and high (> 100 cm^{-1}) in the absorbing region. Also, the spectral emittance of these materials is very high[19], $0.9 < \varepsilon_\lambda < 1.0$, for wavelengths greater than the multiphonon absorption edge. Previous measurements on the extinction coefficient of rare earth doped YAG has shown that for a variety of rare earth dopants $\alpha_b \sim 25$ cm^{-1} in the emission band.[4] The emission band absorption coefficient for other rare earth doped oxide ceramics is similar, at least in the high concentration range, 10 - 40%, necessary for selective emitters. For this analysis, the particle material extinction coefficient is assumed to be a constant in the respective wavelength regions and following values for α_λ will be used; (q_b, $\alpha_{bm} = 25$ cm^{-1}), (q_C, $\alpha_{cm} = 0.1$cm^{-1}), and (q_P, $\alpha_p = 5000$ cm^{-1}). The value $\alpha_p = 5000$ cm^{-1}, used to calculate q_P was derived using Eq (6) to provide a large emittance, $\varepsilon_\lambda = 0.92$, in the multiphonon absorption region. Values for λ_b and Δ_b are estimated from experimental results[20] and are as follows: Yb ($\lambda_b = .955$ μm, $\Delta_b = 0.2$ μm), Er ($\lambda_b = 1.55$ μm, $\Delta_b = 0.2$ μm), Tm ($\lambda_b = 1.8$ μm, $\Delta_b = 0.5$ μm), Ho ($\lambda_b = 2.0$ μm, $\Delta_b = 0.3$ μm), Nd ($\lambda_b = 2.4$ μm, $\Delta_b = 0.4$ μm).

Figures 1 and 2 show the theoretical results for η_E and q_b for emitter temperatures, T_E, ranging from 1000 to 2000 K. For these results, the values for $n_D = 5 \times 10^7$ cm^{-3}, $D = 10$ μm, $L = 5$ cm, $\lambda_p = 5.0$ μm and the volume fraction of particles, $\Phi = .026$, where

$$\Phi = \frac{4}{3}\left(\frac{D}{2}\right)^3 n_D \qquad (20)$$

were chosen. In figure 1 η_E increases with temperature for all rare earth (RE) ions. Differences in η_E for the various RE's at a given temperature can be attributed to several factors. First, η_E for each RE is strongly dependent on the location of the center wavelength, λ_b, of the emission band. For example, η_E for Yb ($\lambda_b = .955$ μm) @ 1200 K is much less than η_E for Nd ($\lambda_b = 2.4$ μm) since $e_{\lambda b}$ for $\lambda = 0.955$ μm, is much less than $e_{\lambda b}$ for $\lambda = 2.4$ μm. The width of the emission band, Δ_b, for a given λ_b also affects the efficiency. This factor is responsible for η_E for Tm (Δ_b

= 0.5) being greater than η_E for Ho (Δ_b = 0.3 μm) at 1000 K even though λ_b (Ho) > λ_b (Tm). The width and center wavelength of the emission band also have a similar effect on the behavior of q_b with respect to T_E as shown in fig. 2. For the temperature range shown in figure 1 the efficiency of all the rare earth elements is monotonically increasing. The analytical model predicts an optimum temperature for maximum efficiency for all the rare earths in excess of 2000 K. The location of the optimum temperature depends on the ratios of out of band emittances to the in band emittance ($\varepsilon_c/\varepsilon_b$ and $\varepsilon_p/\varepsilon_b$).

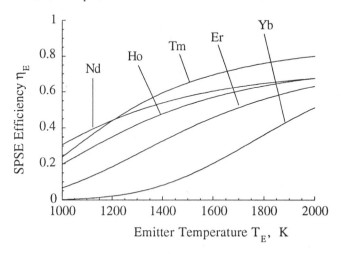

FIGURE 1. Emitter efficiency, η_E, vs emitter temperature, T_E, for an SPSE with Yb, Er, Tm, Ho, or Nd doped particles. Particle density, n_D = 5 x 10^7 cm^3, particle diameter D = 10 μm, volume fraction Φ = .026, emitter thickness L = 5 cm and multiphonon absorption edge, λ_p = 5.0 μm. Extinction coefficients, α_{bm} = 25 cm^{-1}, α_{cm} = 0.1 cm^{-1}, and α_p = 5000 cm^{-1}.

for Yb, figures 1 and 2 show that η_E > 0.5 and q_b > 5 watts/cm^2 can be achieved with a rare earth SPSE.

In figures 3 and 4 the dependence of η_E and q_b for Ho on the rare earth emission band extinction coefficient, α_{bm}, for several values of the particle density, n_D, is shown. Both η_E and q_b reach a maximum rapidly with the rate of increase being the largest for higher values of n_D. The maximum values of η_E and q_b occur at increasingly smaller values of α_{bm} as n_D increases. For maximum power and efficiency it is important to have high particle densities if α_{bm} is relatively low. If α_{bm} is relatively large (> 50 cm^{-1}) high power and efficiency is achieved at all particle densities ($n_D > 5 \times 10^6$).

The dependence of η_E and q_b for Ho on particle diameter, D, is illustrated in figures 5 and 6. Maximum efficiency is achieved for an optimum value of D. For low n_D (< 10^7 cm^{-3}) near maximum η_E can be achieved for a relatively large range of D. At large n_D (> 10^7 cm^{-3}) this range is much less. Figures 5 and 6 also show that η_E reaches a maximum more quickly than q_b as D increases and that for maximum efficiency and power output the particle diameter must be greater than 5 µm.

The dependence of η_E and q_b on emitter thickness, L, shown in figures 7 and 8 is similar to the dependence on the particle diameter. As in the case of the particle diameter there is a maximum efficiency for an optimum value of L. However, the range of L is larger than the range of D where is η_E is a maximum. The existence of a maximum value of η_E for optimum values of L and D is analogous to the optimum film thickness result for the thin film selective emitter.

Figures 9 and 10 show the dependence of η_E and q_b on the emitter temperature, T_E. Similar to the results shown in figure 1, both η_E and q_b are monotonically increasing functions of T_E for T_E < 2000 K. This is the expected result for the power density since $q_b \sim T_E^4$. The maximum value of η_E is achieved for a T_E exceeding 2000 K. In the case of the thin film Ho-YAG emitter the maximum efficiency is achieved at $T_E \sim$ 1650 K.[22]

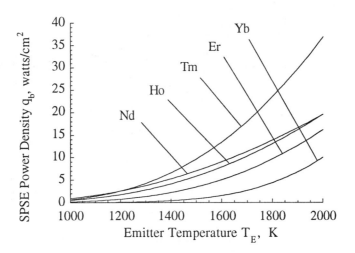

FIGURE 2. Emitter power density, q_b, vs emitter temperature, T_E, for an SPSE with Yb, Er, Tm, Ho, or Nd doped particles. Conditions are the same as in fig. 1

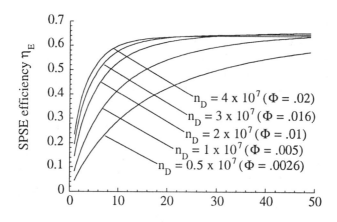

FIGURE 3. Emitter efficiency η_E vs α_{bm} of a Ho SPSE for several particle densities, n_D. $T_E = 1800$ K, $D = 10$ μm, $L = 5$ cm, $\alpha_{cm} = 0.1$ cm^{-1}, $\alpha_p = 5000$ cm^{-1}, $\lambda_p = 5$ μm.

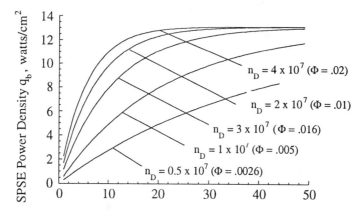

FIGURE 4. Emitter power density q_b vs α_{bm} of a Ho SPSE for several particle densities, n_D. Conditions are the same as in fig. 3.

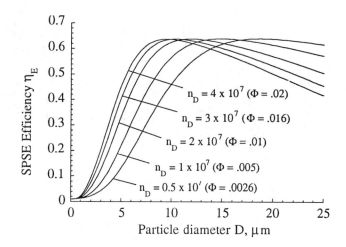

FIGURE 5. Emitter efficiency η_E vs D of a Ho SPSE for several particle densities, n_D. T_E = 1800 K, α_{bm} = 25 cm^{-1}, L = 5 cm, α_{cm} = 0.1 cm^{-1}, α_p = 5000 cm^{-1}, λ_p = 5 µm.

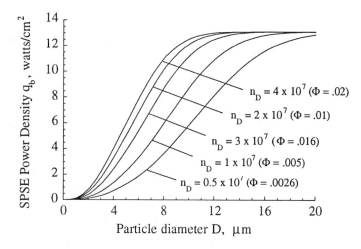

FIGURE 6. Emitter power density q_b vs D of a Ho SPSE for several particle densities, n_D. Conditions are the same as in fig. 5.

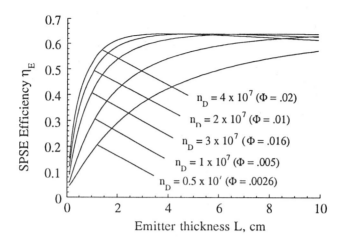

FIGURE 7. Emitter efficiency η_E vs L of a Ho SPSE for several particle densities, n_D. T_E = 1800 K, α_{bm} = 25 cm^{-1}, D = 10 μm, α_{cm} = 0.1 cm^{-1}, α_p = 5000 cm^{-1}, λ_p = 5 μm.

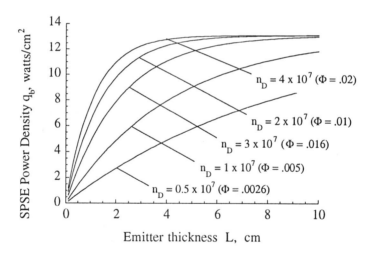

FIGURE 8. Emitter power density q_b vs L of a Ho SPSE for several particle densities, n_D. Conditions are the same as in fig. 7.

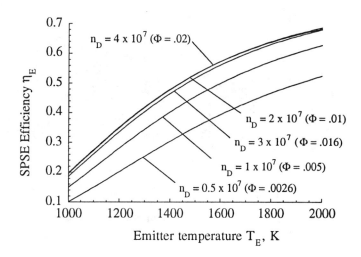

FIGURE 9. Emitter efficiency η_E vs T_E of a Ho SPSE for several particle densities, n_D. L = 5 cm, α_{bm} = 25 cm^{-1}, D = 10 μm, α_{cm} = 0.1 cm^{-1}, α_p = 5000 cm^{-1}, λ_p = 5 μm.

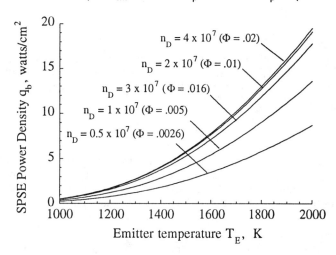

FIGURE 10. Emitter power density q_b vs T_E of a Ho SPSE for several particle densities, n_D. Conditions are the same as in fig. 9.

The dependence of the efficiency on the cutoff wavelength for multiphonon absorption, λ_p, is shown in figure 11. Significant improvement in the efficiency

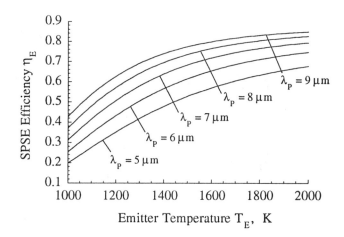

FIGURE 11. Emitter efficiency η_E vs T_E of a Ho SPSE at several multiphonon band edges, λ_p. L = 5 cm, n_D = 5 x 10^7, Φ = .026 α_{bm} = 25 cm^{-1}, D = 10 µm, α_{cm} = 0.1 cm^{-1}, α_p = 5000 cm^{-1}, λ_p = 5 µm.

occurs as λ_p increases. For YAG, $\lambda_p \sim 5$ µm whereas for yttria (Y_2O_3), which can be doped with the rare earth oxides, $\lambda_p \sim 8$ µm. It is expected that the efficiency for both SPSE and thin film selective rare earth doped yttria emitters will exceed the performance of emitters based on other oxide ceramics such as YAG[21,22] with higher values of λ_p.

CONCLUSION

The simplified analysis of the SPSE not including radiation scattering has produced encouraging results for the emitter efficiency and power density. Efficiencies greater than 0.6 and power densities greater than 10 W/cm^2 are predicted for several rare earths (Nd, Ho, Tm, Er, and Yb) for temperatures in the 1500K - 2000K range. The analysis predicts optimum particle diameter (8 - 20 µm) and emitter thickness (2 - 10 cm) to achieve maximum efficiency for an SPSE using Ho doped particles. It is also important to note that these parameters and their effect on the efficiency and power output are heavily dependent on the extinction coefficient of the particle material, α_{bm}, in the emission band. The analysis also predicts that significant increases in efficiency may be achieved by using materials with larger values of λ_p. This is seen in the performance results of an Ho doped SPSE where η_E increases from 0.4 to 0.5 for 1000 < T_E < 2000K as λ_p is increased from 5 µm to 9 µm.

REFERENCES

1. R. L. Bell, Sol. Energy **23**, 203 (1979)
2. W. Spirkl and H. Reis, J Appl. Phys. **57**, 4409 (1985)
3. R. E. Nelson, in *Proceedings of the 32nd International Power Sources Symposium* (Electrochemical Society, Pennington, NJ 1986), pp 95-101
4. R. A. Lowe, D. L. Chubb, S. C. Farmer, and B. S. Good, Appl. Phys. Lett. **64** (26), 3551, (1994)
5. P. L. Adair and M. F. Rose, in *The First NREL Conference on Thermophotovoltaic Generation of Electricity,* (AIP Press, Woodbury, NY 1995) pp 245-262
6. D. M. Wilt et al, Appl Phys. Lett. **64** (18), 2415 (1994)
7. L. Block, P. Daugioda, and M. Goldstein, IEEE Trans. Industry Appl. **IA-28**, 251 (1992)
8. M.D. Morgan, W. E. Horne, and P. R. Brothers, AIP Conf. Proc. **271**, 313 (1993)
9. D. L. Chubb, D. J. Flood, and R. A. Lowe, in *Proceedings of NTSE-92 Nuclear Technologies for Space Exploration,* 1992, p 281
10. K. W. Stone, S. M. Kusek, R. E. Drubka, and T. D. Fay, in *The First NREL Conference on Thermophotovoltaic Generation of Electricity,* (AIP Press, Woodbury, NY 1995) pp 153-162
11. D. C. White and R. J. Schwartz, in *Combustion and Propulsion, 6th AGARD Colloqium on Energy Sources and Energy Combustion,* (Gordon Breach, New York, 1967), pp 897-922
12. G. E. Guazzoni, Appl. Spectra **26**, 60 (1972)
13. C. R. Parent and R. E. Nelson, in *Proceedings of the Twenty-First Intersociety Energy Conversion Engineering Conference* (American Chemical Society, Washington, DC, 1986), Vol. 2, pp 1314-1317
14. D. L. Chubb and R. A. Lowe, J. Appl. Phys. **74**, 5687 (1993)
15. R. Siegel and J. R. Howell, *Thermal Radiation Heat Transfer,* 2nd ed. (Hemisphere, Washington, DC, 1981), Chap. 14, p 96
16. M. E. Thomas, R. I. Joseph, and W. J. Tropf, Appl. Optics **27** No. 2, 239 (1988)
17. W. J. Tropf and D. C. Harris, in *SPIE Vol. 1112 Window and Dome Technologies and Materials ,* (1989)
18. H. C. Van de Hulst, Light Scattering by Small Particles, (Dover Publications, NY, 1981)
19. R. M. Sova, M. J. Linevsky, M. E. Thomas, and F. F. Mark, Johns Hopkins APL Technical Digest **13** (3), 368 (1992)
20. R. E. Nelson and P. A. Iles, ASME/ASES/SOLTECH Joint Solar Energy Conference, 1993
21. R. A. Lowe, B. S. Good, D. L. Chubb, in *Proceedings of the 30th Intersociety Energy Conversion Engineering Conference ,* (ASME International New York, NY, 1995) to be published
22. R. A. Lowe, D. L. Chubb, B. S. Good, in *Proceedings of the First World Conference on Photovoltaic Energy Conversion,* (IEEE Press, Piscataway, NJ, 1994) p 1851

Multiband Spectral Emitters Matched to MBE Grown Photovoltaic Cells

Eva M. Wong, Parvez N. Uppal*, Jeffrey P. Hickey,
Cye H. Waldman**, and Glenn A. Holmquist

Quantum Group Inc., 11211 Sorrento Valley Road, San Diego, CA 92121-1324
*Lockheed-Martin Labs, 1450 South Rolling Road, Baltimore, MD 21227-3898
**Consultant, P.O. Box 231157, San Diego, CA 92023-1157

Abstract. Clearly TPV devices are of considerable interest for power generation. For practical devices it is desirable to have high efficiencies combined with low temperature operation. Photovoltaic cells which can convert the energy at the longer wavelengths of interest are needed to complete such a system. The spectral emission peak of Yb_2O_3 is well matched to the band gap of Si; however, the longer wavelength, spectral emissions of other rare earth oxides can also be exploited through the use of III-V semiconductor compounds such as GaSb or alloys of GaInAsSb. By doping GaSb with InAs, the band gap of the resulting material can be effectively varied depending upon the concentration of InAs in the quaternary alloy. The ability to tailor the emitter materials and, in conjunction, the photovoltaic materials leads to greater efficiencies through spectral matching.

Two binary rare earth oxide combinations, Er_2O_3/Ho_2O_3 and Er_2O_3/Yb_2O_3, were studied. The mixtures were found to give multiple peak spectral emission in the wavelengths of interest. The intensity of the peaks were compositionally dependent though it did not vary in a linear fashion. Photon efficiencies of the molecular beam epitaxially (MBE) grown GaSb cell and GaInAsSb quaternary cell were measured when used in conjunction with the Er_2O_3/Ho_2O_3 emitters in which the concentration of Er_2O_3 and Ho_2O_3 were varied. The results demonstrated promise for further work.

INTRODUCTION

The limitations to high efficiency thermophotovoltaic (TPV) systems are three fold: 1) accommodating higher temperature operation of the emitter without melting, 2) increasing the photon content of peak spectral emitter emissions, and 3) overcoming the photovoltaic (PV) current collection or power density limitations as well as cost. These limitations can be addressed by studying the use of a multiband spectral emitter and the use of matched photovoltaic cells with long wavelength collection capabilities. It has been reported (1,2) that the emitter temperature required for maximum efficiency is lower for erbia and holmia when compared to ytterbia. Quantum Group was granted a patent (3) for the use of doped individual rare earth oxide spectral emitters resulting in strong single and double emitter peaks in the near infrared. With the advent of relatively cost effective III-V single crystal photovoltaic growth techniques, the bandgap match with longer wavelength spectral emission peaks can be achieved and the multiband spectral emitters studied further. The development of these materials offers tremendous flexibility in tailoring the emitter spectral characteristics to those of the photovoltaic bandgap.

MULTIBAND EMITTER THERMODYNAMICS

A principal advantage of the multispectral emitter is its effect upon the operating temperature of the emitter. To begin, compare the performance of a blackbody with that of a single-band selective emitter. Let us assume that the heat input to the radiator due to convection is constant and is balanced by the heat radiated out. The temperature of the radiator can be found by iteration from the emissive power. For a given temperature, the total emissive power of a blackbody is greater than that of a selective emitter, therefore, for a given emissive power, the temperature of the selective emitter must be greater than that of a blackbody. By similar reasoning, a multispectral emitter would have a higher temperature than a blackbody and a lower temperature than a selective emitter. The multispectral emitter allows the system to provide a greater number of photons at a given operating temperature by radiating them in a longer wavelength interval. The overall system efficiency can be greatly enhanced by matching the multispectral emitter to a two-color photovoltaic.

The spectral distribution of emissive power can be advantageous. The emissive power e is given by the integration of the spectral distribution of emissive power as

$$e = \int_{\lambda_1}^{\lambda_2} e_\lambda d\lambda = \int_{\lambda_1}^{\lambda_2} \varepsilon_\lambda e_{\lambda b} d\lambda = \bar{\varepsilon} \int_{\lambda_1}^{\lambda_2} e_{\lambda b} d\lambda \qquad (1)$$

where ε_λ is the spectral emissivity, $\bar{\varepsilon}$ is the average emissivity, and $e_{\lambda b}$ is the Planck spectral distribution of emissive power for a blackbody given by

$$e_{\lambda b}(\lambda, T) = \frac{C_1}{\lambda^5 (e^{C_2/\lambda T} - 1)} \qquad (2)$$

The blackbody emissive power in a small wavelength interval can be expressed as

$$e_b = \int_{\lambda}^{\lambda+d\lambda} e_{\lambda b} d\lambda = e_{\lambda b} d\lambda = \frac{C_1 d\lambda}{\lambda^5 (e^{C_2/\lambda T} - 1)} \approx \frac{C_1 d\lambda}{\lambda^5 e^{C_2/\lambda T}} \qquad (3)$$

The emissive power of multispectral emitters can be expressed as

$$e = \int_0^{\infty} \varepsilon_\lambda e_{\lambda b} d\lambda = N \sum_{i=1}^{N} \overline{\varepsilon}_i \int_{\lambda_i}^{\lambda_i + \Delta\lambda_i} e_{\lambda b} d\lambda \approx \sum_{i=1}^{N} \overline{\varepsilon} e_{\lambda b}(\lambda, T) \Delta\lambda_1 \qquad (4)$$

where N is the number of spectral bands and $\overline{\varepsilon}_i$ is the average emissivity of band i.

It is apparent that the emissive power is dominated by the first or first few terms in the summation. Physically, this is because the shorter wavelength photons are the more energetic ones. Mathematically, it can be seen from the wavelength dependence of the blackbody emissive power in Equation (3). From a TPV point of view, however, at energies above the band gap, the photovoltaic is sensitive to the number of photons rather than their energy. Thus, if photovoltaics which are responsive to longer wavelengths are incorporated, the performance can be significantly improved while producing only a minor impact on the overall system operating conditions.

EMITTER DEVELOPMENT

The relic process was used to fabricate the emitter structures used in this study. The relic process is a means of producing a ceramic material by inbibing an organic material with a metal salt which is subsequently converted to a metal hydroxide and finally a metal oxide. During the conversion from the hydroxide to the oxide, the organic precursor is pyrolized leaving an oxide structure which is a positive image of the original material. The materials thus produced can be compositionally varied to nearly any degree.

The initial approach to determining the optimal oxide composition for multiband emission was empirical in nature. It was demonstrated that binary mixtures of rare earth oxides give selective spectral output at multiple wavelengths and that the radiant emittance at each wavelength band is dependent upon concentration. Because the exact dependence and/or synergistic effects are not well known, the need for empirical studies becomes relevant and important.

Limited data exists in the literature on the spectral characteristics of the rare earth oxides at high temperatures, and even less data exist on the spectral characteristics of mixed rare earth oxides, although it is clear that the data would be of practical interest to thermophotovoltaic systems. Some preliminary studies of mixed rare earth oxides in YAG have been performed more recently (4).

The mixed compounds were found to selectively emit where each component was found to act independently. For example, mixtures of Yb_2O_3 and Ho_2O_3 were found to have spectral emission at all wavelengths unique to the individual oxides (in particular 975, 1200 and 2000nm). The relative intensity of each peak was found to be dependent upon the concentration of each component in the mixture. Mixtures of Yb_2O_3 and Er_2O_3 were found to be particularly interesting in that they both possess a spectral emission peak at 975nm. An example of two mixtures is shown in Figure 1. The feature at 650nm is an artifact of the detector.

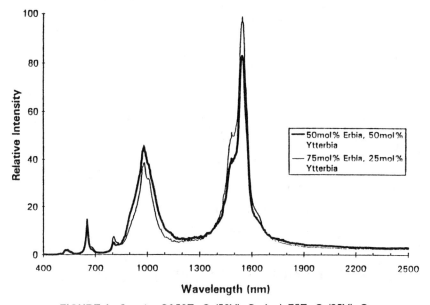

FIGURE 1. Spectra Of $50Er_2O_3/50Yb_2O_3$ And $75Er_2O_3/25Yb_2O_3$.

In Figure 2, an example of a $25Er_2O_3/75Ho_2O_3$ binary compound is compared to a $100Ho_2O_3$ spectra. Note that there are three distinct emission bands: the peaks at 1200 and 2000 nm are attributed to Ho_2O_3 and the peak at 1550 nm is attributed to Er_2O_3. In the binary compound, the peak intensity is nearly equal to that of $100Ho_2O_3$, but there is the additional band at 1550 attributed to Er_2O_3. The fuel input was the same for all spectra however, slight changes in the monochromator detector setups can strongly effect the output. The spectra are therefore plotted as a function of the relative intensity as opposed to an absolute scale.

It is evident that the spectral intensity at any particular wavelength is strongly dependent upon the compositional variation and therefore the relative peak intensity (photon flux) can be adjusted by changing the mol ratio of the mixture. Although these binary compounds have shown behavior that is strongly dependent on concentration, a complete understanding of the effects is needed to be able to predict emissive properties.

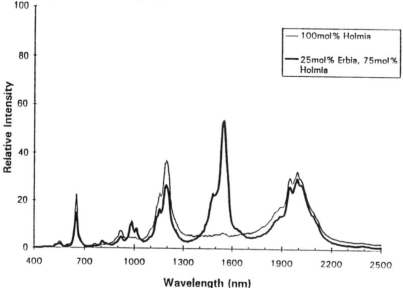

FIGURE 2. Spectra of $25Er_2O_3/75Ho_2O_3$ and $100Ho_2O_3$.

PHOTOVOLTAIC CELL DEVELOPMENT

The selection of the photovoltaic composition is governed by two factors: first, the required band gap for matching to the spectral output and secondly, the ability to match the lattice spacing of the substrate with the first deposited photovoltaic layer and with subsequent photovoltaic layers.

Lower temperature selective emitter TPV applications require PV cells that are sensitive to radiation in the 1.7 to 2.5µm range. In the III-V materials system, there are no binaries and no lattice matched ternaries having a band gap of 2.5 µm. A quaternary system consisting of Ga, In, As and Sb can be made to lattice-match to GaSb for specific compositions. The band gap for the lattice matched alloy compositions can be varied between 1.7 µm and 4.5 µm. For example, the optimum band gaps for the selective emitter peaks at 1.6 µm and 2.0 µm are 0.73 eV and 0.6 eV, and the lattice matched growth of GaSb and the quaternary $Ga_{0.78}In_{0.22}As_{0.2}Sb_{0.8}$ on GaSb is the most promising solution. The absorption wavelength in these quaternary alloys is mainly controlled by the group III elements. In the MBE growth technique, these can be measured accurately in situ. Figure 3 shows how band gap and lattice constant are controlled in the quaternary by changing the ratio of the four elements. The dotted lines show the compositional values for which the material is lattice-matched to GaSb. Note that at E_g=0.5eV, the curvature is quite small in the y-direction of the band gap surface. The cut-off wavelength is mainly determined by the x-value or the Group III composition.

GaSb is a direct gap material and has a high quantum efficiency (as opposed to Ge which has an indirect gap and a low quantum efficiency at 1.6µm). $Ga_xIn_{1-x}As_ySb_{1-y}$ quaternary alloys with specific alloy compositions can be lattice-matched to GaSb substrates with band gaps ranging from 0.73eV to 0.3 eV (1.7µm to 4.2µm).

$In_xGa_{1-x}As$ cells have been developed (5,6), however at the longer wavelengths lower quality can result due to dislocation defects generated to relieve stress caused by lattice mismatch.

GaSb and GaInAsSb Epilayer Development

Two specific band gap energies were identified for the Er_2O_3/Ho_2O_3 system: GaSb with a band gap corresponding to 1.7µm and a quaternary alloy of $Ga_xIn_{1-x}As_ySb_{1-y}$ with nominal composition $Ga_{0.82}In_{0.18}As_{0.16}Sb_{0.84}$ for a band gap corresponding to 2.1µm. The layers were doped p- and n- type to determine the doping efficiencies for the doping levels required for the fabrication of the photovoltaic cells.

The two wavelength bands, 1.7µm and 2.1µm, were selected based upon the spectral response of Er_2O_3/Ho_2O_3 compounds. Er_2O_3 exhibits a distinct emission band centered at 1550nm where the full width at half the maximum peak intensity (FWHM) is approximately 110nm, and Ho_2O_3 has a distinct primary emission band centered at 2000nm, where the FWHM is approximately 180nm, as well as a secondary emission band at 1200nm with a FWHM of approximately 100nm. These spectral emission bands can be matched to the band gap energies of GaSb (0.73eV nominal) and InAs doped GaSb (variable 0.36 - 0.73eV). GaSb is a readily available substrate material. InAs was chosen as the doping material for two reasons: it has a lower band gap energy (i.e., longer wavelength collection capability), and the lattice spacing closely matches that of GaSb. The combination of these factors makes it an excellent choice for multilayer photovoltaic materials because it allows for greater flexibility in designing the collection wavelength(s) and also allows for photovoltaic layers with little lattice strain.

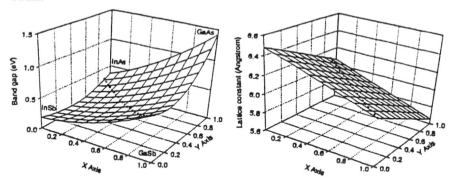

FIGURE 3. Relationship Between Band Gap and Lattice Constant for GaInAsSb.

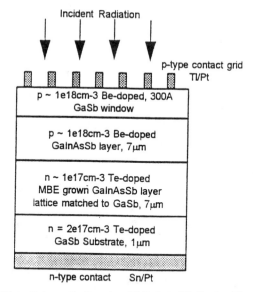

FIGURE 4. Schematic Drawing Of GaInAsSb Device Structure.

A schematic drawing of the cross section of a GaInAsSb device structure is shown in Figure 4. The PV area of each resulting device collector cell was 1 cm^2. After growth, the materials were characterized using x-ray diffraction (XRD) and photoluminescence (PL) techniques. A single crystal x-ray diffraction scan, Figure 5, from a GaInAsSb quaternary layer grown on GaSb shows a near perfect lattice match (+0.18%) to GaSb. The PL results of a GaSb epilayer and a GaInAsSb quaternary layer are shown in Figure 6. Detection of a PL signal and high peak intensity indicate very high material quality and thus the suitability of fabricating high quality devices.

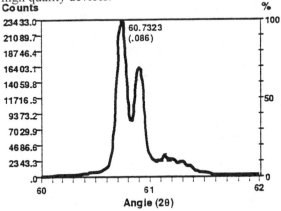

FIGURE 5. XRD Scan From GaInAsSb Layer Grown On GaSb Substrate.

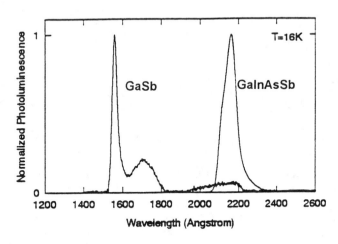

FIGURE 6. 16K Photoluminescence Of GaSb Epilayer And GaInAsSb Epilayer.

285

TEST MEASUREMENTS AND CELL EFFICIENCIES

The spectral emitter cylinders were heated by burning oxygen and methane in a near 2 to 1 stoichiometric ratio. Radiometric measurements were taken at each flow rate and converted to total energy in watts by configuration factor algebra. An IL 1700 thermopile radiometer sensitive between 0.5 to 4 microns was used for these measurements. Each PV cell was tested under a variable resistance to determine the maximum power output in watts for that particular flow rate. PV power output was then compared to total photon output (again using configuration factor algebra) to give a total photon efficiency for each cell.

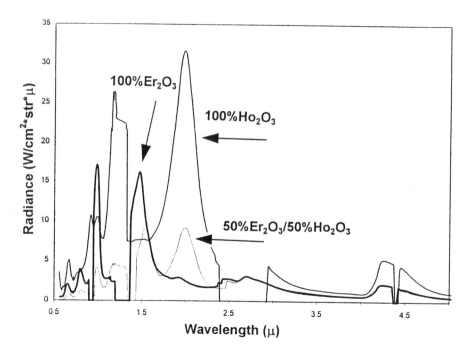

FIGURE 7. Spectra of 100% Er_2O_3, 100% Ho_2O_3, and 50% Er_2O_3/50% Ho_2O_3 showing radiance as a function of wavelength for 30 scfh O_2 and 12 scfh CH_4.

Three selected emitter spectra are shown in Figure 7. Flow rates were consistently oxygen rich being "tuned" at 30/12 standard cubic feet per hour (scfh) oxygen to methane respectively. Spectra taken using a highly accurate spectroradiometer for the 100% Ho_2O_3, 100% Er_2O_3, and a 50/50 combination, covered the range of 0.5 to 14 micron. A limited portion of this range is shown, 0.5-5.0μ, and is the range of interest. It is believed that the combustion products of CO_2, CO, and H_2O are responsible for the features at 0.95, 1.4, 2.5-3.0, and 4.2-4.6 microns. Some data is lost at 1.2 microns due to the spectroradiometer, which uses two dual band detectors where 1.2 microns is a switching point for the Si/PbS detector. Of particular interest is the unusually high intensity of the 1.0 and 1.2 micron features for Er_2O_3 and Ho_2O_3 respectively. These lines are much stronger than would be expected from previous mantle spectra and other examples from the literature. It is possible that CO_2 band emission is responsible for the added radiance.

The open circuit voltage (V_{oc}) and short circuit current (I_{sc}) were measured directly and then multiplied by each other and a fill factor of 0.50 to obtain the watts output. The fill factor for the cells was relatively low due to the quality of the non-optimized metallization and ohmic contacts. The 0.5 to 4.0 micron range is the region of sensitivity for the thermopile radiometer used to determine the configuration factors for the system. This range is not however the narrow range of sensitivity for the two PV cells. With the spectroradiometer data, energy in specific bands was obtained for the regions where the PV cells are sensitive. This put the cell efficiency in terms of band energy that the PV cells can use. The cell efficiency for these regions was found to be between 12% and 20% for the GaSb cell, and 3% to 10% for the quaternary cell as shown in Tables 3 and 4.

As expected for the GaSb cell, efficiencies were highest for the 100% Er_2O_3 spectral emitter. Looking at the efficiencies for the quaternary cell, Table 4, its lowest value is for 100% Ho_2O_3. The primary feature of holmia matches the quaternary cell's sensitivity; just as the primary feature of Er_2O_3 matches the most sensitive part of the GaSb response curve. Therefore it appears that the low value for the quaternary cell is anomalous. These anomalous values may have been related to the series resistance of the cells.

The data taken with the radiometer was divided by its form factor to calculate the total burner output in watts. The PV cells measured power in watts, and these values were divided by their form factors to give a total PV power out. This number was then divided by the total burner output to give the conversion efficiency for the PV cells. The configuration factor equations used for the burner configuration are provided in Leuenberger and Person (7).

It must be noted that a more efficient combustion process improve the total system output in terms of higher energy in the spectral emitter emission lines, and consequently more of the total energy would be available to the PV cells for higher efficiencies and a higher total output. The photovoltaics were early MBE development cells and somewhat series resistance limited. Optimized processing would, of course, improve the cell performance and conversion efficiencies.

TABLE 3. GaSb Cell Efficiency Measurements

Total Radiant Output [a]	GaSb Output Measured	GaSb Output [b]	% of Total Radiant Output in GaSb Band (1.5-1.8µ)	GaSb Conversion Efficiency
$50Er_2O_3/50Ho_2O_3$				
945.65 Watts	2.9E-03 Watts	30.11 Watts	13.7%	23%
100% Ho_2O_3				
969.10 Watts	2.3E-03 Watts	23.26 Watts	9.3%	26%
100% Er_2O_3				
820.61 Watts	3.2E-03 Watts	32.84 Watts	14.2%	28%

TABLE 4. GaInAsSb Cell Efficiency Measurements

Total Radiant Output [a]	GaInAsSb Output Measured	GaInAsSb Output [b]	% of Total Radiant Output in GaInAsSb Band (2.0-2.3µ)	GaInAsSb Conversion Efficiency
$50Er_2O_3/50Ho_2O_3$				
945.65 Watts	4.8E-03 Watts	4.90 Watts	13.1%	5.5%
100% Ho_2O_3				
969.10 Watts	4.0E-03 Watts	4.11 Watts	17.1%	3.1%
100% Er_2O_3				
820.61 Watts	3.3E-03 Watts	3.42 Watts	4.7%	8.8%

[a] from configuration factors & radiometer measurements
[b] from PV cell and configuration factors

CONCLUSIONS

Mixtures of rare earth oxides were made and the materials were found to be selectively emitting in the bands where the individual oxides are known to have spectral emission bands. The intensity of the peaks was found to be concentration dependent, although the exact dependence was not clearly determined. Small changes in composition were found to effect the spectral distribution in the different wavelength bands. These characteristics can be tailored by varying the molar ratios of the two components. The coupling of the tailored multiband spectrum to a photovoltaic cell that can convert both energy bands would yield significantly higher efficiencies where the spectral energy is well matched to the optimal responsivity of the photovoltaic cell. Therefore, multiband emitters which increase the emission band width have clearly been demonstrated, and the ability to tailor the spectral emission by varying the compositional mixture is apparent.

Furthermore, the ability to produce III-V photovoltaic materials with tunable band gaps by molecular beam epitaxy was demonstrated in this work. Cell efficiencies were found to vary depending upon the emitter composition used, however it is expected that with some development work, a significantly optimized cell can be fabricated to produce higher efficiencies. The epitaxially grown quaternary cell demonstrates the ability to tailor the band gap of a photovoltaic by varying the composition of the cell. This ability provides tremendous flexibility in designing TPV systems. With optimization of the process, it is expected the cost drivers for this process could be markedly reduced. The potential for a multilayer, multi-bandgap photovoltaic cell matched to a multispectral emitter should not be ignored as it could provide large efficiency increases in TPV systems without significantly affecting the mechanical device design.

ACKNOWLEDGEMENTS

We are grateful for the funding provided by the U.S. Department of Energy under contract # FG03-94ER81838 and for the guidance provided by our Program Manager, Dr. Cynthia Carter.

REFERENCES

1. Chubb, D.L. and Lowe, R.A., "Thin-Film Selective Emitter," *J. Appl. Phys.* **74** (9), 5687-5698, (1993).
2. Chubb, D.L., Lowe, R.A., and Good, B.S., "Emittance Theory for Thin-Film Selective Emitter," in *AIP Conference Proceedings 321*, 1994, pp. 229-244.
3. Goldstein, M.K., "Multiband Emitter Matched to Multilayered Photovoltaic Collector," U.S. Patent #4,776,895, October 11, 1988.
4. Lowe, R.A., Chubb, D.L., and Good, B.S., "Radiative Performance of Rare Earth Garnet Thin Film Seletive Emitters," in *AIP Conference Proceedings 321*, 1994, pp. 291-297.
5. Wojtczuk, S., Gagnon, E., Geoffroy, L., and Parodos, T., "$In_xGa_{1-x}As$ Thermophotovoltaic Cell Performance vs. Bandgap," in *AIP Conference Proceedings 321*, 1994, pp. 177-187.
6. Wilt, D.M., et al., "InGaAs PV Device Development for TPV Power Systems," in *AIP Conference Proceedings 321*, 1994, pp. 210-220.
7. Leuenburger, H. and Person, R.A., "Compilation of Radiation Shape Factors for Cylindrical Assemblies," paper no. 56-A-144, ASME, November 1956.

Characteristics of Indium Oxide Plasma Filters Deposited by Atmospheric Pressure CVD

S. Dakshina Murthy[*], E. Langlois[*], I. Bhat[*], R. Gutmann[*], E. Brown[†], R. Dzeindziel[†], M. Freeman[†] and N. Choudhury[†]

[*]*Electrical, Computer and Systems Engineering Department and Center for Integrated Electronics & Electronic Manufacturing, Rensselaer Polytechnic Institute, Troy, NY 12180-3590*
and
[†]*Lockheed-Martin Inc., Schenectady, NY 12301-1072*

Abstract. Thin films of undoped and tin-doped In_2O_3 have been investigated for use as plasma filters in spectral control applications for thermal photovoltaic cells. These films are required to exhibit high reflectance at wavelengths longer than the plasma wavelength λ_p, high transmittance at wavelengths shorter than λ_p and low absorption throughout the spectrum. Both types of films were grown via atmospheric pressure chemical vapor deposition (APCVD) on Si (100) and fused silica substrates using trimethylindium (TMI), tetraethyltin (TET), and oxygen as the precursors. The O_2/TMI partial pressure ratio and substrate temperature were systematically varied to control the filter characteristics. The plasma wavelength λ_p was found to be a sensitive function of the O_2 partial pressure and the substrate temperature. Post-growth annealing of the films carried out either in nitrogen or air ambient at elevated temperatures did not have any beneficial effect. Tin-doped In_2O_3 was grown using tetraethyltin (TET) as the dopant. The material properties and consequently the optical response were found to be strongly dependent on the growth conditions such as O_2 and TET partial pressures. Both undoped and tin-doped In_2O_3 grown on fused silica exhibited enhanced transmittance due to the close matching of refractive indices of In_2O_3 and silica. X-ray diffractometer measurements indicated that all these films were polycrystalline and highly textured towards the (111) direction. The best undoped and tin-doped In_2O_3 films had a λ_p around 2.7µm, peak reflectance greater than 75% and residual absorption below 20%. These results indicate the promise of undoped and tin-doped In_2O_3 as a material for plasma filters.

© 1996 American Institute of Physics

Introduction

Photovoltaic (solar) cells are widely used to convert light energy from the sun into usable electrical energy through a direct conversion process. A similar device, the thermal photovoltaic cell or TPV cell, utilizes energy from the near infra-red spectrum to excite charge carriers and create electrical energy. These devices have immense practical value for generating power from thermal radiation sources. In this case, only those wavelengths corresponding to energies equal to or greater than the band gap of the TPV cell material are of use. However, the thermal source radiates energy at a continuum of wavelengths corresponding to black-body radiation. All radiation below the bandgap energy would pass through the cell and be wasted. To increase efficiency, wavelengths corresponding to sub band gap energies should be reflected back into the source in order to reduce the input power to the system. This spectral control technique can be accomplished utilizing a heavily doped semiconductor filter or "plasma filter". These filters have already been developed for use in efficient incandescent and sodium-vapor lamps, solar collectors and energy-efficient windows for houses (1-2). However, these applications require the transmission of radiation in the visible spectrum while reflecting back all infra-red radiation. For use in TPV cells, the filters need to transmit wavelengths shorter than about 2.5µm and reflect all longer wavelengths.

This requirement calls for the use of a highly conductive material with a large enough bandgap whose carrier concentration can be controlled to obtain the required behavior. Transparent conducting oxides such as indium tin oxide or indium oxide are well suited for this purpose (1-2). These films have been synthesized by various methods such as thermal evaporation (3), electron beam evaporation (4), r.f. magnetron sputtering (5), d.c. magnetron sputtering (6), ion beam sputtering (7), spray pyrolysis (8), chemical vapor deposition (9) and pulsed laser deposition (10). These techniques have concentrated on the growth of material transparent to visible light but not to infrared. The optical and electrical properties of these materials have also been well documented (11-12). In general, these methods are used to obtain materials with as high a carrier concentration as possible. However, for TPV applications, it is necessary to tune the "plasma wavelength" of the filter between 2.5-3µm, which requires a finer control of the carrier concentration in the material. We chose to use atmospheric pressure chemical vapor deposition as the growth technique. The use of carefully controlled proportions of the highly pure organometallic reactants allows us to easily regulate the carrier concentration to the required value without resorting to post growth annealing techniques. Also, the use of high deposition temperatures and a chemical vapor deposition technique ensures the growth of inherently good starting material. Another point to note is that the amount of tin incorporated in indium tin oxide layers grown by the other techniques mentioned above (~5%) is higher than we intend to use in our tin-doped layers.

This paper focuses on the development of an In$_2$O$_3$ plasma filter for use in TPV spectral control applications. The growth and characterization of thin layers of undoped and tin-doped In$_2$O$_3$ used as plasma filters is described below. The principle behind the operation of these plasma filters is also explained in brief.

Technical Background

Indium oxide is a solid which most commonly occurs as the sesquioxide, In$_2$O$_3$. As reported in the literature (13), In$_2$O$_3$ is a polycrystalline solid with a cubic unit cell structure whose lattice parameter a$_o$=10.117Å. The structure of In$_2$O$_3$ is related to that of the fluorite structure, from which it can be derived by removing one-quarter of the anions (13).

There are two types of In$_2$O$_3$ plasma filters under consideration in this paper. The first type, commonly referred to as "undoped" Indium Oxide is in fact doped by oxygen vacancy defects, which are always present as a consequence of the growth process. In$_2$O$_3$ is an ionic solid consisting of indium cations (In^{3+}) and oxygen anions (O^{2-}). When an oxygen atom is removed from the normal crystal site, two electrons are left in the vacancy. One or both of these electrons may be excited into the conduction band leaving behind a single or doubly charged vacancy. The presence of these electrons in the conduction band the turns the undoped In$_2$O$_3$ into an n-type conducting material. This behavior may be described as follows (14):

$$O_O \Leftrightarrow V_O^{\cdot} + e' + \tfrac{1}{2}O_2 \qquad (1a)$$

$$O_O \Leftrightarrow V_O^{\cdot\cdot} + 2e' + \tfrac{1}{2}O_2 \qquad (1b)$$

where V ≡ vacancy
O ≡ oxygen
e ≡ electron
Ö/ e' ≡ effective / actual charge

The second type of In$_2$O$_3$ is the tin-doped In$_2$O$_3$ which is extrinsically doped by tin atoms. Tin is a tetravalent atom and becomes a Sn^{4+} cation when fully ionized. If tin is incorporated into an indium (In^{3+}) site, an extra electron is removed to the conduction band leaving behind a positive effective charge. Tin thus behaves as a donor making tin-doped In_2O_3 also an n-type conducting material.

In both types of In_2O_3, the aim is to obtain n-type material with a very high concentration of electrons obtained from the donors. The high carrier concentration is the basis for the working of the plasma filter, which has to do with the collective oscillations of the "plasma" formed by the negatively charged electrons and the positively charged donors. At a particular wavelength of oscillation corresponding to the resonant case, known as the plasma wavelength λ_p, the material undergoes a transition such that at wavelengths $\lambda > \lambda_P$ it is highly reflective and for $\lambda < \lambda_P$ it is highly transmissive in nature. By definition, λ_p is defined as the wavelength where the real part of the refractive index, n, is equal to the complex part, k. λ_p is given by the following equation (1) :

$$\lambda_P = 2\pi c_o \left(\frac{nq^2}{\varepsilon_o \varepsilon_r m^*} - \gamma^2 \right)^{-1/2} \qquad (2)$$

where c_o = speed of light in free space
n = charge carrier concentration
ε_o = permittivity of free space
ε_r = relative dielectric constant of medium
m^* = effective mass of charge carriers
γ = the damping rate
q = electronic charge

In eq.(2), γ^2, which depends on the mobility µ of the electrons is usually quite small compared to the other term and may be neglected. However, the mobility µ does affect the steepness of the reflectance and transmittance curves in the transition region (1).

For our case, the goal is to position λ_P such that shorter wavelengths ($\lambda < \lambda_p$) will be those usable by the TPV cell and longer wavelengths ($\lambda > \lambda_p$) will be reflected back into the source. A nominal λ_p of 2.7µm is chosen for application for a cell with a bandgap around 0.55eV. It is also necessary to make the reflectance and transmittance curves close to unity in their dominating regions and to make them as steep in the transition region as possible. The absorption by the filter also needs to be minimized. This calls for good control of the charge carrier concentration and making the mobility as high as possible to maximize the obtainable power and efficiency of the TPV cell.

Experimental procedure

1) Substrate preparation. The substrates used for the deposition of indium oxide were <100> oriented p-type Si. High resistivity wafers were used, doped to a nominal resistivity of 30 Ωcm in order to minimize free-carrier absorption in the bulk of the wafer. Fused silica substrates, 1mm thick, were also used for some runs. The 1" square Si substrates were immersed in beakers containing trichloroethane (TCA), which were placed in an ultrasonic bath for 2-3 minutes and then transferred to a hot plate for approximately 3 minutes. The wafers were rinsed off with acetone followed by methanol. The cleaning in TCA removes organic contaminants and particulates from the surface and the rinsing steps remove any organic residues left by the TCA. A final dilute HF dip for 30 seconds was performed to etch away the native oxide layer on the Si surface. A similar procedure was used to clean the fused silica substrates, omitting the dilute HF dip. The substrates were then immediately transferred to the growth chamber.

2) Chemical vapor deposition system. All growth runs of indium oxide or tin doped indium oxide were carried out in an atmospheric pressure CVD system. The system consists of a pyrex chamber, a resistively heated susceptor, a glass funnel, and 316SS steel plumbing connecting the chamber to the trimethylindium / tetraethyltin bubblers and to the argon / oxygen tanks. The organometallic indium and tin precursors enter the chamber through the funnel while the oxygen inlet is at the side. This ensures that the reaction occurs at the susceptor, reducing the possibility of gas-phase reactions. Gas flows are controlled using flow meter tubes (a mass flow controller was used for tetraethyltin). A diagram of the CVD system chamber is shown in Fig. 1.

3) Growth Procedure. Growth runs were initiated by flushing all gas lines with ultra-high purity argon. Under a constant flow of 1slpm of argon, the wafer was inserted into the chamber through the inlet port and positioned directly under the glass funnel. The chamber was flushed for 10 minutes with 5 slpm of argon to purge the system of oxygen. Immediately afterwards, the wafer was heated by the susceptor to the growth temperature, typically 450°C.

Precursors used for growing indium oxide or tin doped indium oxide were trimethylindium, tetraethyltin, and extra-dry grade of O_2. Ultra-high purity argon was used as a carrier gas at a total flow of approximately 5 slpm. Reactant flows were started simultaneously at low flows of trimethylindium (75 sccm of Ar was flowed initially through the bubbler) which was subsequently raised to the normal flow rate (300 sccm of Ar through the bubbler) after about ten minutes of growth.

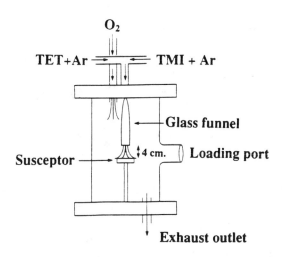

FIGURE 1. Schematic diagram of the Chemical Vapor Deposition system.

This was done to promote uniform nucleation over the wafer surface by growing the initial layer slowly and avoiding the formation of indium droplets caused by excess indium. To achieve a layer thickness of about 0.7 to 0.8 µm, a total growth time of about 65 minutes was necessary. The growth was monitored periodically by observing the color changes due to constructive interference as the film thickness increased. These color changes were then correlated with the thickness values on a color chart. This monitoring helped to observe changes in the growth rate and provided an approximate method for determining when the growth run should be terminated.

An optimum funnel height of 4 cm above the wafer surface was established by trial and error. This was chosen to minimize the possibility of poor nucleation caused by an indium rich surface while maintaining a reasonable growth rate. It was found to be necessary to periodically clean the reaction chamber to prevent dusting of the substrate surface. The procedure detailed above was used for the growth of all the layers discussed below. Systematic variations in growth conditions were used to control the optical properties of the filters.

4) Characterization. Reflectance and transmittance measurements over a spectral range of 1.8-20µm were carried out using a Fourier Transform InfraRed (FTIR) spectrometer. Structural characterization was carried out using a Philips X-Ray diffractometer using Cu Kα radiation. Hall effect measurements were also done to extract electrical properties such as mobility and carrier concentration. However, owing to the highly conducting nature of these films, the parameters extracted from the Hall measurements were not very reliable. Thickness measurements on the films were done using a spectroscopic ellipsometer with a spectral range of 400-740 nm (J. A. Woollam Co.) and also an α-step profilometer.

Experimental Results and Discussion

A number of different studies were carried out on undoped and tin-doped In_2O_3 plasma filters by varying various growth conditions, post-growth treatments and substrates used. Unless explicitly mentioned, the growth temperature was 450°C for all cases. The layers were all grown to a final thickness of about 0.65-0.85μm, which corresponds to the skin depth for absorption of near infrared waves incident on these films.

Undoped Indium Oxide

a) Effect of varying O_2/TMI partial pressure ratio. A very effective method for controlling the carrier concentration and consequently the λ_p in undoped indium oxide was to vary the O_2/TMI partial pressure ratio. A constant partial pressure of 1.4×10^{-4} atm of TMI (corresponding to a flow of 300 sccm through the bubbler) was maintained and the oxygen partial pressure was varied in different runs from 8×10^{-4} atm to 1.5×10^{-3} atm. The growth temperature was maintained at 450°C. The results of these studies are shown in Fig.2. We see that λ_p increases with a rise in the O_2 partial pressure. The native oxygen vacancy concentration is expected to fall at higher partial pressures of O_2, as this would lead to a more stoichiometric material. This drop in vacancy concentration results in a smaller value for the carrier concentration and consequently a larger λ_p. As we can see from eq (2), λ_p is a sensitive function of the carrier concentration and thus small changes in the O_2 partial pressure cause large shifts in λ_p over the range of interest.

b) Effect of varying the growth temperature. Initially, growth of undoped In_2O_3 was carried out at a susceptor temperature of 350°C. Reasonably good optical characteristics were obtained, but the growth rate was found to be very low. In order to increase the growth rate and optimize the optical properties of the film, a series of growth runs were made at temperatures of 400, 450, and 500°C. Here, the partial pressures of TMI and O_2 were maintained constant at 1.4×10^{-4} atm and 1×10^{-3} atm respectively. As seen in Fig. 3, there is an increase in λ_p as the growth temperature is raised. This indicates that the oxygen vacancy concentration drops

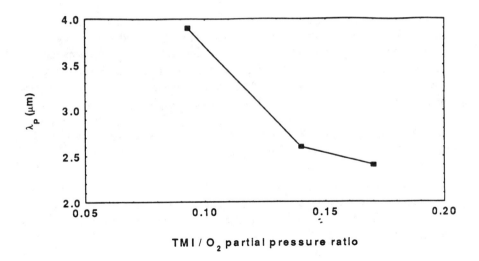

FIGURE 2. Variation of λ_p with O_2 partial pressure used for the growth of undoped In_2O_3 on Si. (TMI partial pressure kept constant at 1.4×10^{-4} atm).

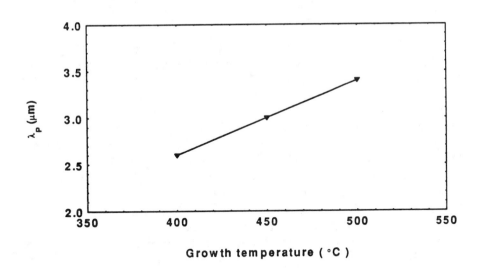

FIGURE 3. Variation of λ_p with the growth temperature used for undoped In_2O_3 growth on Si. (Partial pressures of TMI and O_2 kept constant at 1.4×10^{-4} atm and 1×10^{-3} atm respectively).

at higher growth temperatures, which can be attributed to improved stoichiometry of the film. It appears that the reaction between the indium and oxygen species is more favored at higher temperatures. In short, this provides yet another method of effectively controlling λ_p. Growth at temperatures higher than 500°C was not investigated because of potential problems with increased recirculation and gas-phase reactions, leading to dusting and poor surface morphology.

c) Effect of using fused silica as the substrate material. Fused silica was used as an alternate substrate upon which to grow In_2O_3. The growth was carried out on Si and fused silica substrates at 450°C, under partial pressures of 1.4×10^{-4} atm for TMI and 1×10^{-3} atm for O_2. As we can see from the reflectance and transmittance spectra of Figs. 4 & 5, λ_p is nearly the same on both substrates. However, for the In_2O_3 grown on fused silica, the transmittance is greatly improved and is as high as 80% at a wavelength of 2 μm (Fig. 5). This is a result of the smaller difference in refractive indices of In_2O_3 (n = 2) and fused silica. The refractive index n = 1.45 for fused silica and n = 3.5 for silicon. This also explains the absence of the interference fringe in the reflectance for $\lambda < \lambda_p$. The interesting feature of this study was that similar material was grown on both Si <100>, which is a crystalline substrate and fused silica, which is amorphous. The substrate does not seem to have a significant effect on the growth of undoped In_2O_3. It is likely that the surface of Si is covered by a thin native oxide layer prior to the growth, so that the differences between Si and silica are not significant.

d) Annealing Studies. It has been reported in the literature that post-deposition annealing of indium oxide samples grown by other techniques has helped improve their optical and electrical properties (15-17). However, we discovered that this was not the case for our CVD grown samples. Rapid thermal annealing was carried out in a N_2 ambient at 850°C for 5 min. and 1050°C for 20 seconds and the results are shown in Figs. 6 & 7 respectively. Fig. 8 shows the reflectance spectrum of a sample annealed in an air ambient at 750°C for 5 hours. We can see that annealing in the inert N_2 ambient resulted in a shift of λ_p to lower wavelengths, probably caused by loss of oxygen from the sample, resulting in an increase in the carrier concentration. On the other hand, annealing in an air ambient had the effect of shifting λ_p to longer wavelengths. This may be a consequence of O_2 from the ambient diffusing in and reducing the vacancy concentration. In both cases, *no beneficial effect was observed upon annealing*, in stark contrast to studies by other workers (15-17). This can be attributed to the fact that our material was grown by CVD at a relatively high temperature, which promotes the growth of good stoichiometric polycrystalline material, without the presence of a second phase such as excess indium droplets. The In_2O_3 used by other workers was grown by techniques such as sputtering and reactive evaporation, which yield inherently

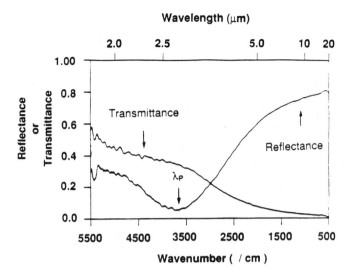

FIGURE 4. Reflectance and transmittance spectra of undoped In_2O_3 grown on Si <100>. The absorption is less than 25%.

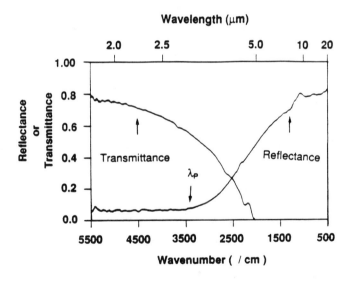

FIGURE 5. Reflectance and transmittance spectra of undoped In_2O_3 grown on fused silica.

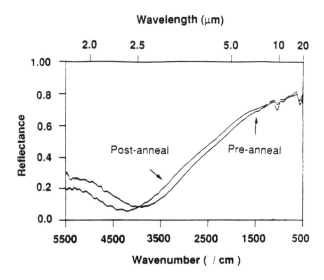

FIGURE 6. Reflectance spectra of undoped In_2O_3 annealed at 850°C for 5 min. (RTA) in N_2.

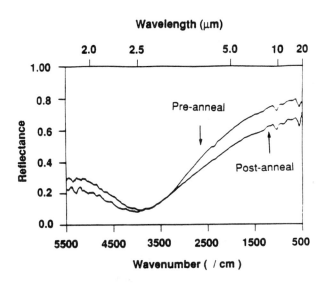

FIGURE 7. Reflectance spectra of undoped In_2O_3 annealed at 1050°C for 20 sec (RTA) in N_2.

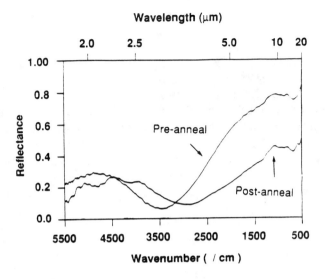

FIGURE 8. Reflectance spectra of undoped In_2O_3 annealed in air ambient at 750°C for 5 hours.

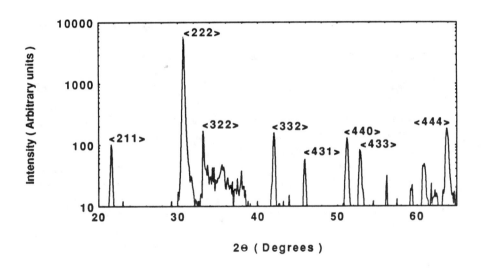

FIGURE 9. Results of X-ray diffraction measurement on undoped In_2O_3 (Cu Kα radiation).

poorer starting material that is far more amenable to improvement upon appropriate annealing.

e) X-ray Studies. X-ray measurements were carried out using a diffractometer in order to study the degree of crystallinity in the In_2O_3 samples. A typical x-ray diffraction spectra of a In_2O_3 / Si sample is shown in Fig. 9. A large number of peaks were obtained with good intensity on all samples, indicating that the In_2O_3 was polycrystalline. On comparison with standard powder diffraction data for In_2O_3, the strong (222) peak showed evidence of high texture in the <111> direction, which agrees with results obtained by other workers on In_2O_3 grown by chemical vapor deposition (9). No significant changes were observed in the relative intensities of the various peaks for measurements on samples grown under a variety of different growth conditions mentioned above: It is also interesting that the x-ray measurements done on indium oxide samples grown on fused silica also exhibit a high degree of texturing in the <111> direction. This shows that the growth temperature is high enough to ensure that the reactant species have enough surface mobility in order to form large grains oriented in a particular direction, irrespective of the substrate material.

Tin Doped Indium Oxide

As seen before (Fig. 4 & 5), good results were obtained with the undoped In_2O_3. However, the transition of the reflectance curve for $\lambda > \lambda_p$ was not as sharp as desired. The absorption also needed to be reduced further. In undoped In_2O_3, the n-type doping arises from the native defects (eq.1), which in turn depends on the oxygen partial pressure during the growth. The oxygen partial pressure has to be kept low during the growth in order to achieve the high carrier concentrations required for these filters. However, the growth of a good film with better surface morphology requires a higher flow of oxygen. In fact, we use a larger O_2 / TMI partial pressure ratio at the beginning of the growth in order to obtain a good nucleation layer without the formation of In droplets. The conflicting requirements mentioned above can be addressed by the use of Sn as an extrinsic dopant. In tin doped films, we can adjust the flow of the Sn precursor to obtain a primary control of the carrier concentration whereas the oxygen flow can be adjusted to optimize the film stoichiometry and morphology. This may lead to reduced scattering in the films and consequently higher mobility, resulting in a sharper transition of the reflectance spectra.

a) Effect of varying TET partial pressure. The tin doping was carried out using very low partial pressures of TET. These ranged from 1-4% of the TMI partial pressure in the *gas phase*. We cannot expect the same ratio for the actual

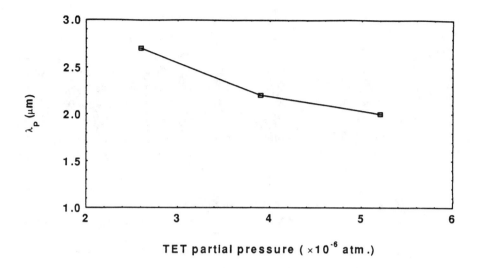

FIGURE 10. Variation of λ_p with TET partial pressure used during the growth of tin-doped In_2O_3. (Partial pressures of TMI and O_2 were kept constant at 1×10^{-4}atm and 3×10^{-3}atm respectively)

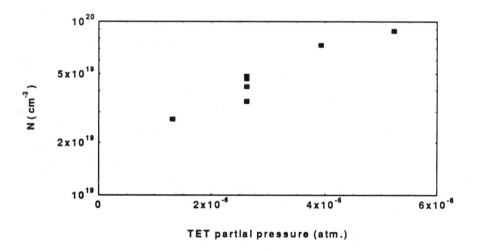

FIGURE 11. Carrier concentration of tin-doped In_2O_3 layers vs. partial pressure of TET used during the growth.

incorporation of Sn and In as the reactivity of TET with O_2 is very low. At percentages less than 2%, no significant change was observed either in the reflectance curve or the position of λ_p, indicating that a negligible amount of active Sn was incorporated as compared to the oxygen vacancies. However, when the tin percentage was increased from 2% to 4%, the result was quite striking (Fig. 10). Here, partial pressures of O_2 and TMI were kept constant at 3×10^{-3} atm and 1.4×10^{-4} atm respectively, while the partial pressure of TET was varied from 2.8×10^{-6} atm. to 5.6×10^{-6} atm. Note that the O_2 partial pressure is maintained at a significantly higher level than for the undoped In_2O_3 films in order to improve the film quality.

Fig.11 sums up the results of varying the tin doping at constant TMI and O_2 partial pressures. Here, the carrier concentrations calculated from eq(2) for λ_p (assuming λ_P to be the minimum of the FTIR reflectance spectrum) have been plotted vs. partial pressures of TET at a constant flow of 15 sccm of O_2 (3×10^{-3} atm). The result indicates that increasing the tin doping results in a direct increase in the carrier concentration. This is yet another way in which to control the position of λ_p around the required value.

b) Effect of varying O_2 partial pressure. Since the carrier concentration in these films was found to be primarily controlled by the Sn precursor partial pressure, the oxygen flow may be varied in order to optimize the crystalline and optical properties. Varying the O_2 partial pressure might also provide a means of separating the contributions from the tin and the O vacancies to the carrier concentration. A series of growth runs were carried out at various partial pressures of O_2 ranging from 3×10^{-3} atm. to 7.2×10^{-3} atm while keeping the partial pressures of TMI and TET constant at 1.4×10^{-4} atm and 3.93×10^{-6} atm respectively. As can be seen from Fig.12, increasing the O_2 flow resulted in λ_p shifting to lower wavelengths. We would expect that the native oxygen defect concentration would fall as the O_2 partial pressure were increased. If the same amount of tin were incorporated in all cases, the carrier concentration should not have been affected by varying O_2. The observed increase in the carrier concentration seems to indicate that tin incorporation is favored at higher O_2 partial pressures. Fig. 13 summarizes the results of this study, where the carrier concentration calculated from λ_p (as above) is plotted versus O_2 partial pressure.

c) Effect of using fused silica as an alternative to silicon < 100>. Good results were obtained with the tin-doped In_2O_3 grown on Si. Fig. 14 shows the transmittance and reflectance spectrum for the best tin-doped In_2O_3 layer grown on Si. The partial pressures of O_2, TMI and TET used were 1×10^{-3} atm, 1.4×10^{-4} atm and 2.8×10^{-6} atm respectively. It was decided to use fused silica as a substrate to

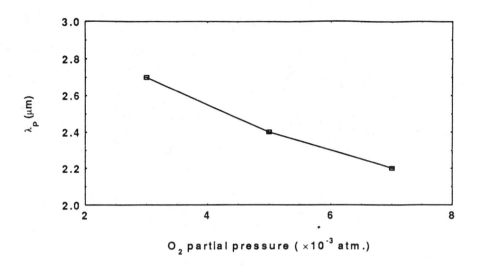

FIGURE 12. Variation of λ_p with O_2 partial pressure used during the growth of tin-doped In_2O_3. (Partial pressures of TMI and TET were kept constant at 1×10^{-4} atm and 3.93×10^{-6} atm respectively).

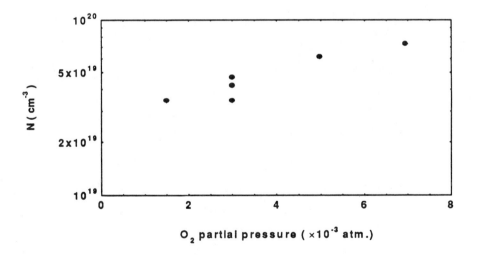

FIGURE 13. Carrier concentration of tin-doped In_2O_3 layers vs. partial pressure of O_2 used during the growth.

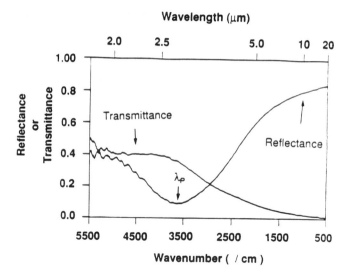

FIGURE 14. Reflectance and transmittance spectra of tin-doped In_2O_3 grown on Si <100>.

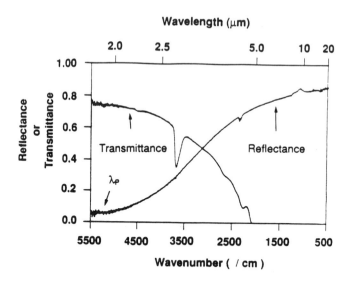

FIGURE 15. Reflectance and transmittance spectra of tin-doped In_2O_3 grown on fused silica.

take advantage of the increased transmittance observed previously for undoped In_2O_3 grown on silica..

Fused silica was used as an alternative to Si <100> on which to grow tin doped In_2O_3. As in the case of undoped In_2O_3 growth on silica, we found that the material was essentially the same as that grown on Si <100>. However, as expected, the reflectance and transmittance spectra were quite different owing to the different optical properties of the substrate material. Fig. 15 shows the reflectance and transmittance spectra for tin-doped In_2O_3 grown on fused silica under partial pressures for O_2, TMI and TET being 3×10^{-3} atm, 1.4×10^{-4} atm and 5.4×10^{-6} atm respectively. We clearly observe the higher transmittance for $\lambda < \lambda_p$ compared to figure 14. λ_p is shifted to a low value of nearly $2\mu m$, but the enhanced transmission makes up for the low value of λ_p. The act of shifting λ_p to a low value also has the beneficial effect of raising the reflectance at high wavelengths. The dip in transmission at $2.74\mu m$ is due to the substrate. The sharp dip in transmittance above $4\mu m$ is again due to the property of the fused silica. This tin doped In_2O_3 film exhibits high transmittance and reflectance in the required regions with low absorption and is one of the best filters grown in the course of this work.

d) Effect of varying layer thickness. As mentioned before, the layers grown in this study had thicknesses ranging from $0.65-0.85\mu m$. This was chosen on the basis of the skin depth for the wavelengths of interest. We would expect to get a sharper transition in the reflectance for thicker layers owing to the larger depth of the plasma through which the wave has to pass before being transmitted through the substrate. On the other hand, the very same reason would lead to an increase in absorption, degrading the filter optical properties. To observe the effect of these conflicting mechanisms, a tin-doped In_2O_3 layer was grown at twice the usual thickness. Partial pressures for O_2, TMI and TET were 3×10^{-3} atm, 1.4×10^{-4} atm and 2.8×10^{-6} atm respectively. As can be seen from Fig. 16, the reflectance does improve compared to that of a layer grown to the usual thickness (Fig.14) with a very sharp transition for $\lambda > \lambda_p$. The larger thickness also leads to the presence of interference oscillations in the reflectance for $\lambda < \lambda_p$. However, the absorption is much higher, leading to a bigger dip in the transmittance at λ_p and a significant decrease in the transmittance for $\lambda < \lambda_p$.

e) X-ray studies. X-ray diffractometer measurements were done as described before. Fig. 17 shows the x-ray diffraction data for a sample of tin-doped In_2O_3 grown on Si. The results indicate that the tin-doped In_2O_3 was also polycrystalline with a <111> texture, similar to the result obtained on undoped In_2O_3. Also, as reported by other workers (13), there was no evidence of a separate SnO_2 peak owing to the high solubility of Sn in the In_2O_3 lattice. These workers also observed

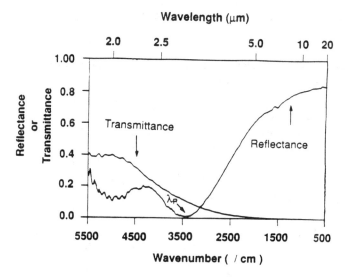

FIGURE 16. Reflectance and transmittance spectra of tin-doped In_2O_3 grown on Si <100> to a thickness of 1.3μm (2 × normal thickness).

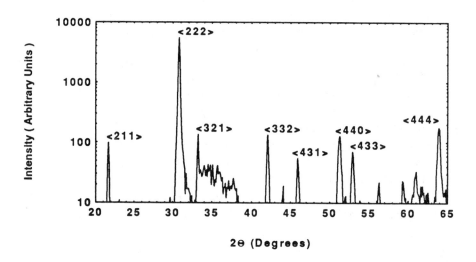

FIGURE 17. Results of X-ray diffraction measurement on tin-doped In_2O_3 (Cu Kα radiation).

that the incorporation of tin in reasonable amounts (~10%) results in an appreciable shift in the peak positions from the undoped to the tin-doped layer. However, we did not observe any noticeable shift in the peak positions of the x-ray diffraction data between undoped and tin-doped In_2O_3. This result suggests that a very small amount of tin (< 1%) was actually incorporated into the lattice, which is not unreasonable since we would expect the incorporation of tin during chemical vapor deposition using a relatively stable species such as TET to be quite low. We intend to confirm this hypothesis by carrying out Secondary Ion Mass Spectrometry (SIMS) on our samples.

Summary and Conclusion

From our results, we see that there are two principal methods of controlling the carrier concentration and consequently λ_p in our plasma filters. The first method is to control the native oxygen vacancy concentration by suitably adjusting the TMI / O_2 partial pressure ratio to obtain the required amount of oxygen vacancy donors. Good results were obtained with this, although in practice it is very difficult to reproducibly control the stoichiometry of the In_2O_3 as well as the surface morphology of the films.

We can also extrinsically dope the indium oxide with tin. This provides us with a more direct method of controlling the carrier concentration, and is easier to control. As discussed earlier, the material grown using tin doping may exhibit better optical properties through improvement in crystalline quality. Fig. 4 & 15 show the results for the best undoped and tin-doped plasma filters. We can see that the absorption is lower and the reflectance higher for the tin-doped material.

Varying the growth temperature is yet another way of shifting λ_p. However, this is not a good technique as low temperatures lead to low growth rates and high temperatures increase the chance of gas-phase reactions. A susceptor temperature of 450°C was found to be appropriate for our system.

Despite the successful results which other researchers have had on post-growth annealing of In_2O_3, annealing seems to be either detrimental or ineffective for our CVD grown filters. This may be attributed to the good inherent crystalline quality obtained directly by CVD techniques.

The choice of substrate does have a significant effect on the optical properties of the plasma filters. X-Ray studies indicate that the same type of <111> oriented polycrystalline material is grown on both substrates. Thus, the change in the filter optical response arises from the optical properties of the substrate material. As expected, the lower refractive index of fused silica results in a much improved transmittance for $\lambda < \lambda_P$. The higher transmittance allows us to shift λ_p to wavelengths as low as 2μm, resulting in an improvement in reflectance at longer wavelengths (Fig.15).

At present, the results on In_2O_3 are very promising but a number of improvements are needed before the filters can be successfully implemented on TPV cells. The uniformity and reproducibility of the layers need to be improved. Changes may be required to the system configuration in order to achieve this. It is also necessary to arrive at an optimum set of growth parameters in terms of reactor conditions, growth time etc. to obtain the best possible response.

The above results indicate that undoped and tin-doped In_2O_3 material is well suited for use as a spectral control filter used in TPV applications.

ACKNOWLEDGMENTS

The authors wish to thank J. Barthel for technical assistance.

REFERENCES

1. Köstlin, H., and Frank, G., " Thin-film reflection filters " , *Philips Tech. Rev.*, **41**, No.7/8 , 225-238 (1983-84).
2. Fan, J. C. C., and Bachner, F. J., " Transparent heat mirrors for solar-energy applications ", *Applied Optics*, **15**, No.4, 1012-1017 (April 1976).
3. Subrahmanyam, A., and Balasubramanian, N., *Semicond. Sci. Tech.*, **7**, 324 (1992).
4. Krokoszinski, H. J., and Oesterlein, R., *Thin Solid Films*, **187**, 179 (1990).
5. Buchanan, M. B., Webb, J. B., and Williams, D. F., *Appl. Phys. Lett.*, **37**, 213 (1980).
6. Maniv, S., Miner, C. J., and Westwood, W. D., *J. Vac. Sci. Tech*, **A1**, 1370 (1983).
7. Fan, J. C. C., *Appl. Phys. Lett.*, **34**, 515 (1979).
8. Groth, R., *Phys. Status. Solidi*, **14**, 69 (1966).
9. Maruyama, T., and Fukui, K., " Indium-oxide thin films prepared by chemical vapor deposition ", *Appl. Phys. Lett.*, **70**, 3848-3851 (1991).
10. Zheng, J. P., and Kwok, H. S., " Low resistivity indium tin oxide films by pulsed laser deposition ", *Appl. Phys. Lett.*, **63**, 1-3 (1993).
11. Köstlin, H., Jost, R., and Lems, W., " Optical and electrical properties of doped In_2O_3 films ", *Phys. Status Solidi*, **a29**, 87 (1975).
12. Müller, H. K., " Electrical and optical properties of sputtered In_2O_3 films ", *Phys. Status Solidi*, **27**, 733 (1968).
13. Parent, Ph., Dexpert, H., and Tourillon, G., " Structural Study of Indium Oxide Thin Films Using X-Ray Absorption Spectroscopy and X-Ray Diffraction ", *J. Electrochem. Soc.*, **139**, No. 1, 277 (Jan 1992).

14. Knofstad, P., *Nonstoichiometry, Diffusion, and Electrical Conductivity in Binary Metal Oxides.*, New York: John Wiley & Sons, Inc., 1972.
15. Haines, W. G., and Bube, R. H., "Effects of heat-treatment on the optical and electrical properties of indium-tin oxide films ", *J. Appl. Phys.*, **49**, 304-307 (1978).
16. Ito, K., Nakazawa, T., and Osaki, K., " Amorphous-to-crystalline transition of indium oxide films deposited by reactive evaporation ", *Thin Solid Films*, **151**, 215-222 (1987).
17. Patel, N. G., and Lashkari, B. H., " Conducting transparent indium-tin oxide films by post-deposition annealing in different humidity environments ", *J. Material. Sci.*, **27**, 3026-3031 (1992).

Characteristics of Degenerately Doped Silicon for Spectral Control in Thermophotovoltaic Systems

H. Ehsani, I. Bhat, J. Borrego and R. Gutmann

*Center for Integrated Electronics and Electronics Manufacturing &
Department of Electrical, Computer and Systems Engineering
Rensselaer Polytechnic Institute, Troy, NY 12180-3590*

and

E. Brown, R. Dzeindziel, M. Freeman and N. Choudhury

Knolls Atomic Power Laboratory, Schenectady, NY 12301-1072

Abstract. Heavily doped Si was investigated for use as spectral control filter in thermal photovoltaic (TPV) system. These filters should reflect radiation at 4 µm and above and transmit radiation at 2 µm and below. Two approaches have been used for introducing impurities into Si to achieve high doping concentration. One was the diffusion technique, using spin-on dopants. The plasma wavelength (λ_p) of these filters could be adjusted by controlling the diffusion conditions. The minimum plasma wavelength achieved was 4.8 µm. In addition, a significant amount of absorption was observed for the wavelength 2 µm and below. The second approach was doping by ion implantation followed by thermal annealing with a capped layer of doped glass. Implantation with high dosage of B and As followed by high temperature annealing (>1000°C) resulted in a plasma wavelength that could be controlled between 3.5 and 6 µm. The high temperature annealing (>1000°C) that was necessary to activate the dopant atoms and to heal the implantation damage, also caused significant absorption at 2 µm. For phosphorous implanted Si, a moderate temperature (800–900°C) was sufficient to activate most of the phosphorous and to heal the implantation damage. The position of the plasma turn–on wavelength for an implantation dose of 2×10^{16} cm^{-2} of P was at 2.9 µm. The absorption at 2 µm was less than 20% and the reflection at 5 µm was about 70%.

INTRODUCTION

Photovoltaic (solar) cells are widely used to convert light energy from the sun into usable electrical energy through a direct conversion process. A similar device, the thermal photovoltaic cell or TPV cell, utilizes energy form the near infrared spectrum to excite charge carriers and create electrical energy. These devices have immense practical value for generating power from sources radiating thermal energy. In this case, only those wavelengths corresponding to energies equal to or greater than the bandgap of the TPV cell material are useful. However, the thermal energy source radiate energy at a continuum of wavelengths corresponding to black body radiation (1). All wavelengths below the bandgap energy would pass through the cell and be wasted. To increase the efficiency, the wavelengths corresponding to the sub band gap energies should be reflected back into the source to reduce the input power to the system. These spectral control technique can be accomplished utilizing a heavily doped semiconductor filter or "plasma filter".

Some semiconductor oxides such as tin oxide (2), indium oxide (3) and indium tin oxide (4) have already been developed for use in efficient incandescent lamps (5) and energy-efficient windows (6) for houses. However, the characteristics of these filters are not suitable for spectral control in TPV systems. The plasma wavelength of these filters are usually in the 0.8-1 µm (7) range whereas, wavelength in the range of 2-4 µm is necessary for efficient TPV systems, depending on the source temperature. Control of the plasma wavelength in semiconductor oxide filters in the range 2-4 µm has been found to be difficult until recently.

This paper focuses the development of degenerately doped Si as plasma filter for TPV applications. The advantages of Si plasma filters (because of the advanced Si technology) are: reproducibility, high doping uniformity on large area wafers, a precise control on the dopant concentration, and the availability of large area wafers. The plasma wavelength can be adjusted over a wide range by varying the dopant concentration in silicon. In this paper, we will describe two methods to introduce impurities in Si, namely, diffusion technique using spin-on-doped glass and ion implantation technique followed by annealing with a cap layer of doped glass. Successful control of the plasma turn-on wavelength in the range 2.5 to 5.5 µm with negligible absorption at lower wavelength was obtained using the later technique.

SIMULATION OF PLASMA FILTER PERFORMANCE

In order to determine the parameters which are important in the design of plasma filters, we have simulated their performance. The model assumed is consists of the plasma filter material of thickness t, in this case, heavily doped silicon. The undoped substrate is ignored in this analysis. The important material parameters of interest are the electron density N, the optical relative dielectric constant ϵ_r and the electron collision frequency ω_c which is determined by the scattering relaxation time.

The simplest case to consider is the one in which there is no free carrier absorption. In this case, the effective electric permittivity of the plasma filter material, ϵ_{eff}, as functions of wavelength is given by:

$$\epsilon_{eff} = \epsilon_r \epsilon_o \left[1 - \left(\frac{\lambda}{\lambda_p}\right)^2\right]$$

where ϵ_r is the optical dielectric constant of the filter material, ϵ_o is the permittivity of free space and λ_p is the plasma wavelength of the material and it is given by (8)

$$\lambda_p^2(\mu m) = \frac{4\pi^2 c_o^2 \epsilon_r \epsilon_o m^*}{q^2 N} = 1.12 \times 10^{21} \times \frac{(m^*/m)\epsilon_r}{N(cm^{-3})}$$

The relationship between the plasma wavelength λ_p and the doping concentration for silicon is shown in Fig. 1. Note that doping level in excess of 1×10^{20} cm^{-3} will be required in order to obtain plasma wavelength below 4 µm. The thickness of the plasma filter material is also an important parameter. In Fig. 2, we show the reflectance characteristics of silicon filters for three different thicknesses normalized to the plasma wavelength. Notice that the figure shows that it is necessary to have a plasma filter material between 0.1 to 1 times the plasma wavelength. For thickness below 0.1, the transition from full transmission to total reflection is very gradual. In the case of a thick plasma frequency filter (t>>λ), Fig. 3 shows the reflection of a silicon plasma filter as a function of normalized wavelength for several values of electron collision frequencies, ω_c. Since collision frequency is inversely related to carrier mobility, high mobility samples will have sharper transition characteristics. An effect which has not been considered is the one of free carrier absorption. Thick plasma filter material has higher total absorption, so a compromise has to be achieved between the free carrier absorption and the sharpness of the transition. Free carrier absorption in the transmission region can be estimated from the absorption coefficient α, given by (9):

$$\alpha = \frac{Nq^3\lambda^2}{4\pi^2\epsilon_o c^3(m*/m_o)^2 m_o^2 \mu n} = 5.23 \times 10^{-17} x \times \frac{N(cm^{-3})\lambda^2(\mu m)}{(m*/m)^2 n\mu(cm^2/Vs)}$$

where c is the speed of light in free space.

EXPERIMENTAL METHODS

The (100) oriented Si wafers used in these experiments were p-type with resistivity in the range of 5-25 ohm-cm, double side polished with the thickness of about 300 µm. Lightly doped silicon was chosen so that residual free carrier

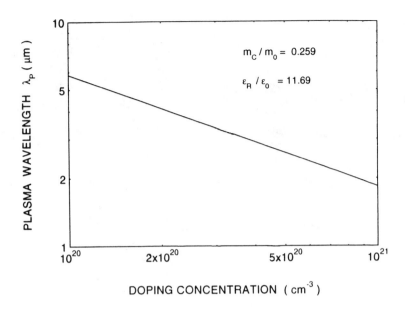

FIGURE 1. Plasma wavelength as a function of electron concentration for n type silicon.

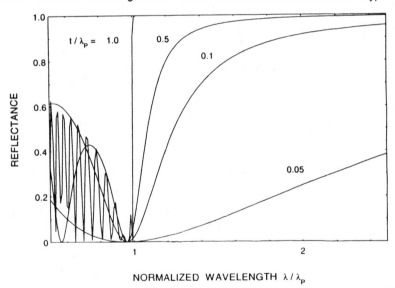

FIGURE 2. Reflectance versus normalized wavelength for silicon plasma filters. t/λ_p refers to the normalized thickness of doped Si.

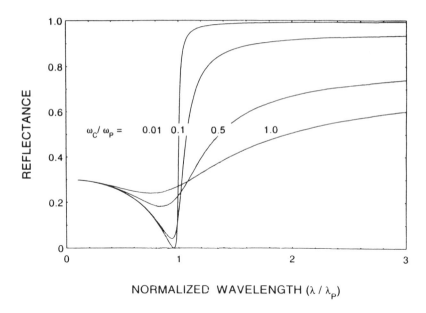

FIGURE 3. Reflectance of Si plasma filter versus normalized wavelength for several values of electron collision frequencies. The thickness of the filter is assumed to be infinite.

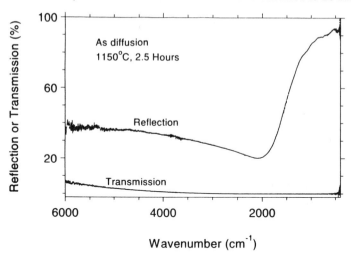

FIGURE 4. Reflection and transmission spectra of a Si wafer diffused with As at 1150°C for 2.5 hours. Spin-on arsenosilicate glass was the dopant source.

absorption in the bulk of the wafer is negligibly small. For doping with diffusion, a spin-on dopant source was first spun on the wafer and the diffusion was carried out in nitrogen atmosphere. For doping with ion-implantation, the wafers were first coated with 200Å thick thermal oxide and the ion implantation was done through this oxide. The thin SiO_2 layer should prevent out-diffusion of dopants during annealing. Some wafers were implanted without this SiO_2 layer, in which case they were coated with the spin-on doped glass prior to annealing. Both conventional furnace annealing and rapid thermal annealing (RTA) was used. Fourier transform infrared spectroscopy (FTIR) was used to determine the reflection and transmission spectra of the Si plasma filters in the range 1.5-20 µm. The percentage absorption at a particular wavelength was calculated by adding the percentage of reflection and transmission at that wavelength and subtracting from 100. Secondary ion mass spectrometric (SIMS) analysis was conducted to obtain the concentration and the distribution of impurities before and after annealing. Hall measurement was carried out using the van der Pauw technique to measure the mobility of the carriers.

RESULTS AND DISCUSSIONS

Diffusion Technique

The first series of experiments were conducted by doping Si with As by the diffusion technique, using spin–on–dopants as the source. Arsenic was used as the dopant source because its diffusion results in an abrupt profile compared to that of phosphorous (10). The long tail in the diffusion profile in the case of phosphorous can significantly increase the free carrier absorption in the plasma filter. The spin–on–dopant (arsenosilicate glass) was applied to the wafers and the diffusion was carried out in the temperature range from 900°C to 1200°C. For temperatures below 1000°C, we did not see any plasma characteristics below 20 µm, indicating insufficient diffusion of dopants into Si. At the diffusion temperatures above 1000°C, the plasma behavior of the wafers was observed within the wavelength range of our measurement, 1.5 to 20 µm. Figure 4 shows the reflection and transmission spectra of a Si wafer diffused with As at 1150°C for 2.5 hours. As can be seen, the position of the plasma wavelength is at 4.8 µm and the maximum reflectivity at 10 µm is about 85%. The carrier concentration calculated from the plasma wavelength is about 1.5×10^{20} cm^{-3}. This value is approximately a factor of 10 lower than the solid solubility of As (1.2×10^{21} cm^3) in Si (11) at 1150 C, indicating that either the diffusion of dopant is limited by the dopant source or a significant amount of As present is electrically inactive. The absorption in these filters was found to be very high (over 60%) as seen in Fig. 4. The calculated absorption coefficient at 2 µm is about 4.5×10^3 cm^{-1}. Based on the time and temperature of the diffusion, the thickness of the doped layer should be about 1 µm. Therefore, the total absorption at 2 µm should be less than 40%. We believe that the extra absorption is caused by the defects generated by As or Si-As precipitates that may be present in the doped region. Various diffusion times and temperatures were investigated, and the minimum value of the plasma wavelength was only 4.8 µm and high absorption was common in all these wafers. Therefore, an alternative to diffusion technique was investigated.

Ion Implantation

Ion implantation is a standard technique being used in silicon microelectronic industry for introducing precise amount of dopants into Si wafers. However, the doping range being used is not high enough so that some modification of the standard technique is necessary in order to get high doping concentration near the surface. The annealing condition need to be optimized to get very abrupt doping profile in order to minimize the free carrier absorption. In this section, the characteristics of plasma filters fabricated using ion implantation technique will be described. As and P were implanted to get n^+ layers on undoped Si, whereas B was implanted to get p^+ layers.

Arsenic Implantation

The high resistivity Si wafers were implanted with 150 kV As with dosage in the range $1\text{-}5 \times 10^{16}$ cm^{-2} and then thermally annealed to activate the dopants and to heal the implantation damage. A 200 Å thick thermal oxide was grown on some of the wafers prior to ion-implantation so that out-diffusion of As can be prevented during post-implantation annealing. Both furnace annealing and rapid thermal annealing (RTA) were used. The results are summarized in Table I. As shown in Table I, furnace annealing at 1000°C for 30 minutes resulted in filters with plasma wavelength at 6 µm, independent of the implantation dose. We believe that 1000°C annealing is not high enough for the complete removal of implantation damage, so that the effective carrier concentration is determined by the residual damage rather than the implantation dose. Table I also shows the results obtained from layers annealed at 1200°C for five minutes in an RTA system. As can be seen, the plasma wavelengths varied with the implantation dose and reached as low as 4.5 µm for samples implanted with a dose of 5×10^{16} cm^{-2}.

Arsenic has high vapor pressure and significant out diffusion of As can occur during the high temperature annealing, even though a 200 Å thick SiO_2 is present. To prevent the possibility of As out diffusion, some of the implanted wafers were etched in HF to remove the thermally grown SiO_2 that was deposited before ion-implantation and then coated with a 0.25 µm thick spin-on As-doped glass (arsenosilicate glass) prior to annealing. This doped glass can also be a source of As for diffusion during annealing. As shown in Table 1, the plasma wavelengths of these filters occurred at a lower value compared to those annealed without the presence of the doped glass. In addition, RTA resulted in filters with plasma wavelength at a lower value compared to those annealed in a furnace. This

TABLE I. A SUMMARY OF ANNEALING CONDITIONS USED AND THE PROPERTIES OF THE DOPED LAYERS.

Exp. Set #	Dose ($\times 10^{16}$ cm^{-2})	Annealing System	Annealing Temp. (°C)	Annealing Time (Min)	n ($\times 10^{19}$ cm^{-3})	λ_p
1	1	Furnace	1000	30	9.6	6
1	2	Furnace	1000	30	9.6	5.8
1	5	Furnace	1000	30	9.6	5.8
2	1	RTA	1200	5	7.7	6.7
2	2	RTA	1200	5	12	5.4
2	5	RTA	1200	5	17	4.5
3	1	Furnace	1150	15	13	5
3	2	Furnace	1150	15	16	4.7
3	5	Furnace	1150	15	17	4.5
4	1	RTA	1200	5	21	4.9
4	2	RTA	1200	5	35	3.9
4	5	RTA	1200	5	63	3.4
5	1	RTA	1200	2	37	3.8
5	2	RTA	1200	2	47	3.6
5	5	RTA	1200	2	73	3.2

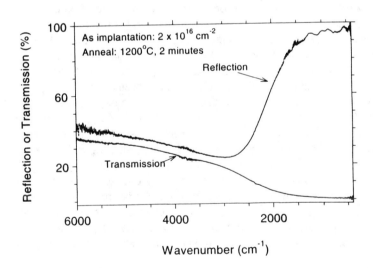

FIGURE 5. Reflection and transmission spectra of a Si wafer implanted with As, and annealed at 1200°C for 2 minutes with a cap layer of As-doped glass. Total absorption at 2 μm is less than 30%.

difference could be caused by the higher redistribution of As in furnace annealing. The plasma wavelength was even lower for samples annealed for two minutes at 1200°C using RTA. Shorter annealing time apparently causes less redistribution of dopants, so that peak carrier concentration is higher in these samples. We have not observed any significant improvement on the plasma characteristics of the wafers annealed between one to two minutes. Figure 5 shows the reflection and transmission spectra of As-implanted layer annealed at 1200°C for two minutes in RTA with a cap layer of doped glass. Note that the reflectivity at 10 µm reaches 90%, but the absorption at 2 µm is about 28%. These results indicate that the redistribution and out diffusion of dopant impurities should be minimized to obtain low value for λ_p, with minimum free carrier absorption. All further studies were conducted using RTA system with the doped glass as the cap layer.

The plasma wavelength and the total absorption at 2 µm of As doped layer as a function of the implantation dose was carried out and the results are summarized in Fig. 6. All the wafers were annealed for two minutes at 1200°C. This figure indicates that the plasma wavelength shifts to lower value as the implantation dose increases from 1×10^{16} cm^{-2} to 5×10^{16} cm^{-2}. With a further increase of the dose, the plasma wavelength moves to higher wavelengths.

Low absorption by the plasma filters is also important for efficient performance of TPV systems. As seen in Fig. 6, the absorption increases as the As concentration increases, with significant absorption observed for implantation doses above 5×10^{16} cm^{-3}. The increase of both the absorption and the plasma wavelength beyond 5×10^{16} cm^3 dose could be caused by the defects generated by the precipitation and local chemical reaction of As and Si. This reaction can decrease the carrier concentration and can generate a high density of defects. Therefore, the absorption is not due to just the free carriers, but also due to defects.

In order to investigate the concentration and distribution of dopant impurities before and after annealing, SIMS measurements have been conducted. SIMS analysis was performed on wafers implanted with 2×10^{16} cm^{-2} dose of As. Figure 7(a) shows the SIMS data before annealing and Fig. 7(b) is the SIMS profile after the wafer was annealed at 1200°C for two minutes. It is interesting to observe that the total amount of As in the wafer increased after the short annealing and that As diffused deep up to 0.9 µm. The As concentration was about 6×10^{20} cm^3 with the peak concentration occurring near the surface indicating that the excess As comes from the doped glass. Since implanted region is highly defective or even amorphous, high diffusion rate for As is expected in the ion implanted layer. If we assume 100% activation, the plasma wavelength should be lower and the percentage absorption should be higher than the values obtained by experiments, indicating incomplete As activation.

The Hall mobility of 27 cm^2/Vsec was obtained on a Si wafer implanted with As with the dose of 2×10^{16} cm^{-2}, annealed at 1200°C for two minutes.

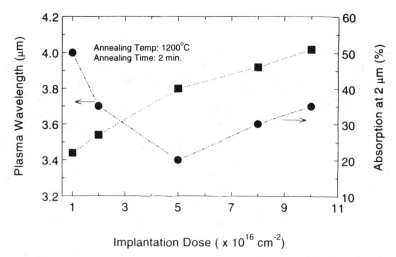

FIGURE 6. Plasma wavelength and absorption as a function of implantation dose. Note that both the absorption and the plasma wavelength increases with the increase in the implantation dose beyond 5×10^{16} cm^{-2}.

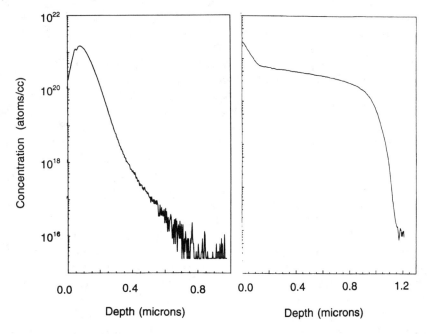

FIGURE 7. SIMS profile of As in Si (a) before and (b) after annealing at 1200°C for 2 minutes. Implantation dose = 2×10^{16} cm^{-2}. Significant redistribution of As is observed.

Phosphorus Implantation

Since As implanted layers required high temperature annealing for significant activation and had relatively high absorption at 2 µm, P was investigated for doping. Phosphorous implantation was carried out at 120 kV, with the dose in the range $1\text{-}10 \times 10^{16}$ cm^{-2}, followed by annealing in RTA system. Prior to annealing, the implanted wafers were etched in dilute HF solution to remove any oxide and a 0.24 µm thick layer of P-containing doped glass was applied by the spin-on technique.

Figure 8 show the plasma wavelength and the total absorption at 2 µm as the annealing temperature was varied over a wide range. These annealings were carried out using RTA for two to five minutes in nitrogen ambient. Both the plasma wavelength and the total absorption monotonically increases as the annealing temperature is increased. Unlike Si implanted with As, a moderate temperature of about 800-900°C is sufficient to activate implanted P. Phosphorous implanted Si may have less damage to start with because P is lighter than As and the diffusion rate of P is higher than that of As. Therefore annealing at lower temperature may be sufficient to activate P. As the annealing temperature is increased, the plasma wavelength shifted to higher wavelength, caused by the redistribution of P. The absorption also increases with the annealing temperature because it increases exponentially with the thickness. Figure 9 shows a typical reflection and transmission spectra of a Si wafer implanted with P with the dose of 2×10^{16} cm^{-2}, annealed at 850°C for five minutes. Absorption at 2 µm is less than 25% (without anti-reflection coating) and the total reflection at 10 µm is more than 85%.

Figure 10 show the effect of implantation dose on the plasma wavelength and the total absorption at 2 µm, for wafers annealed at 850°C for five minutes in an RTA system. As the implantation dose is increased from $1\text{-}3 \times 10^{16}$ cm^{-2}, the plasma wavelength decreases and the total absorption increases, as expected. Beyond 3×10^{16} cm^{-2}, the data indicate that the low annealing temperature is not sufficient to fully activate the implanted phosphorous. The variation of both the plasma wavelength and the absorption in the range of implantation dose studied is not large, so that either of these wafers will be a good filter for TPV systems.

Figures 11(a) and 11(b) show the SIMS profile of P in Si before and after rapid thermal annealing, respectively. Unlike Si implanted with As, the Gaussian distribution of the P dopant remain Gaussian after the annealing, with a small increase of the width of the profile. This indicates that annealing at 850°C for a short period does not redistribute phosphorous atoms significantly, but is high enough to activate phosphorous. On the other hand, high temperature annealing is required to activate implanted As, which also redistribute As by diffusion.

The effect of implantation energy in the range 90-150 kV on the filter characteristics was also investigated. But, we have not seen a significant difference on the reflection and transmission spectra of these wafers after they have annealed at 850°C for five minutes. This results are expected because the changes in energy simply places P at different depths from the surface, keeping the profile essentially

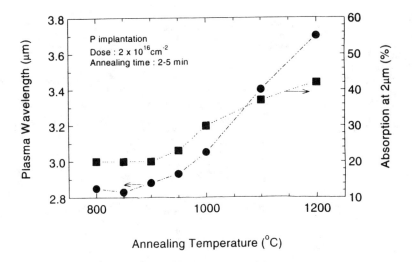

FIGURE 8. Plasma wavelength and absorption as a function of annealing temperature for Si implanted with phosphorous. Note that annealing at low temperature is sufficient to activate phosphorous.

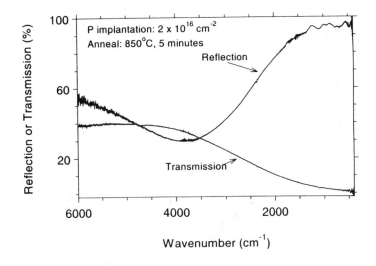

FIGURE 9. Reflection and transmission spectra of a Si wafer implanted with P, and annealed at 850°C for 5 minutes with a cap layer of P-doped glass. Total absorption at 2μm is less than 25%.

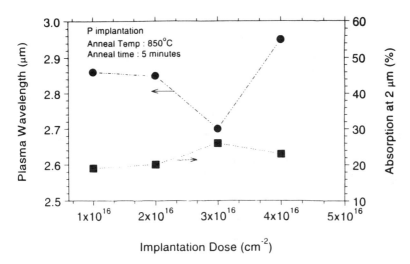

FIGURE 10. Plasma wavelength and absorption as a function of implantation dose for phosphorous.

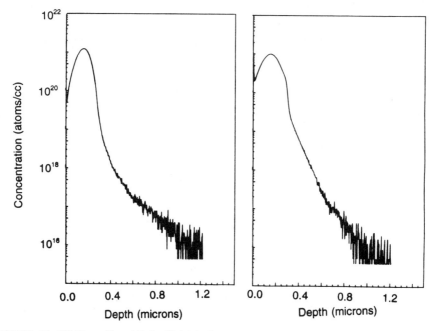

FIGURE 11. SIMS profile of P in Si (a) before and (b) after annealing at 850°C for 5 minutes. Implantation dose = 2×10^{16} cm^{-2}.

the same. In order to obtain a more uniform P distribution, double implantation at two energies (100 kV and 120 kV) was performed. An improvement was seen on the reflection spectra. Since activation-anneal of P does not cause redistribution, improvements in the filter characteristics can be achieved using multiple implantation. These studies are underway and will be reported later.

Hall measurement showed a mobility of 32 cm^2/Vsec for a P-doped sample annealed at 850°C for five minutes. This value is slightly larger than the electron mobility obtained on the wafers implanted with As.

Boron Implantation

Because holes have larger effective mass than that of electrons, p-type Si will have less free carrier absorption. Therefore, boron was also investigated as the dopant source. Boron implantation was carried out on bare silicon, and the wafers were annealed in RTA system to activate boron. Prior to annealing, the wafers were coated with spin-on dopant glass to a thickness of 0.24 µm to prevent any possible out-diffusion.

Figure 12 shows the plasma wavelength and the total absorption at 2 µm, as a function of the annealing temperature for wafers implanted with 2×10^{16} cm^{-2} of boron. Similar to As, high temperature annealing was found to be necessary to activate boron. Since B is a light element, the mechanism for ion stopping during ion implantation is due mostly to electronic stopping. Therefore, the implanted area may

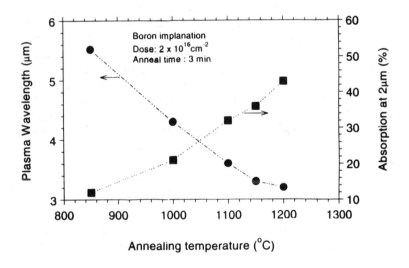

FIGURE 12. Plasma wavelength and absorption as a function of annealing temperature for Si implanted with boron. Note that annealing at high temperature is required to activate boron.

not become amorphous, but may remain polycrystalline. Generally, high temperature annealing is necessary to recrystallize polycrystalline materials. Figure 13 shows the effect of the implantation dose on the filter characteristics. These and other results show that activation of B is limited by the solid solubility at high temperatures. If the implantation dose is very high, boron will precipitate during high temperature annealing. These precipitated centers could be the source of defects and high absorption. Figures 14(a) and 14(b) show the SIMS profile of boron in Si before and after 1200°C two minute anneal. As can be seen, significant redistribution of boron is observed after the annealing. The redistribution of B should be negligible if the implanted region were single crystal.

The Hall measurement showed that the mobility of holes in p-type Si is about 28 cm^2/Vsec, for the samples implanted with a dose of 2×10^{16} cm^{-2} and annealed at 1150°C for two minutes. Note that similar values were obtained for both p-type and n-type layers, when the doping concentration is very high.

CONCLUSIONS

The plasma behavior of Si heavily doped with As, P, and B was investigated. We have seen that high doping concentration can be achieved by the implantation technique, followed by rapid thermal annealing with a cap layer of spin-on doped glass. High temperature annealing (>1100°C) is necessary to activate significant percentage of As and B. This results in significant redistribution of the implanted species, causing high absorption at 2 µm and below. The high absorption is probably caused by both the free carriers and the defects generated by As and B precipitates. On the other hand, a moderate annealing temperature (800–900°C) was sufficient to activate P in Si. This causes negligible redistribution and hence results in low absorption because the thickness of the doped layer is low. In addition, low temperature annealing doesn't cause P precipitates so that absorption caused by defects is also low. Therefore, P doped silicon shows high promise as a candidate for plasma filter in TPV application. The plasma wavelength for an implantation dose of 2×10^{16} cm^{-2} of P was at 2.9 µm. The absorption at 2 µm was less than 20% and the reflection at 5 µm was about 70%. Further improvement in filter characteristics can be achieved by optimizing the doping profile using multiple implantation technique at different energies.

ACKNOWLEDGMENTS

The authors wish to thank John Barthel for technical assistance and Dr. Greg Charache for making the arrangements for the SIMS measurements.

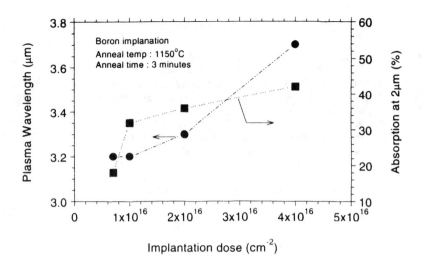

FIGURE 13. Plasma wavelength and absorption as a function of implantation dose for boron. As the implantation dose is increased, boron activation is actually reduced.

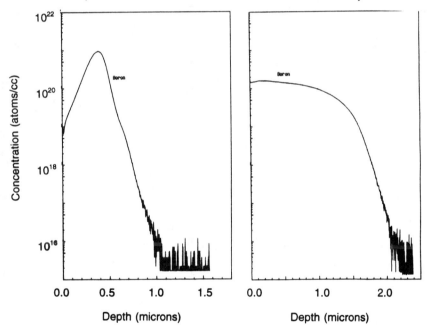

FIGURE 14. SIMS profile of B in Si (a) before and (b) after annealing at 1150°C for 5 minutes. Implantation dose = 2×10^{16} cm^{-2}.

REFERENCES

1. Zissis, G.J., *Optical Engineering* **15**, 484-497 (1976).
2. Grath, R. and Kauer, E., *Philips Tech. Rev.* **26**, 105-111 (1965).
3. Kosilin, H., Jost, R. and Lems, W., *Phys. Stat. Sol.* **29**, 87-93 (1975).
6. Manivannan, P. and Subrahmanyan, A., *J. Phy. D: Appl. Phys.* **26**, 1510-1515 (1993).
5. Kauer, E., *Philips Tech. Rev.* **26**, 33-47 (1965).
6. Kosilin, H., *Philips Tech. Rev.* **34**, 242-243 (1974).
7. Kosilin, H. and Frank, G., *Philips Tech. Rev.* **41**, 225-238 (1983).
8. Jackson, J.D., *Classical Electrodynamics*, 2nd Edition, 1975.
9. Hunsperyer, R.G., *Integrated Optics*, 1982.
10. Ghandhi, S.K., *VLSI Fabrication Principles*, John Wiley, 1983.
11. Sze, S.M., *High Speed Semiconductor Devices*, 1990.

TPV Plasma Filters Based on Cadmium Stannate

X. Wu, W.P. Mulligan, J.D. Webb, and T.J. Coutts

National Renewable Energy Laboratory, 1617 Cole Blvd., Golden, CO 80401

Abstract: A selective filter is an important component in a high-efficiency thermophotovoltaic (TPV) system. Compared to dielectric stack interference filters, semiconductor plasma filters have the potential for higher performance at lower cost. Conventional transparent conductive oxides (TCOs), such as ITO, SnO_2 and ZnO, are inadequate for low temperature (800-1200°C) TPV system applications, because of their low mobility (~20 $cm^2V^{-1}s^{-1}$) and high carrier concentration (>5 x 10^{20} cm^{-3}). A cadmium stannate (Cd_2SnO_4) based selective filter has been developed in this study. We will report experimental results on Cd_2SnO_4 deposited by r.f. magnetron sputtering. The principle variables investigated were the composition of the sputtering gas, the substrate temperature, and the conditions of post-deposition thermal treatment. The electrical, optical and compositional properties of the films have been characterized using Hall effect measurement, optical and infrared spectroscopy, X-ray diffraction, scanning electron microscopy, and atomic force microscopy. Mobilities as high as 65 $cm^2V^{-1}s^{-1}$ with a carrier concentration 2-3 x 10^{20} cm^{-3} have been obtained. The results indicate the ability to control the short-wavelength transmittance, the long-wavelength reflectance, and the position and abruptness of the plasma edge. The plasma edge can be controlled between 1.5 and 3.0 μm.

INTRODUCTION

Due to the broad spectra nature of blackbody radiation, spectral control is necessary for high efficiency, high power density thermophotovoltaic (TPV) systems. Possible methods to achieve spectral control include selective thermal emitters, rear surface reflectors on the TPV converters, or selective filters between the IR source and the TPV converter. A selective filter should be highly reflective to radiation with wavelengths longer than the cutoff wavelength, and highly transparent (with minimal absorption) to radiation with wavelengths shorter than the cutoff, with an abrupt transition between the two regions. Depending on the temperature of the thermal emitter and the band gap of the converter, the optimal filter should have a cutoff wavelength (or filter edge) at about 2 to 3 μm.

The propagation of electromagnetic waves in metals and degenerate semiconductors can be well described by considering the dielectric function of the free electron gas, or 'plasma.' When the dielectric function of the free electron gas is positive and real, waves can propagate without attenuation, leading to

© 1996 American Institute of Physics

transparent behavior in this frequency range. When the dielectric function is negative, waves are reflected. The wavelength where the transition from transparent to reflective behavior occurs, strictly defined as the wavelength where the dielectric function equals zero, is know as the plasma wavelength (or plasma frequency). For metals this transition occurs in the ultraviolet, but for degenerate semiconductors it occurs in the near IR, because the free electron concentration is typically one to two orders of magnitude lower than in metals. For this reason degenerate semiconductors are of interest for TPV selective filter applications, and are known as plasma filters.

The Drude theory of free electrons was used to model the IR reflectivity of plasma filters. The plasma wavelength can be to be given by (1):

$$\lambda_p = \sqrt{\frac{\pi \varepsilon_\infty m^*}{ne^2}} \quad \text{(CGS)} \quad (1)$$

where ε_∞ is the high frequency limit of permittivity, m* is the effective mass of the free electrons, n is the electron concentration, and e is the fundamental unit of charge. For the modeling results presented below, carrier concentration was chosen such the plasma wavelength occurs at 3 µm. Assuming reasonable values for carrier effective mass (m* = 0.20 m_e) and the high frequency limit of permittivity (ε_∞ = 8), the complex dielectric function can be calculated numerically, taking electron mobility as a parameter. The carrier relaxation time was assumed to be constant and independent of frequency. The frequency dependent refractive index and extinction coefficient are then determined easily. Finally the Fresnel coefficients and the reflectance can be calculated. The model results are presented in Figure 1.

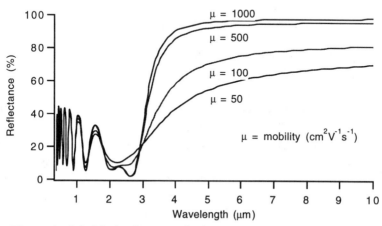

Figure 1. Modeled reflectance for λ_p = 3 µm, mobility as a parameter.

It can be seen that both the long wavelength reflectance and the sharpness of the transition from transparent to reflecting behavior are strongly dependent on electron mobility. The long wavelength reflectance can also be increased by increasing the carrier concentration, but carrier concentration is not a free parameter if the plasma wavelength is to be kept at a fixed value. Near the plasma wavelength, electromagnetic radiation strongly resonates with and is absorbed by the free electrons. Our modeling results suggest that both the width and magnitude of this absorption region are reduced as mobility increases. Thus the optimal plasma filter operating at a wavelength of about 3 μm should have a carrier concentration of about 10^{20} cm^{-3}, and a mobility as high as possible.

Transparent conductive oxides (TCOs) are degenerate wide bandgap semiconductors which are extraordinary in that they can be easily doped to the 10^{20} to 10^{21} cm^{-3} level. Cadmium stannate (Cd_2SnO_4 or CTO) is a transparent conductor reported to have unusually high mobility and high carrier concentration, was selected as the material for this study. The properties of cadmium stannate films were first reported by Nozik (2), who prepared amorphous films by RF sputtering and reported mobilities as high as 100 cm^{2}V^{-1}s^{-1} at a carrier concentration of 5 x 10^{18} cm^{-3}. Nozik attributed this unusually high mobility to a low electron effective mass (m* = 0.04 me). Haacke (3,4) investigated polycrystalline films prepared by RF sputtering onto heated substrates. Both carrier concentration and mobility were found to increase after a post-deposition anneal at high temperature. Haacke was not able to obtain films which were completely free of secondary phases.

Possible means of increasing the mobility in cadmium stannate films include growth of single phase polycrystalline films, increasing the grain size, growth of single crystal films on lattice matched substrates (for example, bulk single crystal TCOs (5) have exhibited mobility in excess of 300 cm^{2}V^{-1}s^{-1}), understanding the defect mechanisms and ensuring all donors are ionized, and exploring low scattering cross-section dopants. The focus of this work was on the first item.

EXPERIMENTAL

Cd_2SnO_4 films were prepared by RF magnetron sputtering of 2 inch diameter planar oxide targets prepared by Cerac Inc. The component oxides (33 mol% SnO_2 and 67 mol% CdO) were pre-reacted following a procedure outlined by Nozik.[2] X-ray diffraction (XRD) showed that the target was single-phase orthorhombic Cd_2SnO_4. Sputtering was carried out in a CVC SC-3000 system, evacuated to a background pressure of 2x10^{-7} torr and then back-filled with UHP oxygen (99.993% purity) or UHP argon (99.999% purity) or a mixture of both.

Corning 7059 glass and polished crystalline silicon (c-Si) wafer substrates were placed on either a water-cooled or a heated sample holder parallel to the target surface. The distance between the substrate and the target was varied from 6 cm to 9 cm. Deposition was performed at a pressure of 5-20 mTorr, with an RF power between 100 and 140 watts. The average deposition rate was about 7 nm/min in pure oxygen, and about 25 nm/min in pure argon. After deposition, some samples were annealed in a tube furnace containing an ambient of flowing (1500 sccm) UHP argon or UHP oxygen or a mixture of both, at temperatures ranging 580 to 700°C, for 10-30 minutes. Some samples were also annealed in CdS vapor in argon at the same temperatures, as previous studies have shown this can increase carrier concentration.[4]

A Cary 2300 Spectrophotometer was used to acquire the UV/visible transmittance and reflectance spectra. A Nicolet Magna 550 Fourier transform infrared (FTIR) spectrophotometer was used to acquire the infrared transmittance and reflectance data at 16 cm^{-1} resolution. A gold-coated polished silicon wafer was used as the reflectance standard. The IR reflectance data were not corrected for the reflectance of gold. Hall effect measurements were made using the Van der Pauw technique with indium dot contacts. Sheet resistance was measured using a four-point probe. Film thickness was determined from the position of neighboring interference maxima in optical transmittance curves, and checked using a Dektak 3 step height profilometer and by SEM cross-section. XRD patterns were acquired using a Rigaku diffractometer, with Cu K_α radiation. Surface roughness was measured by atomic force microscopy.

RESULTS

The primary goal this work was to obtain a reproducible process to produce high mobility cadmium stannate film. Several key deposition and post-deposition annealing parameters were studied in detail. Deposition parameters studied included substrate temperature (20-700°C), sputtering gas (argon, oxygen, or mixtures of both), chamber pressure (5-20 mTorr), and film thickness. Post-deposition annealing parameters studied include annealing temperature, annealing time, annealing gas (argon, oxygen, or mixtures of both), and gas flow rate through the annealing tube furnace.

The optimal process was found be sputtering in a pure oxygen atmosphere onto room temperature substrates, followed by an anneal at 680°C in pure argon for 20 minutes. Optimal optical and electrical parameters were obtained for films with thickness greater than 5000 Å. XRD patterns for a film as-deposited and after anneal are shown in figure 2. The as-deposition film is amorphous, but the film crystallizes after annealing. The peaks have been indexed as spinel phase cadmium stannate, which is the known stable crystalline phase of this material in thin film form (6). There is no indication of secondary phases in the post-annealed polycrystalline film above the XRD detection limits (~1%). This process is very repeatable, and has the advantage of only one high temperature process step, compared to Haacke's best process, which required deposition on substrates heated above 500°C, and a subsequent high temperature anneal.

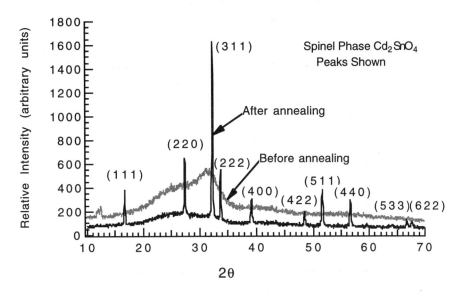

Figure 2. XRD Patterns for a typical TCO film, before and after anneal.

Films deposited onto heated substrates were mostly amorphous in the as-grown state, without evidence for the Cd_2SnO_4 spinel phase, even for substrate temperatures as high as 700°C. This is inconsistent with previous reports by Haacke, who obtained crystalline films for substrate temperatures greater than 500°C. We suspect the difference may be related to the lower power environment of magnetron sputtering compared to conventional RF diode sputtering, which was used by Haacke (4).

Electron mobilities as high as 65 $cm^2V^{-1}s^{-1}$ at a carrier concentration of 2 x 10^{20} cm^{-3} have been obtained. Using our optimal process, mobilities consistently exceed 50 $cm^2V^{-1}s^{-1}$. Even at carrier concentrations as high as 9 x 10^{20} cm^{-3} the mobility is still as high as 55 $cm^2V^{-1}s^{-1}$. The resistivity of this film is very low: 1.28 x 10^{-4} ohm-cm. To the best of our knowledge, these are the highest mobility and highest conductivity transparent conductive films ever attained by sputtering.

By controlling both deposition and annealing parameters, we have been able to vary the carrier concentration in these films from about 1×10^{20} to 9×10^{20} cm^{-3}. As mentioned, the plasma wavelength is proportional to the reciprocal of the square root of carrier concentration. Hence we have been able to vary the minimum of the reflectance curve (which occurs at a somewhat shorter wavelength than the true plasma frequency) between about 3.0 to 1.3 microns. Films with a carrier concentration of 1×10^{20} cm^{-3} and a reflectance minimum at 3.0 μm were obtained by sputter deposition in a pure argon atmosphere, followed by a 680°C anneal in argon. The experimentally measured specular IR reflectance for this film is presented in figure 3.

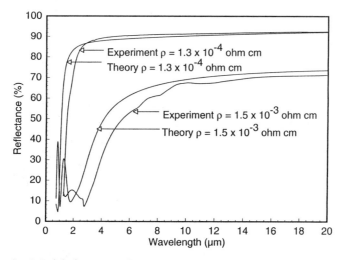

Figure 3. Modeled vs. experimental reflectance, at high and low carrier concentration.

The agreement between experiment and theory (model) is good. Our highest carrier concentration films (9×10^{20} cm^{-3}) are obtained by deposition in pure oxygen on room temperature substrates, followed by a 680°C anneal in argon and CdS vapor atmosphere. The measured specular IR reflectance for such a film is also presented in figure 3. The reflectance minimum occurs at about 1.3 μm, and the specular reflectance exceeds 90% at 5 μm. We expect the total reflectance will be an additional 2-3% higher than the specular reflectance. Agreement between theory and experiment is excellent.

We have also been successful in controlling the carrier concentration and plasma wavelength simply by varying the percentage of oxygen in the annealing atmosphere. Figure 4 shows IR reflectance data for four samples from the same deposition run (pure oxygen sputtering gas, room temperature substrate), annealed in different percentages of oxygen.

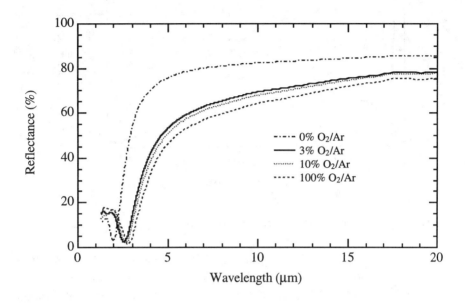

Figure 4. Control of the plasma wavelength by varying O_2 content during anneal.

For this series, the reflectance minimum shifts from about 2.8 µm for a pure oxygen annealing atmosphere, to about 2.0 µm for annealing in pure argon. Other films annealed in pure argon have shown reflectance minima as low as 1.5 µm. The reduction in carrier concentration and shifting of the plasma wavelength to longer wavelengths is possibly due to a reduction in the concentration of oxygen vacancies as the partial pressure of oxygen increases. Oxygen vacancies are known to be donors in cadmium stannate (4). The variation of electrical parameters with percentage of oxygen in the annealing atmosphere is varied is shown in figure 5. Carrier concentration drops rapidly as oxygen content is increased, while mobility increases slowly. The net result is a decrease in conductivity for films annealed in high oxygen content atmospheres.

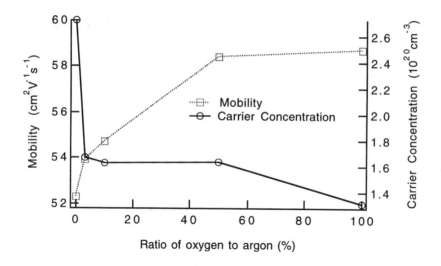

Figure 5. Electrical parameters vs. ratio of oxygen to argon during anneal.

The optical properties of a typical CTO film are shown in figure 6. The films show very low absorptance in the visible. Absorptance increases in the near IR as the plasma wavelength is approached. The fundamental absorption edge occurs in the UV, and shifts to shorter wavelengths as the carrier concentration is increased, a phenomenon known as the Burstein-Moss shift (7).

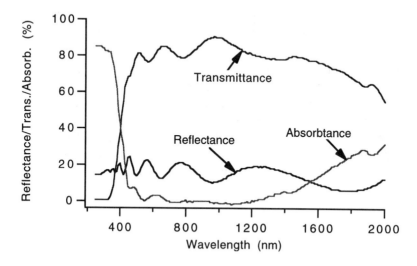

Figure 6. U.V. and visible properties of a typical CTO film.

An atomic force micrograph of a typical film is shown in figure 7. In general these films are very smooth, with average roughness on the order of 20 Å over an area of 1 µm². Because of their relatively smooth surfaces, the difference between specular and total reflectance for CTO films is small. Cadmium stannate films have also been found to be very durable, and are stable at temperatures up to 700°C in argon.

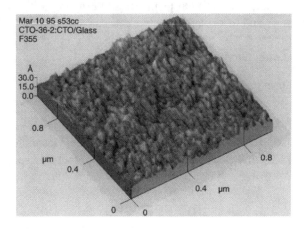

Figure 7. Atomic force micrograph of a typcial CTO film.

SUMMARY AND PLANS

A free electron model has been used to show that both high carrier concentration and high mobility are required for plasma filters to be practical in TPV applications. Cadmium stannate is a good potential candidate for plasma filter applications because it can be doped to exceptionally high carrier concentrations (~10^{21} cm^{-3}), and because its carriers exhibit relatively high mobility at high concentration.

We have developed a process consisting of room temperature sputtering and a single high temperature anneal which can reproducibly produce single phase polycrystalline cadmium stannate films of very high quality. To the best of our knowledge, our mobility and conductivity results are the highest ever reported for a transparent conductive films prepared by sputtering. We have demonstrated the ability to control the plasma wavelength between about 1.3 to 3.0 µm. For films with a reflectance minimum at 1.3 µm, the specular reflectance reaches over

90% at 5 µm. Agreement between experimental reflectance data and theory and is excellent.

We believe further increases in the electron mobility of cadmium stannate thin films are highly likely. There is still room to further optimize the growth and annealing processes. The effects of grain size have not yet been systematically studied, and the defect chemistry and donor mechanisms are not well understood or controlled. In addition, we plan to investigate growth on high quality lattice matched substrates, and to explore possible extrinsic dopants. Finally, we plan to investigate other TCOs potentially low electron effective mass and high mobility, such as cadmium indate.

ACKNOWLEDGMENTS

We would like to thank Helio Moutinho and Alice Mason for atomic force microscopy and scanning electron microscopy measurements, respectively.

REFERENCES

1. J.I. Pankove, *Optical Processes in Semiconductors*, Dover, New York (1971).
2. A.J. Nozik, *Phys. Rev. B*, **6** (1972), 453.
3. G. Haacke, *Appl Phys Lett*, **28** (1976), 622.
4. G. Haacke, W.E. Mealmaker and L.A. Siegel, *Thin Solid Films*, **55** (1978), 67.
5. S. Shimada, I.Sato, I., and K.Kodaira, *J. Electrochem. Soc.*, **135** (1988), 3165.
6. L.A. Siegel, *J. Appl. Crystallogr.*, **11**(1978), 284.
7. E. Burstein, *Phys. Rev.*, **93** (1954), 632.

Thermophotovoltaic Devices Utilizing a Back Surface Reflector for Spectral Control

GW Charache, DM DePoy, PF Baldasaro, and BC Campbell

Lockheed-Martin, Inc.
Schenectady, NY 12301

Abstract: The back surface reflector (BSR) represents the spectral-control technology that offers the highest spectral utilization factor, F_u, where F_u is defined as the fraction of the total absorbed radiation with energy greater than the semiconductor bandgap [1]. In order for this technology to succeed, an integrated photovoltaic cell - spectral control thermophotovoltaic device design is required which simultaneously minimizes free carrier absorption and series resistance losses. For this study, BSR technology was developed for GaSb, InAs and InP substrate systems. Reflection and contact resistance results will be presented for the above material systems. To date, $F_u > 80\%$ have been obtained for all three material systems, with potential for $F_u > 90\%$.

INTRODUCTION

Thermophotovoltaic (TPV) systems have recently re-emerged as a viable technology for a number of commercial and military applications [2]. This is due, in part, to the continued development of solar cells and advanced III-V semiconductor epitaxial technologies. However, significant differences remain between photovoltaic and thermophotovoltaic technologies, foremost of which is recuperation of low-energy photons. Since the heat source is in close proximity to the TPV devices, reflection of below-bandgap radiation and its reabsorption by the radiator is possible. This recuperation is crucial to the efficiency of TPV systems since approximately 75% of the energy in a blackbody source is below bandgap in an optimally matched system [3].

Both front-surface and back-surface spectral-control devices have been proposed to act as below-bandgap recuperators. Front-surface spectral control devices consist of an interference filter or an interference/plasma tandem filter, which are designed to reflect the below-bandgap radiation and transmit the above-bandgap radiation [1]. Utilizing these techniques, the voltaic diode and spectral-control portions of an overall TPV device can essentially be designed independently. Interference filters alone exhibit very high reflection for a limited

Material	Application	Reference
Germanium	TPV-recuperator	[4-6]
Amorphous-Silicon	PV- absorption enhancement	[7-10]
Amorphous-Silicon	PV- temperature reduction	[11,12]
Silicon	TPV - recuperator	[13-18]
Silicon	PV- absorption enhancement	[19]
GaAs	PV- temperature reduction	[20]
InGaAs	Detector - absorption enhancement	[21]

Table 1. - Summary of back surface reflector applications.

bandwidth. The range of high reflection is limited by the requirement for high above bandgap transmission. It has been shown that in order to keep above bandgap transmission > 90%, the range of high reflection can only extend to approximately $2\times\lambda_g$, where λ_g is the wavelength corresponding to the TPV cell bandgap [1]. The use of an interference/plasma tandem filter improves the long wavelength reflection compared to an interference filter used alone. However, parasitic absorption of above-bandgap radiation by the plasma filter reduces both the efficiency and power density of the TPV system. In addition, tandem filters have inherently higher reflection of above bandgap photons compared to interference filters alone. Conversely, the back-surface spectral-control device has no parasitic absorption of the above-bandgap radiation, a wide bandwidth, and low reflection of above bandgap photons. However, the design of the voltaic diode must accommodate spectral-control limitations due to the parasitic absorption of below-bandgap light by free carriers during its transit through the TPV device. Thus, an integrated view of voltaic diode / spectral control must be taken.

Back-surface spectral-control devices have been utilized by researchers for a number of optoelectronic devices [Table 1]. For all these applications, there are a number of criteria that must simultaneously be satisfied for a back surface reflector (BSR) technology to succeed. These include:

•High reflectivity

•Low specific ohmic contact resistance

•Strong adhesion

•Thermal stability

The quantitative values of these criteria, however, vary depending on application. For TPV systems, the requirements are the most severe due to the large fraction of below bandgap radiation that must be recycled, the strong wavelength dependence of free-carrier absorption, and the high operational current levels.

This paper presents a simple model that will serve as a guide to the feasibility and design of a BSR-TPV device. Experimental results will be presented for all candidate semiconductor substrates (GaSb, InAs, InP) for low temperature (800 - 1000 °C) radiator TPV systems.

TPV-BSR DEVICE DESIGN

The primary goal in designing a TPV-BSR system is to maximize the spectral utilization factor, F_u, without degrading the fill factor. These two goals are conflicting due to an increase in free carrier absorption and a decrease in series resistance as the number of electrical carriers increases in the TPV device. The spectral utilization factor quantifies the efficiency with which the below-bandgap energy absorption is minimized and is defined as the fraction of the total absorbed radiation with energy greater than the semiconductor bandgap [1]. F_u is calculated from

$$F_u = \frac{absorption > E_g \, in \, cell}{total \, absorption \, in \, cell \, and \, filter},$$

where the absorption of below bandgap photons through a layer of thickness t, can be expressed as

$$absorption < E_g = 1 - \exp[-\alpha \, t],$$

where α is the free carrier absorption cross-section. The free carrier absorption cross-section α for n-type GaSb at 9 µm can be approximated by,

$$\alpha \approx 6 \times 10^{-17} \cdot N,$$

where, N is the free carrier doping level (cm^{-3}) [22]. Experimentally it has been observed that the absorption coefficient of p-type GaSb is an order of magnitude greater than in n-type material. This necessitates the use of a p-on-n device to limit the free carrier absorption by using a thin, p-type emitter. Figure 1 illustrates the net absorption versus free electron concentration for a number of different substrate thicknesses (assuming a single pass). Thus, in order to obtain a net substrate absorption of less than 5%, the substrate doping-thickness product (areal density of free carriers) must be less than 1×10^{15} cm^{-2}. Similarly, Figure 2 illustrates the net absorption vs. hole concentration for the p-type emitter. In order to obtain a net emitter absorption of less than 5%, the emitter areal density of free carriers must be less than 1×10^{14} cm^{-2}.

The series resistance of the emitter, base and substrate must be evaluated with the constraints imposed by free carrier absorption. For this evaluation, the top contact grid design assumes a 1×1 cm^2 cell active area, a single 500 µm wide central busbar, and 10 µm wide grid fingers with a 100 µm pitch. The net emitter sheet series resistance is given by the net power loss in the emitter divided by the square of the short circuit current density:

$$R_{emitter} = \frac{L \cdot S^2 \cdot W}{6 \cdot q \mu_e \cdot N_e \cdot t_e \cdot A^2}$$

where L is the grid length, S is the grid pitch, W is the width of the cell active area, μ_e is the hole mobility in the emitter, N_e is the emitter doping level, t_e is the emitter thickness and A is the device active area. Figure 3 plots the emitter resistance versus doping for various emitter thicknesses assuming a hole mobility of 300 cm^2/V-s. Finally, the substrate series resistance is given by,

$$R_{substrate} = \frac{t_s}{q \mu_s \cdot N_s \cdot A}$$

where t_s, μ_s, and N_s are the substrate thickness, electron mobility, and doping level respectively, and A is the device area. Assuming an electron mobility of 3000 cm^2/V-s and a substrate thickness of 500 microns, the series resistance is negligible for dopings above 5×10^{16} cm^{-3}. As can be seen from this analysis, the control of the emitter areal density is more crucial than the substrate.

Figure 4 plots the net reflection at 9 µm for a GaSb TPV cell for various emitter dopings and thicknesses. The data was calculated using a simple multiple bounce program assuming normal incidence, substrate doping and thickness of 1×10^{17} and 100 microns, a back surface reflectivity of 90%, a GaSb front surface reflectivity of 40%, a front-contact top-side grid reflectivity of 95%, a front-contact bottom-side grid reflectivity of 80% and front contact area coverage of 10%.

BSR EXPERIMENTAL RESULTS

As mentioned in the introduction, a back surface reflector must simultaneously have high reflectivity and form a low resistance ohmic contact to the back of the device. These two seemingly conflicting requirements can by accomplished a number of ways [Fig. 5]. Figure 5A illustrates a non-alloyed BSR that would be suitable for n-InAs (Fermi-level pins in the conduction band) or a heavily doped surface layer formed by diffusion or ion-implantation on GaSb or InP (tunneling contact).

An alternative to a non-alloyed contact is an alloyed/grided back contact. Here, a compromise is reached between the reflective (non-ohmic) and non-reflective (ohmic) areas. The relative areas of each are determined by the value of the back contact resistance. For example, to achieve a net resistance of < 1.0 mΩ, a contact resistance of < 5×10^{-5} Ω-cm^2 is required for 5% ohmic grid coverage on a 1×1 cm^2 device.

In the first grided concept [Fig. 5B], the reflective metal is deposited directly on top of the alloyed grids. This concept, along with the non-alloyed concept, suffer from potential low temperature metal-semiconductor reactions which may degrade long-term reflectivity. In order to prevent potential reactions, a dielectric spacer may be added. Figure 5C demonstrates this option where the reflective and ohmic regions are formed by the same metallization layer. This requires that the alloyed-reflective metal forms an adequate ohmic contact. If this can not be achieved, two separate metallization steps may be employed [Fig. 5D].

Experimental specular reflection (11°) versus wavelength results for n-GaSb, n-InAs and n-InP are shown in Figures 6-8, respectively. The doping concentration and thickness for all samples is approximately 5×10^{16} cm^{-3} and 350-400 μm, respectively. Reflection measurements are only carried out to 10 μm because < 5 % of the total energy from a 800 °C black body has a wavelength beyond 10 μm.

GaSb utilized a Sn-Au (3 nm, 200 nm) 5% grid coverage BSR with a 100 nm PECVD SiN dielectric spacer [Fig. 5C]. Ohmic contacts were formed using a 350 °C, 5 second rapid thermal anneal. The tin layer serves the dual purpose as an adhesion layer between the silicon nitride and the gold and also as a back contact resistance reducer in the alloyed areas. InAs utilized a Ti-Au (3 nm, 200 nm) non-alloyed reflective contact [Fig. 5A]. Here, the titanium provides an adhesion layer that is sufficiently thin to prevent absorption. InP also utilized a Sn-Au (3 nm, 200 nm) 5% grid coverage BSR with a 100 nm PECVD SiN dielectric spacer [Fig. 5C].

Figure 9 illustrates the measured ohmic contact resistance using the Cox-Strack method for a number of n-GaSb (1×10^{17} cm^{-3}), n-InAs (5×10^{16} cm^{-3}) and n-InP (6×10^{16} cm^{-3}) contacts. As is readily evident, a number of contact metallizations exist which satisfy the 5×10^{-5} Ω-cm^2 requirement.

CONCLUSION

Back Surface Reflector (BSR) technology has been utilized throughout the optoelectronics community for a number of applications, including TPV devices. This paper has outlined a TPV-BSR design procedure that demonstrates its feasibility for low-temperature (800-1000 °C) radiator applications. Finally, initial experimental results were presented demonstrating proof-of-concept for all potential binary substrates (GaSb, InAs, InP) for TPV applications.

REFERENCES

1. PF Baldasaro, EJ Brown, DM DePoy, BC Campbell, and JR Parrington, "Experimental Assessment of Low-Temperature Voltaic Energy Conversion," 1st NREL Conference on Thermophotovoltaic Generation of Electricity, AIP Conference Proceedings **321**, 29 (1994).

2. The First NREL Conference on the Thermophotovoltaic Generation of Electricity, Copper Mountain, CO, AIP Conf. Proc. **321** (1994).

3. LD Woolf, "Optimum Efficiency of Single and Multiple Bandgap Cells in Thermophotovoltaic Energy Conversion," Solar Cells **19**, 19 (1986).

4. SZ Jamagidze, et al., "The Efficiency of the Long-Wavelength Spectrum Energy Return to the Radiator Using Back Reflecting Contacts of Germanium Thermophotovoltaic Convertors," Soobshcheniya Akademii Nauk Gruzinskoy SSR, **138**, NR 1, 45 (1990).

5. J Werth, "Thermophotovoltaic Energy Conversion," Proc. 17th Power Sources Conf., 23 (1963).

6. J Werth, "Design Study of a Thermophotovoltaic Converter," Proc. 18th Power Sources Conf., 153 (1964).

7. A Banerjee and S Guha, "Study of Back Reflectors for Amorphous Silicon Alloy Solar Cell Application, Jl. Appl. Phys., **69**, 1030 (1991).

8. HW Deckman et al., "Optically Enhanced Amorphous Silicon Solar Cells," Appl. Phys. Lett., **42**, 968 (1983).

9. J. Morris, " Absorption Enhancement in Hydrogenated Amorphous Silicon-based Solar Cells, Jl. Appl. Phys., **67**, 1079 (1990).

10. S. Wiedeman and FR Jackson, "Photovoltaic Device having Enhanced Rear Reflecting Contact," U.S. Patent 5,230,746 (1993).

11. AT Chai, "Back Surface Reflectors for Solar Cells," Proc. 14th IEEE PVSC, 156 (1980).

12. AT Chai, "Solar Cell Having Improved Back Surface Reflector," U.S. Patent 4,355,196 (1982).

13. R.W. Swanson, "Recent Developments in Thermophotovoltaic Conversion," Proc. 14th IEEE PVSC, 186 (1980).

14. R.W. Swanson, "Silicon Photovoltaic Cells in TPV Conversion," EPRI Research Project ER-1272 (1979).

15. R.W. Swanson, "Silicon Photovoltaic Cells in ThermophotovoltaicEnergy Conversion," Proc. 13th IEEE PVSC, 70 (1979).

16. LD Woolf, "Thermophotovoltaic Converter and Cell for Use Therein," U.S. Patent 4,234,352 (1980).

17. AM Vasilev, et al., "Optical Characteristics of Silicon Photocells and the Efficiency of a Thermophotovoltaic Converter,"Teplofizika Vysokikh Temperatur, 5, 1079 (1967).

18. LD Woolf, "Photothermophotovoltaic Converter," U.S. Patent 4,746,370 (1988).

19. R Hezel and R Ziegler, "Ultrathin Self-Supporting Crystalline Silicon Solar Cells with Light Trapping," Proc. 23rd IEEE PVSC, 260 (1993).

20. LD Woolf and JC Bass, "Solar Cell with Low-Infrared Absorption and Method of Manufacture," U.S. Patent 4,773,945 (1988).

21. H Fukano et al., "High-Speed InP-InGaAs Heterojunction Phototransistors Employing a Nonalloyed Electrode as a Reflector," IEEE Jl. Quantum Electronics, 30, 2889 (1994).

22. JI Pankove, "Optical Process in Semiconductors," Dover Publications Inc. (1971).

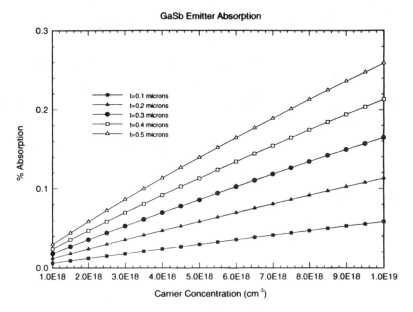

Fig. 1 - Modeled GaSb emitter free-carrier absorption vs. carrier concentration for various thicknesses.

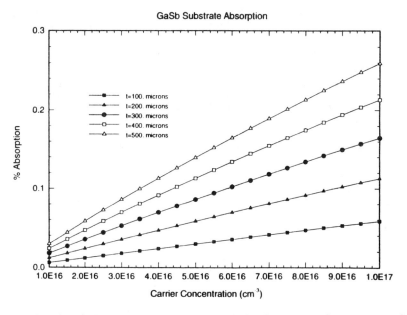

Fig. 2 - Modeled GaSb substrate free-carrier absorption vs. carrier concentration for various thicknesses.

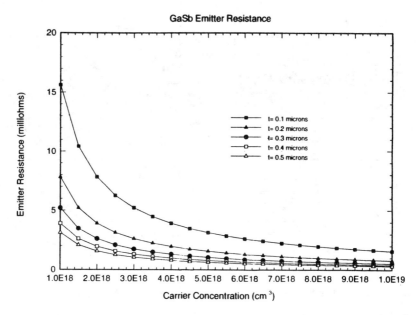

Fig. 3 - Modeled GaSb emitter series resistance vs. carrier concentration for various thicknesses.

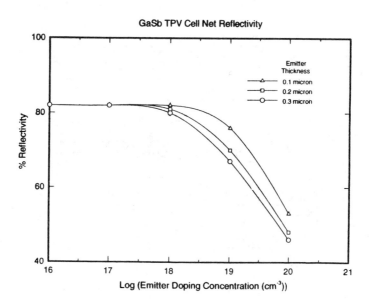

Fig. 4 - Modeled GaSb TPV cell net reflectivity vs. emitter carrier concentration for various emitter thicknesses.

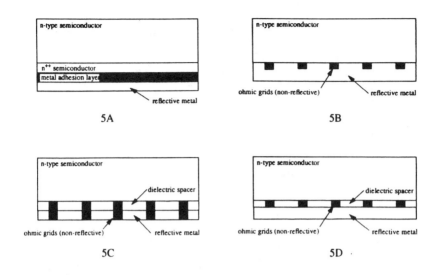

Fig. 5 - Thermophotovoltaic back-surface reflector (BSR) concepts.

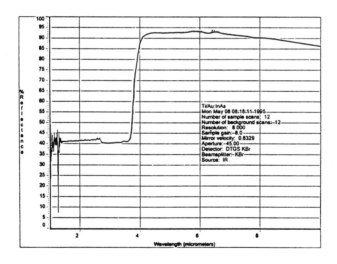

Fig. 6 - Measured below-bandgap reflectivity of an n-InAs substrate with a BSR.

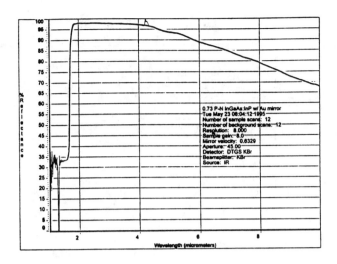

Fig. 7 - Measured below-bandgap reflectivity of an n-InP substrate with a BSR.

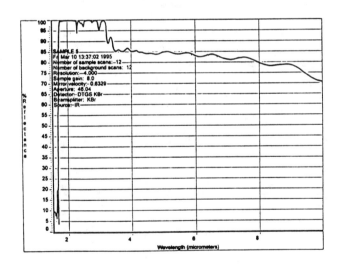

Fig. 8 - Measured below-bandgap reflectivity of an n-GaSb substrate with a BSR.

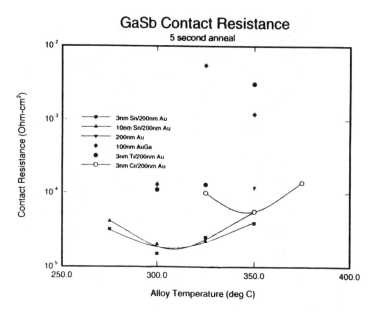

Fig. 9 - Measured n-GaSb contact resistance for various metalizations and alloy temperatures.

Measurement of Conversion Efficiency of Thermophotovoltaic Devices

GW Charache[†], DM DePoy[†], M. Zierak[*], JM Borrego[*],
PF Baldasaro[†], JR Parrington[†], MJ Freeman[†], EJ Brown[†],
MA Postlethwait[†] and GJ Nichols[†]

[†]*Lockheed-Martin, Inc., Schenectady, NY 12301*

[*]*Center for Integrated Electronics, Department of Electrical, Computer and Systems Engineering, Rensselaer Polytechnic Institute, Troy, NY 12180*

Abstract: In this paper we present two methods for determining the conversion efficiency of TPV devices. In the first, the conversion efficiency is calculated from measurements of the external quantum efficiency and reflection as a function of wavelength, and from the I-V characteristics under high-level illumination. This is an indirect method based on separate differential measurements. In the second method, a novel heat transfer technique is utilized to combine both the voltaic diode and spectral control efficiency into a single measurement.

INTRODUCTION

One of the most important characteristics of a thermophotovoltaic (TPV) device is the conversion efficiency. For this discussion, a TPV device refers to a combined voltaic diode and spectral control device. TPV device efficiency is defined as the ratio of the electrical power output from the voltaic diode to the power absorbed by the TPV device from a heat source [1].

$$\eta_{TPV} = \frac{PO_{electric}}{P_{abs.}}.$$

When calculating or measuring solar photovoltaic device efficiency, $P_{abs.}$ is replaced with the incident power, since any incident power not absorbed is not utilized. This is not the case for TPV systems due to the ability to recuperate both above-bandgap and below-bandgap photons. Thus, reflection of above bandgap light does not decrease efficiency, as in solar cells, but does decrease power density. Therefore, proper calculation or measurement of efficiency must be made of the total absorbed power and not the incident power.

There are several methods in which TPV device efficiency can be determined and proper measurement and reporting are crucial in establishing a credible knowledge base for TPV system development. A number of difficulties exist in comparison to measured solar photovoltaic efficiencies due to non-standard heat sources; efficiency dependence on radiator temperature, cell temperature, and cell bandgap; and the ability to recuperate below bandgap photons.

The most direct measurement involves exposing a TPV device to a heat source and measuring the current-voltage (I-V) characteristic. The TPV device conversion efficiency is the maximum electric power produced divided by the net power delivered to the heat source. The main limitation of this measurement is the inability to form a closed cavity that prevents parasitic absorption or escape of photons in non-cell areas. Because of these difficulties, other approaches must be applied to TPV devices.

One alternative method consists of first measuring the external quantum efficiency and both the above- and below-bandgap reflection of the TPV device as a function of wavelength. The absorbed power and photocurrent are then calculated by numerical integration over the known power spectrum of the source. With the photocurrent known, the I-V characteristic of the cell is measured with the cell exposed to a light source which produces the same photocurrent. From the measured I-V characteristic, the maximum output power is determined. The conversion efficiency is then calculated by dividing the maximum output power by the calculated absorbed power from the heat or light source to be used in the application.

The advantage of this technique is that the light source which is used for measuring the illuminated I-V characteristics does not have to be calibrated, nor does its photon flux spectrum have to be known. It is just necessary that it produce the same photocurrent as the one produced by the heat or light source in a particular application. In addition, this method is not limited to black body sources. For any known spectrum, the calculations can be made. The disadvantage of this technique is that three separate measurements must be made and the efficiency is calculated. Thus, this is an indirect method of efficiency determination.

The second efficiency measurement overcomes the disadvantage of the above technique; however, now a proto-typical radiator spectrum is required. For this measurement a novel heat transfer technique is utilized to measure the ratio of the generated electric power of a TPV device to the total radiative energy absorbed by the TPV device from a proto-typic radiator. This technique provides a single direct measurement that can be used to compare the relative efficiencies of competing cell and spectral control technologies and a quantitative measure of TPV device efficiency.

EFFICIENCY CALCULATION BASED ON DIFFERENTIAL MEASUREMENTS

For the first method, the calculation of TPV device efficiency begins with the calculation of the power absorbed by the TPV device. For a source with a power spectrum, $N_2(\lambda)$, the absorbed power is given by,

$$P_{abs.} = A_{cell} \cdot FR \cdot \int_0^\infty [1-R(\lambda)] \cdot N_2(\lambda) \, d\lambda$$

where A_{cell} is the total area of the device, FR is the flux ratio of the source power spectrum to an ideal power spectrum, and $R(\lambda)$ is the reflection spectrum of the TPV device. $1-R(\lambda)$ represents the absorption spectrum of photons within the TPV device. The photocurrent (I_{ph}) produced by the incident radiation can be calculated using the external quantum efficiency of the voltaic diode, $QE_{ext}(\lambda)$, the photon flux of the spectrum of the source, $N_1(\lambda)$ and the active area of the device, A_{active}:

$$I_{ph} = qA_{active} \cdot FR \cdot \int_0^\infty QE_{ext}(\lambda) \cdot N_1(\lambda) \, d\lambda \ .$$

where q is the charge on an electron.

Using any light source, the above photocurrent is induced and the maximum electrical output power is measured. In our measurement setup, the I-V curve in only swept in the 4th quadrant; therefore a determination is made of the short circuit current which corresponds to the given photocurrent. These two currents are identical if the series resistance is zero; however, when the series resistance is non-zero, the short circuit current can be significantly lower than the photocurrent. We overcome this obstacle by using the dark I-V parameters to calculate the short circuit current using:

$$I_{ph} = I_{sat} [\exp(\frac{-I_{sc}R_s}{nk_bT/q}) - 1] - I_{sc},$$

where, I_{sat} is the dark current, I_{sc} is the short circuit current, R_s is the series resistance, n is the ideality factor, k_b is Boltzmann's constant, and T is temperature. This equation is solved by an iterative process to determine I_{sc}. Note, that I_{ph} is positive and I_{sc} is negative when generating power.

With the dark characteristics of a cell previously measured, one can predict the illuminated I-V characteristic using:

$$I = I_{sat} [\exp(\frac{V-IR_s}{nk_bT/q}) - 1] - I_{ph},$$

where, V is the voltage across the cell and I is the current through the cell. From this curve the maximum output power can be determined, and then the efficiency calculated. This predicted efficiency can be calculated for a number of different flux ratios and then plotted as a function of any parameter.

A variety of measurements were taken for an n-p InGaAs (E_g = 0.73 eV) TPV device, grown lattice matched to an InP substrate. This device has approximate emitter and base thicknesses of 0.2 and 3.0 microns, respectively, and does not have an AR coating.

The measured external quantum efficiency of this device is shown in Figure 1. This measurement was calibrated with another 0.73 eV InGaAs cell measured at NREL. For wavelengths greater than 1.8 μm, a pyroelectric detector was used in conjunction with the calibrated cell, assuming a constant responsivity of the pyroelectric detector.

This device's dark I-V characteristics are shown in Figures 2 and 3. The parameters (n, R_s, and I_{sat}) were extracted from a linear fit of the raw data between 0.5V and 0.6V [2]. This gives values of the ideality factor of 1.09, the series resistance of 6.3 mΩ, and a saturation current of 4×10^{-8} Amps. Note, that the change in slope of the log I-V curve between 0.4V and 0.5V indicates a non-constant ideality factor as a function of injection level, which is assumed in our simple model. The dark I-V data was then used to predict an illuminated I-V curve. A comparison of this predicted curve to a measured illuminated curve are shown in Figure 4. The discrepancies between these two curves are presumably due to injection level dependent effects.

Figure 5 plots the calculated TPV device conversion efficiency as a function of short circuit current for this device assuming a 1000 °C blackbody radiator and a cell temperature of 50 °C. Included on this plot are the efficiencies for the bare cell and a cell with assumed perfect spectral control using the dark I-V data. The perfect spectral control curve represents the upper limit of this cell's particular performance for this radiator temperature.

Figure 6 plots the calculated peak TPV device conversion efficiency as a function of radiator temperature assuming a constant flux ratio of 1. Included on this curve are the efficiencies of a bare cell and a cell with a perfect filter (assumed no absorption of below bandgap light). Again, the last curve represents the upper limit of this particular cell's performance and emphasizes the need for an excellent spectral control technology.

EFFICIENCY MEASUREMENT

In the second measurement, the efficiency is determined by directly measuring the ratio of the electrical power out ($PO_{electric}$) of a TPV diode to the total radiative power absorbed by the TPV device,

$$\eta = \frac{PO_{electric}}{Q_{radiative}}.$$

$Q_{radiative}$ is determined by measuring the heat-up rate of a TPV device mounted to a copper block. A schematic diagram of this heat transfer measurement is shown in Figure 7. For this measurement a TPV device is mounted to an insulated copper block of mass, m. A conductive paste is used to make a good electrical and thermal contact between the TPV device and the copper block. A near blackbody heat source (SiC rod) is used to illuminate the TPV device. Initially, an I-V curve is taken of the illuminated diode, which yields the maximum power point. Next, with the diode open circuited, the rate of temperature rise of the copper block is measured. The radiative energy absorbed by the TPV device is then given by,

$$Q_{radiative} = Q_{total} - Q_{cond,conv} = mC_p \frac{dT}{dt}\bigg|_{total} - mC_p \frac{dT}{dt}\bigg|_{cond,conv}$$

where C_p is the specific heat of copper, and dT/dt is the temperature rise of the copper block per unit time. Q_{total} represents the measured total power absorbed by the TPV device from the silicon carbide rod. $Q_{cond,conv}$ represents the combined conductive and convective heat transfer from the silicon carbide rod to the copper block. This is determined by measuring the heat-up rate of the copper block with a gold coated silicon wafer in place of the TPV device. (Note: the heat capacity of the TPV device was calculated to be ~2% of the heat capacity of the copper block and thus considered negligible.)

Due to the low flux ratio of the experimental setup, a method of extrapolating efficiency to higher current levels is required. This only involves scaling the open circuit voltage and adjusting the fill factor to account for the higher current levels since the current and the heat-up rate are both proportional to the flux ratio.

Figure 8 compares the measured efficiency, utilizing the copper block, and calculated efficiency, based on the method presented in the previous section, for a 0.56 eV InGaAs TPV cell. For these measurements an operational current density of 4 Amps/cm^2 and a cell operating temperature of 100 °F are assumed. Included in Figure 8 are the efficiencies of a bare cell, a cell with an interference filter for spectral control and a cell with a "blocking" filter. The "blocking" filter is an idealized filter, not in thermal contact with the measured cell, that absorbs all below bandgap light before it can be absorbed by the TPV cell. This curve only represents the upper limit of conversion efficiency with this TPV device, since it is not a real system that could be implemented. This figure illustrates the good agreement obtained with these two efficiency measurement techniques. Differences are attributed to the non-ideal blackbody characteristics of the SiC heater used in these experiments (calculations assumed an ideal blackbody).

CONCLUSION

We have shown two methods by which to calculate the conversion efficiency of TPV devices. The first requires measurements of the external quantum efficiency, below bandgap reflection, and an illuminated I-V curve. The advantage of this technique is that it does not require a calibrated blackbody radiator. The disadvantage is that it is an indirect method of acquiring the efficiency of a TPV device. It was also shown that the efficiency could also be estimated within 10% using a dark I-V characteristic in place of an illuminated I-V curve.

The second conversion efficiency measurement combines both the voltaic diode and spectral control efficiency into a single integrated measurement. In this technique, the measured electric power out of a TPV diode is divided by the total radiative energy absorbed by the TPV device. This technique also requires the measurement of an illuminated I-V curve to extrapolate the efficiency to higher illumination levels.

REFERENCES

1. L.D. Woolf, "Optimum Efficiency of Single and Multiple Bandgap Cells in Thermophotovoltaic Energy Conversion," Solar Cells, **19**, 19 (1986).

2. M. Zierak, et. al., "Characterization of InGaAs TPV Cells," Proc. of 1st NREL Conf. on TPV Gen. of Elec., 473 (1994).

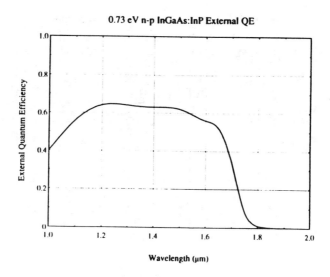

Fig. 1 - Measured 0.73 eV n-p InGaAs:InP TPV cell external quantum efficiency.

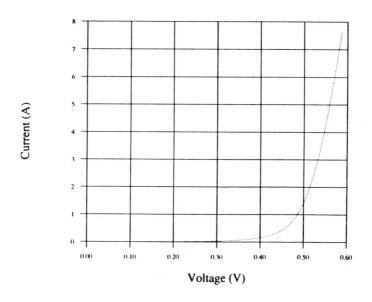

Fig. 2 - Measured 0.73 eV n-p InGaAs:InP TPV cell dark current vs. voltage curve.

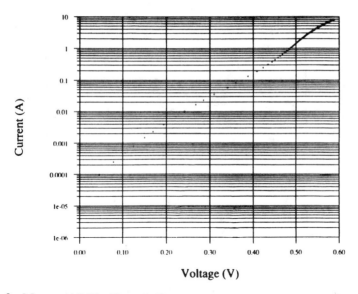

Fig. 3 - Measured 0.73 eV n-p InGaAs:InP TPV cell log current vs. voltage curve.

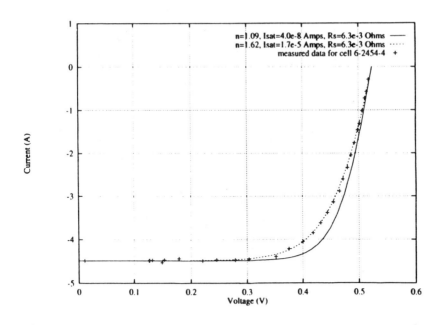

Fig. 4 - Comparison of measured light and predicted light I-V for a 0.73 eV n-p InGaAs:InP TPV cell.

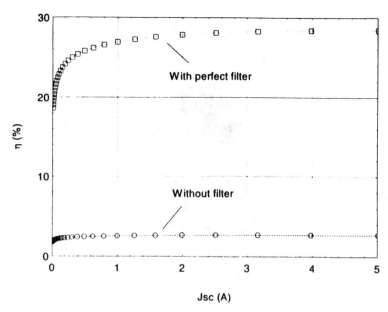

Fig. 5 - Calculated efficiency vs. short circuit current density for a 0.73 eV n-p InGaAs:InP TPV cell and a 1000 °C radiator.

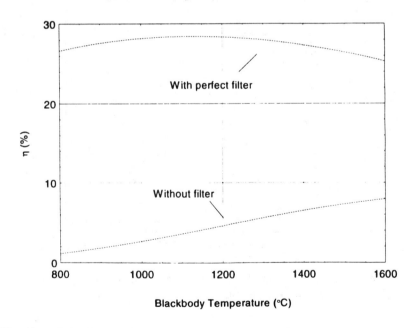

Fig. 6 - Calculated efficiency vs. radiator temperature for a 0.73 eV n-p InGaAs:InP TPV cell with and without a "perfect filter".

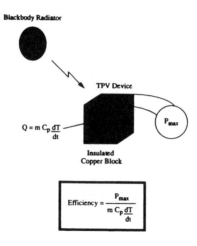

Fig. 7 - Integral TPV device efficiency measurement.

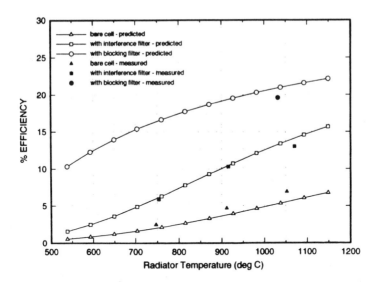

Fig. 8 - Measured and predicted efficiency measurements for a 0.55 eV n-p InGaAs:InP TPV cell with various filters.

TPV CELLS WITH HIGH BSR

P.A. Iles and C.L. Chu

Applied Solar Energy Corporation
City of Industry, CA 91745-1002, USA

Abstract. This paper reviews the use of back surface reflectance (BSR) in a variety of PV cells. The major controlling factors are illustrated by plots of reflectance versus wavelength. Possible application to TPV cells is discussed.

INTRODUCTION

Solar cells with high reflectance at the back surface have been used extensively. In space, BSR can reduce the operating temperature of cells. Reflection of near-bandgap wavelengths can increase cell currents by increasing the absorption length for these lightly absorbed wavelengths. In combination with surface texturing, high BSR can also provide light trapping, in which multiple reflections can increase the absorption lengths to 10-20 times the cell thickness. In direct-gap materials, BSR can increase photon recycling, where the bandedge radiation from charge carrier recombination can be re-absorbed to generate more carriers.

For TPV cells, addition of high BSR appears to provide the ideal selective optical filter, allowing full use of the absorbed wavelengths, and reflecting back to the emitter a high fraction of the not-absorbed spectrum.

Cell Reflectance

Figure 1 shows the reflectance features of a PV cell. Part of the incident radiation is reflected at the front surface (reflectance r). For a cell with no AR coating, r is usually around 30-35% and does not vary much with wavelength. Radiation is absorbed up to the cut-off wavelength for the particular semiconductor, and the longer wavelengths are transmitted to the back surface, where they are reflected or absorbed. The value of r_b, the back surface reflectance depends on the surface finish of the semiconductor, on the smoothness of the back contact metallization interface, and on the reflectance of the metal in contact with the semiconductor.

© 1996 American Institute of Physics

Figure 1: Optical Reflections in PV Cell

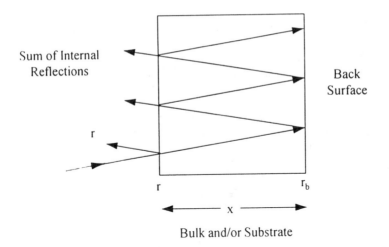

Bulk and/or Substrate

If the PV cell is grown on a substrate, there may also be a small reflection at the interface between the cell and the substrate. The properties of the substrate dictate the beyond-bandgap transmission and the back surface reflection.

If an AR coating is used on the front surface, r depends on the wavelength, and the r-values at different wavelengths are incorporated in the measured values of reflectance.

The reflected rays undergo multiple reflections and the total reflectance at each wavelength is given by (1)

$$R_T = r + \frac{(1-r)^2 r_b}{1-r.r_b} \qquad (1)$$

R_T is the value measured at each wavelength. Knowing the values of r at a given wavelength, r_b can be calculated. If the cell and/or the substrate introduce beyond bandgap absorption (from free carrier absorption), R_T is given by equation (2).

$$R_T = r + \frac{(1-r)^2 r_b \exp(-2\alpha_f \cdot x)}{1 - r \cdot r_b \exp(-2\alpha_f \cdot x)} \qquad (2)$$

In (2) x is the free carrier absorption length, usually the thickness of the cell or substrate, or in some cases the thickness of a highly doped back surface field layer. α_f is the free carrier absorption coefficient, dependent on the physical properties of the semiconductors involved, and proportional to the doping density and usually varying with (wavelength)^2. Experimental values for α_f usually lie between 1 and 100 cm^{-1}. The metals with best infrared reflectance are the noble metals, and Figure 2 shows that high R-values can be obtained out to wavelengths near the end of BB emission. The R-λ plot in Figure 3 for Au deposited on Ge (or other polished substrates) confirms the handbook results in Figure 2.

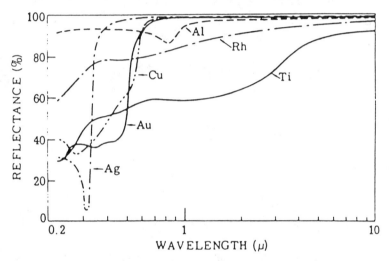

Fig. 2 Reflectance of various films of silver (Ag), gold (Au), aluminum (Al), copper (Cu), rhodium (Rh), and titanium (Ti) as a function of wavelength.

Illustrative R-λ Plots

We will present a series of R-λ plots, taken using an integrating sphere attached to a monochromator. The detectors measure the hemispherical reflectance from a sample placed at the exit port of the integrating sphere.

A variety of semiconductor samples is presented, to illustrate the main factors controlling the measured R-λ curves.

Figure 3: Reflectance vs. Wavelength for Gold Film Deposited on Germanium

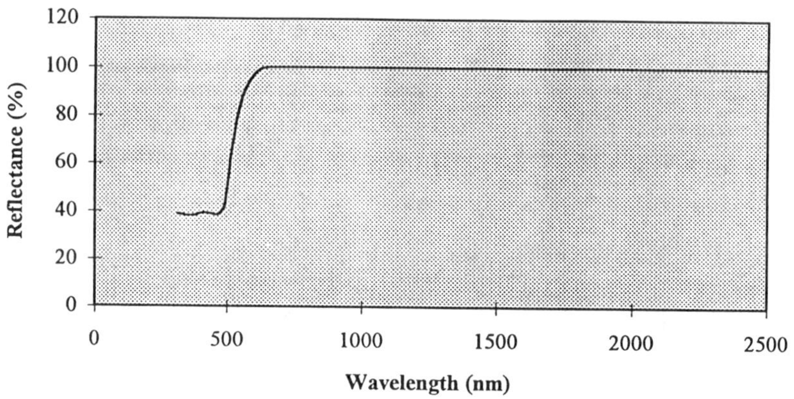

Factors Controlling BSR

Figure 4 shows Si cells with different back surface metallization conditions.

The cells all had polished (planar) front surfaces, on which a double layer AR coating was deposited. The BSR (the additional term in (1)) can be seen by the rise in reflectance beyond the bandedge of Si (~1.15μm). The bottom three curves had TiPdAg back surface metallization, with the Ti directly in contact with silicon. Ti has low IR reflectance (see Figure 2) and the low R-values also depend on the surface finish. When Al is directly in contact with Si and heated above the eutectic alloying temperature, the BSR-value is only slightly higher than for Ti, because the refreezing of the Si-Al alloy causes uneven growth features which reduce the polished surface reflectance. When the Al-Si contact was heated to lower temperatures, higher BSR (lower α_s) values were measured.

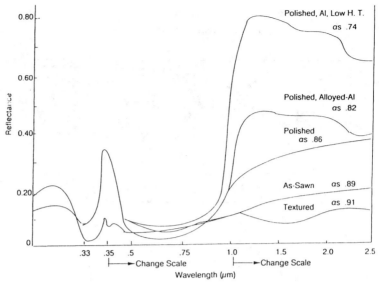

FIGURE 4 Reflectance versus wavelength for silicon cells with various back surface conditions (αs values shown on curves).

These cells were space-cells, and the solar absorbtance (α_s) values are shown. For an oriented array, cells with αs values around 0.74 run about 17°C cooler than cells with $\alpha_s = 0.91$. Some Si space-cells have shown α_s values around 0.64.

Figure 5 shows R-λ for a Si slice with a highly polished back surface, no AR coating, before and after a gold BS contact was deposited. Curve a (before BS metallization) shows that r_b is around 0.35, the value typical of uncoated Si.

Figure 6 shows the R-λ for a polished GaAs slice with Au as BS metallization.

Figure 7 uses a polished Ge wafer. The front surface reflectance was higher than for Si (higher refractive index) and the bandedge around 1.6 μm can be seen. The BSR with Au deposited was lower than the BSR for Si with the same metallization.

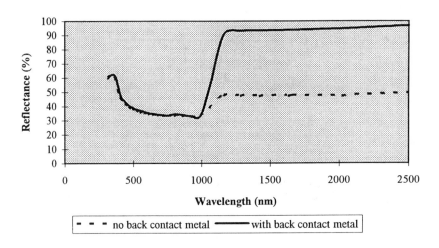

Figure 5: Reflectance vs. Wavelength for a Polished Silicon Slice

- - - no back contact metal ——— with back contact metal

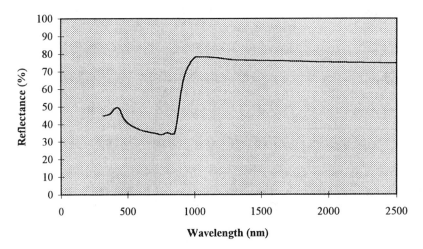

Figure 6: Reflectance vs. Wavelength for a Polished GaAs Slice With Back Contact Metallization

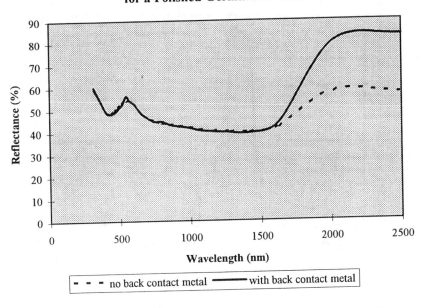

Figure 7: Reflectance vs. Wavelength for a Polished Germanium Slice

- - - no back contact metal ——— with back contact metal

Figure 8 shows a Si cell with AR coating on the front, and a BSR consisting of a thin SiO_2 layer with Al forming a back surface mirror. This structure allows high temperature heating of the Al contact, and if grids are etched in the SiO_2 before Al evaporation, the contact resistance can be low while the optical reflectance remains high where Al is in contact with SiO_2. Two curves are shown. (a) for a high resistivity Si wafer, and (b) for low resistivity Si (0.2 ohm-cm, around 10^{17} boron atoms cm^{-3}). Curve (b) shows the decreasing R values (with λ), caused by free carrier absorption. This BSR technology, combined with a textured front surface can provide light trapping and enhanced current from near-bandedge absorption. The trapping also reduces the escape rays, so that the measured BSR is lower.

Figure 9 shows R-λ for an Si cell, with DLAR coating on the front surface. All three cells had polished back surfaces, and cells (b) and (c) had a boron diffused BS field layer which preserved the BS optical finish. Two different boron diffusion schedules were used as shown for (b) and (c), and the fall-off in R with λ shows the increased effects of free carrier absorption when the diffused layer was more heavily doped (lower $R_\square$).

Figure 8   R-λ for Si Cells (with dielectric BSR)

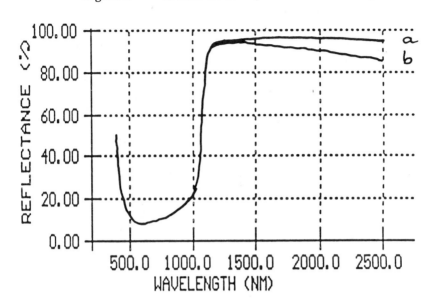

Figure 9   R-λ for Si Cells (with/without BSF)

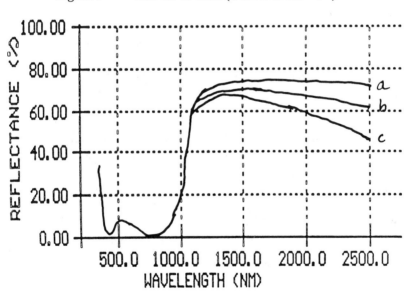

Current GaAs space cells are grown on Ge substrates and there is sufficient difference in refractive index between the two materials, that when the GaAs layer is only a few microns thick, the R-λ plot has interference fringes in the infrared region, superimposed on the normal Ge back surface R-λ plot. If high reflectance metals are deposited on the Ge back surface, enhanced BSR is observed beyond ~1.6μm.

Tests on TPV Cells

A few measurements have confirmed that the InP-based cells behave generally as expected from the above results.

Figure 10 show results for a P-type InP slice, 10a is for an N/P InGaAs cell on a P-InP substrate, 10b is before BSR metallization, 10c is after metallization. The lack of enhanced R above the InP bandedge shows significant free carrier absorption. 10d shows the effect of adding a DLAR, and 10e is for an infared rejection optical filter added to the cell in 10d. The IR rejection filter used was designed for use with GaAs/Ge cells to decrease αs without relying on reflection of wavelengths which are heavily absorbed in Ge.

Figure 11 shows a sequence similar to that in Figure 10, for an P/N InGaAs cell (around 0.6 eV), grown on an N-type InP substrate. All four curves shown free carrier absorption above 2 μm and the R-value increased when the Au back-surface metal was added.

Comments On The Use of BSR in TPV Cells

In many ways high BSR is the ideal method to reflect the unabsorbed wavelengths back to the thermal source. The reflectance begins at the appropriate cut off wavelength for each TPV converter semiconductor. The high BSR is maintained to wavelengths well beyond the long wavelength limit of BB radiation from the low temperature sources used for TPV. Use of BSR can either supplement or even replace the use of selective optical filters placed in front of the cell, giving large cost reduction, particularly to avoid reflectance which is high over only a limited range of the beyond-bandgap BB radiation. As we saw above, the basic design rules for TPV converter materials are the same as for the PV materials namely high back surface finish, and use of highly reflecting metals with a smooth interface with the semiconductor.

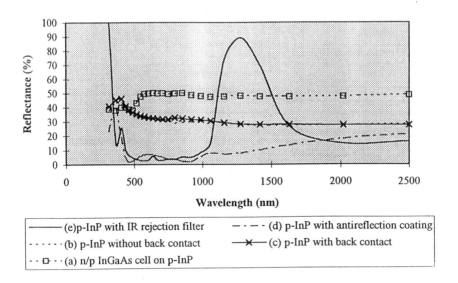

Figure 10: Reflectance vs. Wavelength for P-In-P

——— (e) p-InP with IR rejection filter — - - — (d) p-InP with antireflection coating
- - - - - (b) p-InP without back contact —×— (c) p-InP with back contact
- - ◻ - - (a) n/p InGaAs cell on p-InP

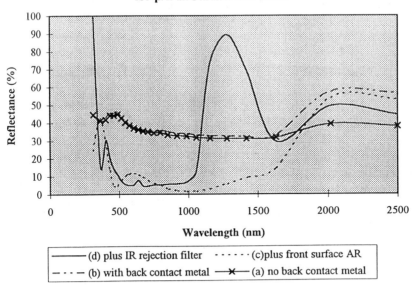

Figure 11: Reflectance vs. Wavelength for p/n InGaAs/n-InP Slice

——— (d) plus IR rejection filter - - - - - (c) plus front surface AR
— - - - — (b) with back contact metal —×— (a) no back contact metal

For the BSR technique which uses a thin insulator between the semiconductor and the BS metal, it is necessary to provide a grid contact to reduce the series resistance at the back surface. In practice this can be achieved by using about 5-10% contact area for the grids; although the BSR is not as high in this contact area, the overall BSR is not lowered much. Preliminary estimates show that our current grid to chrology can provide suitable back grid contacts (and low series resistance) with the dielectric layer for the high current levels typical of TPV. If high BSR can be obtained without the dielectric layer, the heat treatment of the BS metallization can give low series resistance, and the heat sinking to remove excess heat will also be easier. The major problem in using the BSR is the possibility of high free carrier absorption in the bulk of the TPV cell, or more often in the substrate on which the cell is grown. Our tests on N/P InGaAs cells on P-InP showed serious absorption losses, with no BSR enhancement, and the P/N InGaAs cells on N-InP substrates gave only moderate BSR values. These cells were not optimized for BS conditions. Even with high polish and Au metallization, P-InP slices showed low BSR. With the same BS conditions, cells grown on N-InP substrates may have acceptably high BSR, especially if the N-doping concentration and/or thickness of the substrate can be reduced.

This suggests that the possibility of obtaining high BSR should be another criterion in selecting the material to be used in the TPV cell. Already Si and Ge have been shown to be capable of high BSR values, and possibly GaSb is also good.

The only remaining factor is the possibility that at the high injection levels typical of TPV, the injected free carriers may provide sufficient absorption to offset the BSR, and necessitate use of optical filters. Preliminary tests using $0.25 A/cm^2$ ($=4 \times 10^{19}$ injected carriers per cm^3) for Si cells showed no significant reduction of BSR.

In conclusion, we have shown that the incorporation of a high BSR is possible for TPV cells, and should be evaluated, to reduce or relax the added complexity of the optical filters. In either case, the beyond-bandgap radiation must be reflected to increase the conversion efficiency, and to increase the output power density.

SESSION IV:
TPV CELLS I—InGaAs CELLS

Lattice-Matched and Strained InGaAs Solar Cells for Thermophotovoltaic Use

Raj K. Jain[*], David M. Wilt[^], Rakesh Jain[+],
Geoffrey A. Landis[x], and Dennis J. Flood[^]

[*]*University of Toledo*, [^]*NASA Lewis Research Center*
[+]*Cleveland State University*, [x]*Ohio Aerospace Institute*
Photovoltaic Branch, MS 302-1
NASA Lewis Research Center
Cleveland, OH 44135

Abstract. Lattice-matched and strained indium gallium arsenide solar cells can be used effectively and efficiently for thermophotovoltaic applications. A 0.75 eV bandgap InGaAs solar cell is well matched to a 2000 K blackbody source with a emission peak around 1.5 μm. A 0.60 eV bandgap InGaAs cell is well suited to a Ho-YAG selective emitter and a blackbody at 1500 K which have emission peak around 2.0 μm. Modeling results predict that the cell efficiencies in excess of 30% are possible for the 1500 K Ho-YAG selective emitter (with strained InGaAs) and for the 2000 K blackbody (with lattice-matched InGaAs) sources.

INTRODUCTION

Power generation by thermophotovoltaics (TPV) offers several advantages. Photovoltaic devices can be used effectively and efficiently for solar as well as non-solar based TPV applications. The recent advancements made in material and devices will further strengthen and expand the use of this new technology in a cost-effective way. Today's advanced growth techniques allow bandgap engineering of materials suitably matched to the emission spectra of the radiation source.

Indium gallium arsenide (InGaAs) is a material in which the bandgap energy can be changed by varying the In/Ga ratio. Compared to other ternary materials, the InGaAs growth technology on indium phosphide substrates is well advanced. The flexibility in growth and better understanding of processing/etching makes InGaAs technology very attractive for several electronic and opto-electronic devices.

InGaAs cells offer a wide choice for the various broad-band and narrow-band emission sources for TPV applications. The bandgap energy of InGaAs can be varied between 1.42 and 0.36 eV. The InGaAs lattice-matched to InP has a In/Ga ratio of 0.53/0.47 and has a bandgap energy of 0.75 eV. As the In concentration increases the InGaAs bandgap energy decreases and the lattice constant increases, and the material becomes more and more lattice mismatched. This results in compressively strained InGaAs structures, which is relieved by the generation of misfit dislocations if the thickness is more than the critical layer thickness. These defects caused by lattice mismatch greatly affect the performance of the device (1). InGaAs cells of 0.75 eV and lower bandgap energy are of great interest for TPV use. The ideal solar cell bandgap will depend on the trade-off between the device efficiency and the source input power, which in turn depends on the temperature. The specific TPV application will dictate the selection of the photovoltaic cell as well as the thermal source. Very high temperature sources offer a large power density, but source heat management, temperature limitations of materials and the cooling of the cell pose a serious problem. Sources between 1000 to 2000 K offer a good possibility for terrestrial as well as space power needs.

This work is in continuation of the work presented at the previous TPV conference, where the calculated performance of lattice-matched (0.75 eV) InGaAs cell under an erbium-doped yttrium-aluminum-garnet (Er-YAG) selective emitter at 1500 K was reported (2). The 0.75 eV InGaAs cell is an ideal choice for Er-YAG selective emitters (emission peak 1.5 µm). This work discusses the suitability of the lattice-matched and strained (lattice mismatched) InGaAs solar cells for TPV applications. The lattice-matched cell has a bandgap energy of 0.75 eV with a cut-off wavelength of 1.65 µm. This makes it a suitable choice for a blackbody at 2000 K, which has a emission peak around 1.5 µm. The strained InGaAs cell has a bandgap energy of 0.6 eV with a cut-off wavelength of 2.1 µm. This makes the 0.6 eV cell well matched to 1500 K blackbody and holmium doped YAG (Ho-YAG) selective emitter. Both of these sources have emission peak around 2.0 µm. Modeling calculations have been performed for the high AM0 efficiency lattice-matched and mismatched InGaAs cells under narrow-band and broad-band emission sources. Numerical results indicate that InGaAs cell efficiencies exceeding 30% are achievable if the series resistance is reduced and the un-utilized radiation is reflected back to the source by the filter and recycled. It is to be noted that optimized InGaAs cells designed for specific TPV applications may lead to higher efficiencies.

InGaAs SOLAR CELL

TPV applications require specially designed solar cells which can operate at high power densities as well as convert the source radiation efficiently. To date, most of the InGaAs solar cell developmental work has been for photovoltaic applications and the efforts were concentrated mainly on the 0.75 eV bandgap InGaAs cell that is lattice-matched to indium phosphide (see 2 and references therein). First results on the development of lattice mismatched InGaAs cells were reported at the First NREL TPV Conference by NASA, RTI, and Spire.

Wilt and coworkers at the NASA Lewis Research Center have developed high efficiency lattice-matched and mismatched InGaAs cells (3,4). Table 1 shows the measured AM0 performance parameters of the InGaAs solar cells. Cell results were measured at NASA Lewis. It is to be noted that the data presented in Table 1 for 0.75 eV cell are with anti-reflection (AR) coatings, while 0.60 eV cells are without AR coatings. We typically expect a roughly 35% improvement in performance with AR coating. One can easily notice the effect of lattice mismatch on the performance of strained InGaAs cell. The degradation is caused by the generation of dislocations and increases as the mismatch (or In/Ga ratio) increases. Figure 1 shows the structure of the two InGaAs cells. More cell details are available elsewhere (3,4). We have used the measured current-voltage characteristics and the processing data of the 0.75 eV and 0.60 eV cells for our modeling calculations and predicted the performance under a blackbody and a Ho-YAG selective emitter.

Table 1. Measured Performance Data of Lattice-Matched and Strained InGaAs Solar Cells under AM0 Spectrum and at 25°C. (Cell Size 0.6 cm x 0.8 cm)

	Lattice-Matched 0.75 eV InGaAs Cell With AR Coating	Strained 0.60 eV InGaAs Cell Without AR Coating
Short-Circuit Current	27.1 mA	23.23 mA
Open-Circuit Voltage	376.4 mV	199.1 mV
Fill Factor	0.72	0.62
Efficiency	11.21%	4.34%

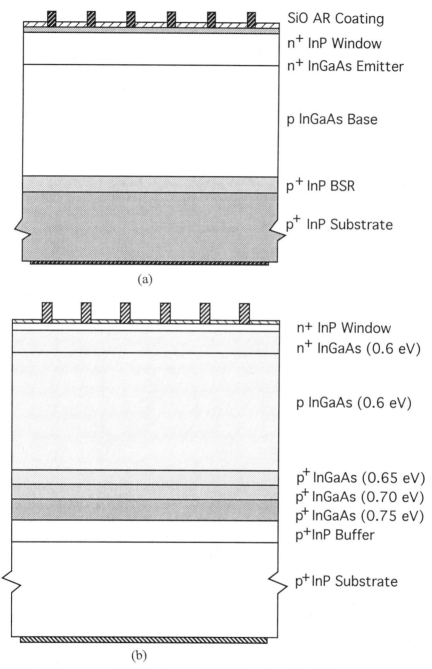

FIGURE 1. Structure of the lattice-matched (a) and strained (b) InGaAs cell.

RADIATION SOURCES

A wide variety of narrow-band as well as broad-band radiation sources can be utilized for TPV applications. Selective emitters offer advantages over blackbody sources by narrowing the emission band, but offer lower power densities. However, blackbody sources along with suitable filters can provide high power densities and offer a possibility to operate with a wide number of cell materials.

Recently Lowe et al. reported the development of a thin film Ho-YAG selective emitter (5). The Ho-YAG selective emitter has an emission peak around 2.0 μm, which makes it well suited for the 0.60 eV InGaAs cell considered in this work. The 0.60 eV and 0.75 eV InGaAs cells are also suitable for 1500 K and 2000 K blackbody sources respectively. In this work Ho-YAG selective emitter and blackbody radiation sources have been considered.

MODELING APPROACH

In the present work, we have considered the measured data of the highest efficiency 0.75 eV and 0.6 eV InGaAs solar cells fabricated at the NASA Lewis. The cell processing details (thickness, dopings, etc.) and performance results served as the baseline for predicting the performance of InGaAs cells for TPV applications. Computer simulations were performed using the PC-1D computer code (6) to match the measured I-V characteristics as well as the external quantum efficiency of InGaAs cells.

The device modeling requires an accurate knowledge about the material physical parameters. Although most of the InGaAs work has been on lattice-matched devices, little or no experimental information is available on the various physical parameters. Similar information on lattice mismatched InGaAs material is almost nonexistent. The optical absorption coefficient and the intrinsic carrier concentration are very important parameters, which greatly influence the device performance. We have calculated these important material parameters for strained InGaAs based on the available and our estimated data for lattice-matched InGaAs (2).

Figure 2 shows the optical absorption coefficient (α) versus wavelength (λ) for the 0.75 eV InGaAs at 300 K. The absorption coefficient values up to 800 nm wavelength were calculated by the authors, using the relation:

$$\alpha = 4\pi\kappa/\lambda \tag{1}$$

where κ is the extinction coefficient. The InGaAs extinction coefficient results up to 800 nm were reported in reference 7, based on the work of four references. The 0.75 eV InGaAs optical absorption coefficient data for 1100 nm and higher wavelengths were taken from reference 8. α versus λ results between 800 nm and 1100 nm for the 0.75 eV InGaAs were extrapolated by the authors. These absorption coefficient versus wavelength values were used in the present and earlier reported work (2) for predicting the performance of lattice-matched InGaAs (0.75 eV) for TPV applications.

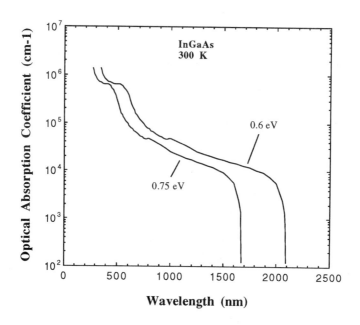

FIGURE 2. Optical absorption coefficient of lattice-matched (0.75 eV) and strained (0.60 eV) InGaAs material.

Figure 2 also shows the optical absorption coefficient versus wavelength results for the 0.60 eV InGaAs at 300 K. These results have been calculated by the authors by assuming the same absorption coefficient as of 0.75 eV InGaAs but shifted in wavelength by the bandgap ratio (0.75/0.60).

The intrinsic carrier concentration n_i for InGaAs was calculated by using the scaling relation described by Jain et al. (2):

$$n_i(\text{InGaAs}) = n_i(\text{GaAs})\exp[-\Delta E_g/(2kT)] \tag{2}$$

where ΔE_g is the bandgap difference, k is the Boltzman's constant, and T is the temperature in K. Figure 3 shows the plot of the intrinsic carrier concentration versus InGaAs bandgap energy at 300 K. An n_i value of 2.59×10^6 cm^{-3} for GaAs at 300 K was used.

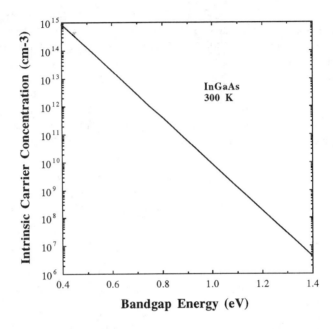

FIGURE 3. Plot of the intrinsic carrier concentration versus InGaAs bandgap energy.

It should be noted that in the absence of measured data, our calculated results based on sound approach offer the best estimate of the important parameters (α and n_i) and are the first reported results (2). However, experimental measurements of the important material parameters of InGaAs and their temperature dependence should be pursued. This study will be very valuable in accurate prediction and understanding of the physics and technology of the devices made from this promising bandgap engineered material for TPV and other use.

RESULTS AND DISCUSSION

Although the lattice-matched and strained InGaAs cells considered in the present work are not optimized for TPV applications, they are the high efficiency cells reported and therefore used as the baseline cells for our modeling calculations as well as for predicting the performance under the selective emitter as well as the blackbody at 1500 K and 2000 K. Cell temperature is 25°C.

Figure 4 shows the calculated 0.75 eV (lattice-matched) InGaAs cell efficiency versus input power for the blackbody at 2000 K. As mentioned earlier, the 0.75 eV cell is ideally suited to a 2000 K blackbody emission spectrum which has a peak around 1.5 µm. The efficiency of the existing cell decreases with increase in the power density (curve 1), due to the cell series resistance. It should be noted that in TPV applications, cells will be operating under high intensities, and therefore should have minimum possible series resistance to reduce the losses. If the cell series resistance is reduced to zero, power losses are reduced and the cell efficiency increases (curve 2), but still it is low for TPV applications.

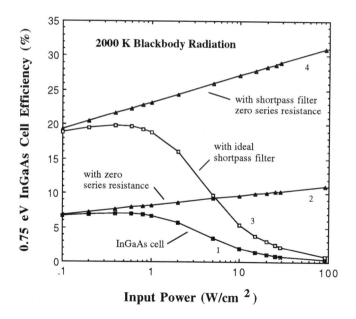

FIGURE 4. Calculated 0.75 eV InGaAs cell efficiency versus input power for various conditions under a blackbody at 2000 K.

Use of a reflecting filter could improve this efficiency. Curves 3 and 4 show cell performance if an ideal reflective filter is used between the blackbody and the cell. The filter reflects back to the source all radiation longer than 1.6 µm. The cell efficiency improves significantly, but is still dominated by the cell resistance at high intensities (curve 3). The cell efficiency improves significantly and increases with intensity if the cell series resistance can be reduced to zero (curve 4), approaching 31% for input power of 91 W/cm². In comparision, under a 1500 K Er-YAG selective emitter and ideal filter conditions, the 0.75 eV InGaAs cell has shown efficiencies approaching 30% for a relatively low input power of 6 W/cm² (2).

Figure 5 shows the calculated 0.60 eV strained InGaAs cell efficiency versus input power for the blackbody at 1500 K. As mentioned earlier, the 0.60 eV cell is well suited to a 1500 K blackbody emission spectrum which has a peak around 2.0 µm. The results are similar in behavior (see Fig. 4) but quantitatively different. In calculations, it is assumed that an ideal filter reflects back all radiation longer than 2.0 µm. The 0.60 eV cell efficiency improves to 19% for input power of 30 W/cm² with ideal shortpass filter and zero series resistance. The efficiencies are lower compared to the 0.75 eV cell under 2000 K blackbody (see Fig. 4). The lower 0.6 eV cell efficiency results can be explained by the higher dark current at lower bandgap and also the lack of an AR coating.

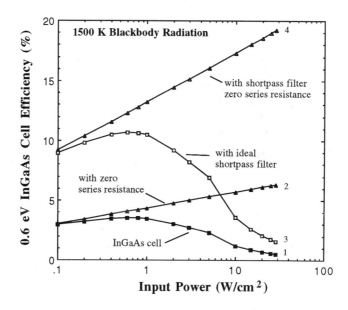

FIGURE 5. Calculated 0.60 eV InGaAs cell (no AR coating) efficiency versus input power for various conditions under a blackbody at 1500 K.

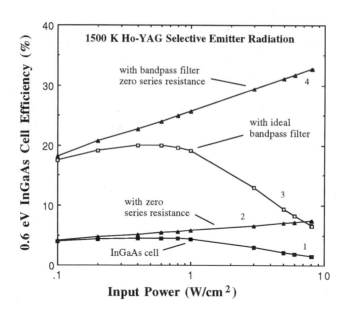

FIGURE 6. Calculated 0.60 eV InGaAs cell (no AR coating) efficiency versus input power for various conditions under a Ho-YAG selective emitter at 1500 K.

Figure 6 shows the calculated 0.60 eV strained InGaAs cell efficiency versus input power for the Ho-YAG selective emitter at 1500 K. As mentioned earlier, the 0.60 eV cell is well suited to a Ho-YAG selective emitter emission spectrum which has a peak at 2.0 μm. The results are similar in behavior (see Fig. 5) but quantitatively different. The efficiencies are higher compared to the 1500 K blackbody case (see Fig. 5). In case of the blackbody, a large portion of the emitted radiation is un-utilized. For the case of an ideal filter, we restrict the emission of the Ho-YAG selective emitter into a band between 1.9 and 2.1 μm. The rest of the emission is reflected back to the source. The efficiency of the current cell improves almost three times. It is still dominated by the cell resistance at high intensities (curve 3). The cell efficiency improves significantly and increases with intensity if the cell series resistance is reduced to zero. InGaAs cell efficiencies in excess of 32% (curve 4) are predicted for the Ho-YAG selective emitter but at a lower power density (8 W/cm^2) compared to the blackbody case.

In all of these results, series resistance has been a major effect, especially at high intensities. Figure 7 shows the effect of series resistance on efficiency for the 0.6 eV InGaAs cell (no AR coating), with the Ho-YAG selective emitter.

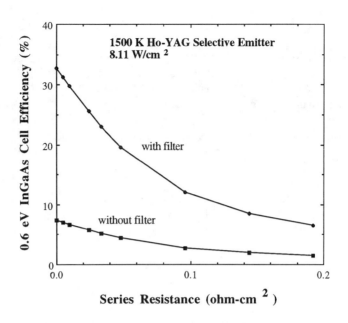

FIGURE 7. Effect of series resistance on the performance of 0.6 eV InGaAs cell (no AR coating) with the 1500 K Ho-YAG selective emitter.

The TPV cell efficiencies calculated are for the unoptimized InGaAs cells. Preliminary modeling studies suggest that efficiencies in excess of 40% are achievable for the optimally designed InGaAs cells.

CONCLUSIONS

The high AM0 efficiency lattice-matched and strained InGaAs solar cells have been modeled and their performance under blackbody and Ho-YAG selective emitter radiation predicted for various conditions.

The 0.75 eV InGaAs cell efficiencies of 31% are predicted under a 2000 K blackbody source with ideal filter and zero cell series resistance. The lattice-matched InGaAs cell is an ideal choice for the 2000 K blackbody source.

The 0.60 eV InGaAs cell efficiencies in excess of 32% are calculated under a Ho-YAG selective emitter at 1500 K. Strained cells exhibit lower efficiencies under 1500 K blackbody source. The strained 0.60 eV InGaAs cell is well matched to the spectrum of Ho-YAG selective emitter as well as the 1500 K blackbody sources.

Cell series resistance is a very important parameter and must be reduced to minimum. Cells also have to be optimized to match the radiation spectrum which will require more work leading to TPV application specific cell designs as well as advanced antireflection coatings.

Achievement of high cell efficiencies requires effective thermal control to keep the cell temperature low. Efficiencies exceeding 40% are predicted for the optimized devices.

High cell efficiencies are desired to achieve high total TPV system efficiencies, and are dependent on the development of suitable filters as well as efficient selective emitters.

ACKNOWLEDGMENTS

Helpful discussions with Donald L. Chubb and Roland Lowe on selective emitters are greatly appreciated.

REFERENCES

1. Jain, R. K., and Flood, D. J., "Influence of the Dislocation Density on the Performance of Heteroepitaxial Indium Phosphide Solar Cells," IEEE Transactions on Electron Devices **40**, 1928-1934 (1993).
2. Jain, R. K., Wilt, D. M., Landis, G. A., Jain, R., Weinberg, I., and Flood, D. J., "Modeling of Low-Bandgap Solar Cells for Thermophotovoltaic Applications," in the First NREL Conference on Thermophotovoltaic Generation of Electricity, AIP Conference Proceedings 321, 1994, pp. 202-209.
3. Wilt, D. M., Fatemi, N. S., Hoffman, R. W., Jenkins, P. P., Brinker, D. J., Scheiman, D., Lowe, R., and Jain, R. K., Appl. Phys. Lett. **64**, 2415-2417 (1994).
4. Wilt, D. M., Fatemi, N. S., Hoffman, R. W., Jenkins, P. P., Scheiman, D., Lowe, R., and Landis, G. A., in the First NREL Conference on Thermophotovoltaic Generation of Electricity, AIP Conference Proceedings 321, 1994, pp. 210-220.
5. Lowe, R., Chubb, D. L., Farmer, S. C., and Good, B. S., "Rare-Earth Garnet Selective Emitter," Appl. Phys. Lett. **64**, 3551-3553 (1994).
6. Basore, P. A., "PC-1D version 3: Improved Speed and Convergence," in Conference Record of the 21st IEEE Photovoltaic Specialists Conference, 1991, pp. 299-302.
7. Alterovitz, S. A. in Properties of Lattice-Matched and Strained Indium Gallium Arsenide, Edited by Bhattacharya, P., INSPEC, 1993, pp. 188 and references therein.
8. Escher, J. S. in Semiconductors and Semimetals, Volume 15, Edited by Willardson, R. K., and Beer, A. C., Academic Press, 1981, pp. 208.

$In_xGa_{1-x}As$ TPV Experiment-based Performance Models

Steven Wojtczuk

Spire Corporation
One Patriots Park
Bedford, MA 01730-2396

Abstract: Indium gallium arsenide ($In_xGa_{1-x}As$) cell models extracted from measured data on thermophotovoltaic (TPV) cells with bandgaps of 0.75 to 0.55 eV are presented. The dark current model is based on a fit to values extracted from open-circuit voltages at high photocurrents (where the ideality factor is close to unity) for the various bandgap cells. Quantum efficiency models of Hovel and a very simple base model are compared with measured data. A standard model for the series resistance of the cell is presented and agrees with measured data. The quantum efficiency model is used with the standard blackbody equations to predict the cell photocurrent at 800 and 1200C over the 0.5 to 0.75 eV bandgap range. The dark current, series resistance, and photocurrent are used to numerically determine maximum output power for $In_xGa_{1-x}As$ cells over the above bandgap range at the these two blackbody temperatures.

INTRODUCTION

Low-bandgap indium gallium arsenide ($In_xGa_{1-x}As$) cells are of interest for the electrical conversion of thermal photons from heat sources of ~ 800C (2.7 μm blackbody peak) up to ~ 2400C (1.1 μm peak). This paper describes a model method for $In_xGa_{1-x}As$ cells in the 0.5 eV (2.5 μm cutoff wavelength) to 0.75 eV (1.65 μm). The $In_xGa_{1-x}As$ bandgap is set during epitaxial growth by the adjusting indium composition. Examples in this paper use 800 and 1200C blackbodies since many thermal sources used with $In_xGa_{1-x}As$ cells are in this range. The models are applicable to sources at any temperature. Performance predictions require a model of the cell I-V curve at each bandgap. This requires a model of the cell dark current and series resistance. Maximum power points are calculated by shifting the I-V curve by the amount of photocurrent calculated by weighting the power spectrum of the blackbody radiation with the wavelength-dependent quantum efficiency model.

CELL STRUCTURE

The $In_xGa_{1-x}As$ cell epilayer structure considered in this study is shown in Figure 1. An N-on-P cell is used so that the higher mobility majority carrier electrons allow a low sheet resistance in the relatively thin, heavily-doped N emitter layer. The N-on-P design also allows for longer minority-carrier electron diffusion lengths in the more lightly doped P base layer, of great importance since most of the longer-wavelength blackbody irradiation is near the bandgap and is absorbed in the base layer. The $In_xGa_{1-x}As$ material is only lattice-matched to the InP wafer at the $In_{0.53}Ga_{0.47}As$ (0.74 eV) composition. When grown at compositions away from lattice match (e.g. 0.5 eV $In_{0.8}Ga_{0.2}As$), dislocation defects are created which can thread through the cell junction, increasing the dark current. A compositionally-graded layer is normally used to reduce the threading dislocation density. At high photocurrent levels, very good cell performance for mismatch up to 1.3% (down to the 0.55 eV $In_xGa_{1-x}As$ composition) has been achieved with these grading layers. Some additional details on similar cells can be found in (1).

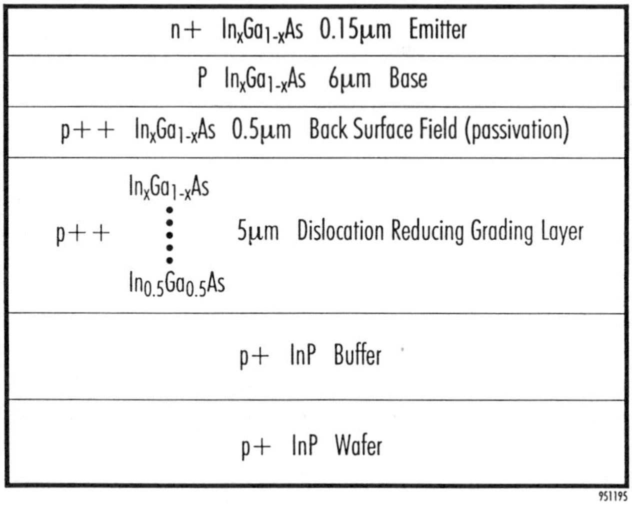

n+ $In_xGa_{1-x}As$ 0.15μm Emitter
P $In_xGa_{1-x}As$ 6μm Base
p++ $In_xGa_{1-x}As$ 0.5μm Back Surface Field (passivation)
p++ $In_xGa_{1-x}As$ ⋮ $In_{0.5}Ga_{0.5}As$ 5μm Dislocation Reducing Grading Layer
p+ InP Buffer
p+ InP Wafer

Figure 1. *Typical $In_xGa_{1-x}As$ TPV cell structure.*

QUANTUM EFFICIENCY

The complete quantum efficiency model of Hovel (2) is compared in Figure 2 with measured data for a Spire 0.74 eV $In_{0.53}Ga_{0.47}As$ cell. Accurate wavelength-dependent absorption coefficient data (3) is crucial for good quantum efficiency models. However, data is only readily available for the 0.74 eV lattice-matched composition. Other $In_xGa_{1-x}As$ compositions were modeled in this work by shifting the absorption edge of the 0.74 eV data to the correct bandedge for the particular $In_xGa_{1-x}As$ composition. The contribution from the emitter, base, and space-charge depletion region (scr) are shown in Figure 2 from Hovel's model. The contributions from the emitter and scr are only significant at shorter wavelengths, where the light is absorbed closer to the surface. Although accurate, the Hovel model is relatively complex to manipulate. A simple model for the quantum efficiency, shown in Figure 2, does almost as well in an empirical sense, although physically this model is inaccurate in describing the cause of the quantum efficiency at shorter wavelengths (even though the actual fit at short wavelengths is surprisingly good). The base diffusion length (~ 10 μm for most bandgaps) was the key fit parameter.

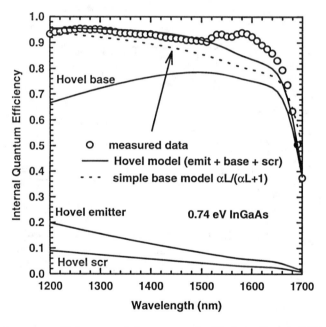

Figure 2. *Quantum efficiency models compared with data from a 0.74 eV cell.*

DARK CURRENT MODEL

The radiative dark current model of Henry (4,5) sets a lower ultimate limit for the dark current. However, real $In_xGa_{1-x}As$ cells typically have dark currents above this level due to material and junction defects, even with grading layers to help accommodate lattice-mismatch. Figure 3 shows this radiative limit, a few measured data points of particular interest, and a semi-empirical model non-linear least-squares fit to all the data. The dark current data points were obtained at each bandgap by measuring Voc and Isc for that cell at several illumination levels, plotting this data (Voc-Isc method avoids series resistance effects) on a log scale, finding the region where the cell ideality factor was close to unity, and using that Voc and Isc to extract the dark current from the standard equation.

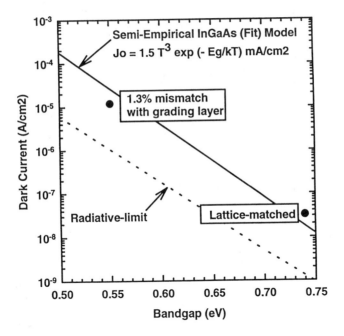

Figure 3. $In_xGa_{1-x}As$ dark current models versus bandgap.

SERIES RESISTANCE MODEL

The series resistance model used for this study was a standard one, similar to that described by many authors (6,7). The results are shown in Table I for a 1 cm by 1cm cell with a single central busbar. The largest contribution to the series resistance in these cells was the alloyed back contact. The back contact resistivities achieved (10^{-2} - 10^{-3} Ω-cm^2) could be much improved. The emitter sheet resistance and front contact resistivity was measured with standard transmission line test patterns and used as inputs to the model. Emitter loss is negligible because of the large amount of front grid metal that is feasible on TPV cells. For TPV, the light reflected off the large (~15% of area) front grid is "recycled" back to the close proximity heat source. This is entirely unlike regular solar cells, where such light is not recycled and such a shadow loss would not be acceptable. The excellent front contact is a tunneling-transport contact between the heavily-doped low-bandgap InGaAs and the grid metal.

Table I Results of $In_xGa_{1-x}As$ Cell Series Resistance Model

Alloyed AuZn/P+ InP back contact	~ 10-20 mΩ
Gold 3µm thick 10 µm wide front contact grid	~ 6 mΩ
25 mil thick P+ InP bulk wafer	~ 3 mΩ
Gold 3 µm thick 1mm wide busbar	~ 3 mΩ
Emitter sheet resistance (15 Ω/square)	~ 0.2 mΩ
Non-alloyed AuCr/ N+ InGaAs front contact	~0.1 mΩ

POWER OUTPUT vs. BANDGAP

The quantum efficiency model was used with the standard blackbody equations to predict the cell photocurrent at 800 and 1200C over the 0.5 to 0.75 eV bandgap range. The dark current, series resistance, and photocurrent models determined illuminated I-V curves used to numerically calculate maximum output power and efficiency (Figure 4) for $In_xGa_{1-x}As$ cells over the above bandgap range at the these two blackbody temperatures. Figure 5 shows the open-circuit voltage and short-circuit current density for the same cases.

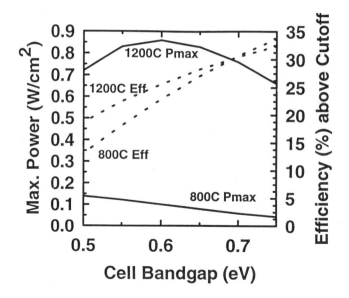

Figure 4. $In_xGa_{1-x}As$ power output and efficiency vs. bandgap.

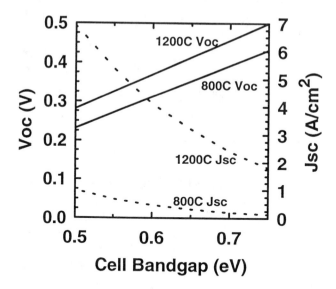

Figure 5. $In_xGa_{1-x}As$ Voc and Jsc vs. bandgap.

Figures 4 and 5 each assume the cell temperature was at 25C. Also, for the calculation, we assumed only half the power from the blackbody was incident on the cells for each temperature. This means that the power density incident on the cells was 3.8 W/cm^2 for the 800C blackbody and 13.3 W/cm^2 for the 1200C blackbody. The efficiency on Figure 4 is defined as the ratio of the cell electrical power output to the blackbody optical power from photons with energy above the cell's bandgap incident on the cell. The power from photons below the cells bandgap is assumed to be returned to the blackbody by a spectral filter on the cell.

SUMMARY

Simple models for the dark current, quantum efficiency, and series resistance of $In_xGa_{1-x}As$ cells are discussed and compared to measured data. The models were used to determine I-V curves for $In_xGa_{1-x}As$ cells in the 0.5 to 0.75 eV bandgap range. Photocurrents were calculated for 800 and 1200C blackbody spectra using the quantum efficiency model. These photocurrents and the model I-V curves were used to numerically determine maximum power points for $In_xGa_{1-x}As$ cells over the above bandgap range. The author hopes these plots may be useful as more quantitative indicators of the trends of $In_xGa_{1-x}As$ cell power output, efficiency, photovoltage, and photocurrent versus $In_xGa_{1-x}As$ bandgap and blackbody temperature.

REFERENCES

1. Wojtczuk, S., Gagnon, E., Geoffroy, L., and Parodos, T., "$In_xGa_{1-x}As$ Thermophotovoltaic Cell Performance vs. Bandgap," in *Proc. of 1st NREL TPV Conf.*, AIP Conf. Proc. 321, 1994, pp. 177-187.

2. Hovel, H.J., *Solar Cells, Semiconductors and Semimetals,* Vol. 11, Academic Press, 1975, pp. 17-20.

3. Borrego, J., Zierak, M., and Charache, G., "Parameter Extraction for TPV Cell Development," in *Proc. of 1st NREL TPV Conf.,* AIP Conf. Proc. 321, 1994, pp. 371-378.

4. Henry, C.H., "Limiting efficiencies of ideal and multiple energy gap terrestrial solar cells," *J. Appl. Phys.* **51**(8), August 1980, pp. 4494-4500.

5. Baldasaro, P.F., Brown, E.J., Depoy, D.M., Campbell, B.C., and Parrington, J.R., "Experimental Assessment of Low Temperature Voltaic Energy Conversion," in *Proc. of 1st NREL TPV Conf.,* AIP Conf. Proc. 321, 1994, pp. 29-43.

6. Basore, P.A. "Optimum Grid-Line Patterns for Concentrator Solar Cells Under Non-Uniform Illumination", *Solar Cells,* **14**, 1985, pp. 249-260.

7. Gessert, T.A., and Coutts, T.J. "Grid metallization and antireflection coating optimization for concentrator and one-sun photovoltaic cells," *J. Vac. Sci. Technol. A,* **10**(4), 1992, pp. 2013-2024.

Molecular Beam Epitaxy of $In_{0.74}Ga_{0.26}As$ on InP for Low Temperature TPV Generator Applications

T. S. Mayer, W. Hwang, R. Kochhar, M. Micovic, and D. L. Miller

Electronic Materials and Processing Research Laboratory
Department of Electrical Engineering
Penn State University, University Park, PA 16802

S. M. Lord

Department of Electrical Engineering
Bucknell University, Lewisburg, PA 17837

Abstract: The growth by molecular beam epitaxy of $In_{0.74}Ga_{0.26}As$ is investigated because of its importance as a PV converter for a variety low temperature TPV system configurations. In this work, a linearly graded buffer layer is used to grow high quality $In_{0.74}Ga_{0.26}As$ layers on a lattice mismatched InP substrate. The thickness of the buffer layer and the substrate temperature during the growth of the buffer and active layers were varied in order to optimize the active layer material quality. The resulting p^+-i-n^+ epitaxial layers were compared using double crystal x-ray diffraction, spectral response, and current-voltage measurements. A more conventional PV cell structure was also evaluated using current-voltage measurements.

INTRODUCTION

Thermophotovoltaic (TPV) generators designed with low temperature emitters can offer advantages over high temperature systems in the areas of reliability, ruggedness, and fuel selection. Unfortunately, existing photovoltaic cell technology is not suitable for use in these systems due to their inherent short-wavelength cutoffs. In order to achieve high conversion efficiencies at low temperatures, cell cutoff wavelengths must be extended beyond 2.0 µm. This can be accomplished by fabricating photovoltaic (PV) cells using Indium Gallium Arsenide ($In_xGa_{1-x}As$). By controlling the indium composition of the active device layers, the cutoff wavelength of the PV cell can be varied from 1.7 to 2.6 µm.

The growth of $In_{0.74}Ga_{0.26}As$ was investigated because of its importance in a variety of system configurations. In particular, $In_{0.74}Ga_{0.26}As$, which has a cutoff wavelength of 2.2 µm, is ideally suited for use as a PV converter in filtered broadband and holmia-based selective emitter systems operating in the temperature range of 1000°C to 1200°C [1,2]. In both cases, the 2.2 µm cutoff is slightly beyond the peak in their emission spectra. Therefore, coupling $In_{0.74}Ga_{0.26}As$ PV cells with these broadband or selective emitters will result in a system design that maximizes its output power density and cell conversion efficiency.

In this work, molecular beam epitaxy (MBE) was used to grow $In_{0.74}Ga_{0.26}As$ p^+–i–n^+ and p^+–n devices on InP substrates. To permit the growth of high quality $In_{0.74}Ga_{0.26}As$ with such a large mismatch to the InP substrate (1.4%), a thin linearly-graded buffer layer (LGBL) was inserted between the substrate and the active layers. Such buffer layers have been used to minimize the number of active layer threading dislocations [3] that often behave like electrically and optically active defects [4]. Moreover, in contrast to other buffer layers that have been used for the growth of lattice mismatched materials by VPE and MOCVD [5], these LGBL's are typically less than 2 µm thick making them compatible with the relatively slow growth rate (1-1.5 µm/h) of MBE.

In order to optimize the active layer material quality, the effect of varying the thickness of the buffer layer and the substrate temperature during the growth of the buffer and active layers was investigated. A series of p^+–i–n^+ devices were grown with buffer layer thicknesses of 0.5, 1.0, and 2.0 µm and substrate temperatures ranging from 300°C to 450°C. The samples were compared using double crystal x-ray diffraction, spectral response, and current-voltage measurements. A more conventional PV cell structure was also grown to evaluate the $In_{0.74}Ga_{0.26}As$ material for use as a converter in low temperature TPV systems. The results of this work are presented in the following sections.

MATERIALS GROWTH AND DEVICE FABRICATION

The samples that were used to compare the material quality for various growth conditions consisted of p^+–i–n^+ diodes with active layer structures as given in Table 1. All of the epitaxial layers were grown on n^+–InP substrates in a Varian GEN-II Molecular Beam Epitaxy (MBE) system. The substrate and the active layers were separated with a graded buffer region were the indium composition was increased linearly from 53% (lattice matched to InP) to 74%. The active layers consisted of a 1 µm thick undoped layer of $In_{0.74}Ga_{0.26}As$ followed by a 100 nm p^+–$In_{0.74}Ga_{0.26}As$ layer and a 50 nm p^+–$In_{0.74}Al_{0.26}As$ window layer. Moreover, a 50 nm p^+–$In_{0.74}Ga_{0.26}As$ cap layer was included to facilitate the fabrication of nonalloyed ohmic contacts.

Three sets of samples were prepared in order to study the effect of substrate temperature and buffer layer thickness on the quality of the active layer material. For the first set, the linearly graded buffer layer (LGBL) thickness was fixed at

TABLE 1. Baseline p^+-i-n^+ device structure used to compare the quality of the active layer material grown at substrate temperatures ranging from 300°C to 400°C and buffer layer thickness of 0.5, 1.0, and 2.0 µm.

Layer	Material	Doping (cm^{-3})	Thickness (nm)	Growth Temperature
Contact	In$_{0.74}$Ga$_{0.26}$As:Be	5x10^{18}	50	300°C, 350°C 400°C, 450°C
Window	In$_{0.74}$Al$_{0.26}$As:Be	5x10^{18}	50	
Cap	In$_{0.74}$Ga$_{0.26}$As:Be	6x10^{17}	100	
Active	In$_{0.74}$Ga$_{0.26}$As	undoped	1000	
n$^+$–Buffer	In$_{0.74}$Ga$_{0.26}$As:Si	5x10^{18}	50 – 1000	
Linearly Graded Buffer	In$_x$Ga$_{1-x}$As:Si (0.53 < x < 0.74)	5x10^{18}	500, 1000, 2000	300°C, 350°C 400°C, 450°C
Lattice Matched Buffer	In$_{0.74}$Ga$_{0.26}$As:Si	5x10^{18}	300	450°C
n$^+$–Substrate	InP:S	2-5x10^{18}		

1 µm and the substrate temperature was increased from 300°C to 450°C in increments of 50°C. In this case, the substrate temperature was held at a constant value during the growth of the LGBL and the active layers. Next, the substrate temperature was held at 400°C and the growths were repeated with LGBL thickness of 0.5 and 2.0 µm. The final set was comprised of samples were the substrate temperature of the LGBL was 350°C while that of the active layer was 400°C. The growths were performed for LGBL thicknesses of 0.5, 1.0, and 2.0 µm.

The growths were conducted at a rate of approximately 1.5 µm/h, and a group V molecule to Ga beam equivalent pressure of 15. Moreover, in order to obtain good surface morphology at relatively low substrate temperatures, the arsenic species used in all growths was the dimer As$_2$ [3]. The linear increase in the indium composition with position was approximated by reducing the grade to a series of small steps. In particular, for each case described previously, the LGBL was divided into 50 equal segments and the concentration of indium was increased an equal amount at each step. Because the indium composition increased from 53% to 74% in the buffer layer, the gradient of the indium concentration varied from 40% In/µm (increase of 0.4% In/step with step size of 100 Å) for the 0.5 µm LGBL to 10% In/µm (increase of 0.4% In/step with step size of 400 Å) for the 2.0 µm LGBL.

Following the epitaxial growth, the samples were divided and part of each sample was processed into diodes ranging in size from 25×10^{-6} to 1.6×10^{-3} cm^2. First, an electrical contact was made to the n^+-substrate by alloying indium. A liftoff procedure was then used to form the top ohmic contacts. In this process, 250 nm of Ti/Au was deposited in a thermal deposition system. Following the metal liftoff, the devices were completed with a mesa isolation etch in H$_3$PO$_4$:H$_2$O$_2$:DI to a depth of 500 nm.

MATERIALS CHARACTERIZATION

In order to assess the indium composition and the material quality, double crystal x-ray diffraction was performed on the as-grown samples. A typical rocking curve for the sample with a 1μm thick LGBL grown at a buffer temperature of 350°C and an active temperature of 400°C is shown in Fig. 1. In this figure, the substrate peak, linearly graded layer, and active layer are clearly discernible. However, as expected for the growth of lattice mismatched materials, the active layer peak is relatively broad.

For the purpose of analysis, x-ray rocking curves were obtained from each sample for reflections from the (004) and (115) (high and low angles) planes. Following the initial set of measurements, the samples were rotated by 180° and the measurements were repeated in order to account for sample tilt. The average of the

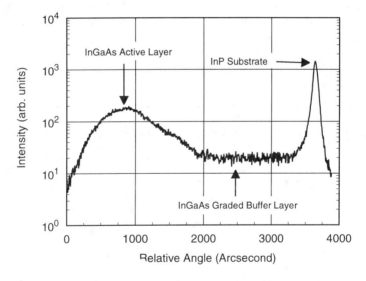

FIGURE 1. Typical double crystal x-ray (115) diffraction rocking curve.

two measurements was used to calculate the indium composition, lattice relaxation, and full width at half maximum (FWHM) of the active later (a uniform bi-axial compression of the action layer was assumed and the in-plane lattice constants were taken to be the same in both the <110> and <110> directions). The results of these calculations are summarized in Table 2 for two sets of samples.

From the table, it is evident that the actual indium composition of the active layers was close to the targeted value of 74% for all of the samples. For the growth parameters listed in the table, the amount of lattice relaxation ranged from approximately 86% to nearly complete relaxation. Moreover, the trend in the data suggests that the samples grown at lower LGBL temperatures are slightly more relaxed than those grown at higher LGBL temperatures. Finally, the FWHM varied from 605 to 1100 arc-seconds. These values are comparable to those quoted for the epitaxial growth of materials with a similar active to substrate lattice mismatch [6]. Unlike the lattice relaxation, however, the FWHM showed no apparent dependence on the parameters of buffer growth temperature or thickness.

The spectral response of mesa isolated p^+–i–n^+ devices was measured in order to verify that the material is photo-sensitive at wavelengths beyond 2 µm. The devices that were used for these measurements consisted of a "snowman" structure that had a 100 µm diameter optically active area and a 25 µm diameter ohmic contact. The room temperature spectral response of the sample with a 2 µm thick LGBL grown at a buffer temperature of 400°C and an active temperature of 400°C is shown in Fig. 2. This data demonstrates that the targeted cutoff wavelength of 2.2 µm was achieved. The gradual rolloff of the spectral response at long wavelengths is due to the very thin 1 µm active layer. In this device, many of

TABLE 2. Results of the double crystal x-ray diffraction for the samples grown at a buffer temperature of 350°C and 400°C and an active temperature of 400°C.

T_{sub} Buffer (°C)	T_{sub} Active (°C)	Buffer Thickness (µm)	Indium Composition	Relaxation	FWHM (004) (Arc-seconds)
400	400	0.5	73	90	620
400	400	1.0	71	86	720
400	400	2.0	71	86	775
350	400	0.5	73	88	665
350	400	1.0	74	92	605
350	400	2.0	73	100	1100

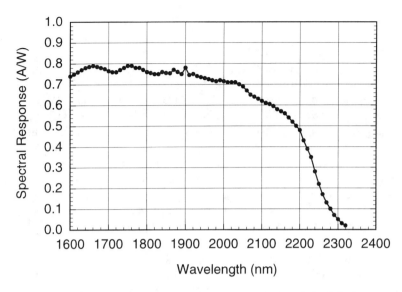

FIGURE 2. Typical room temperature spectral response of a p^+–i–n^+ device.

the long-wavelength photons striking the sample are absorbed in the buffer layer and are not collected as current. Therefore, the long wavelength response could be improved by increasing the thickness of the active layer.

ELECTRICAL CHARACTERIZATION

The effect of the buffer thickness and the growth temperature on the electrical performance of the p^+–i–n^+ devices was evaluated by measuring the reverse leakage current (dark current) as a function of reverse bias voltage. The reverse leakage current is comprised of components due to diffusion, thermal generation, and tunneling [7]. At low-to-moderate reverse biases, thermal generation in the bulk neutral and depletion regions dominates the characteristics. This generation current depends on the density and type of traps in the semiconductor making it extremely sensitive to the quality of the active layer material. Therefore, a measure of the reverse leakage current allows rapid evaluation of a sample's material quality.

Initially, devices from the first set of samples that consisted of material grown with a fixed buffer thickness of 1 µm and a substrate temperature that was varied from 300°C to 450°C were characterized. Because there are few published reports of lattice mismatched growth of InGaAs by MBE [6], these samples were used to determine the range of substrate temperatures that warranted further investigation. The room temperature reverse leakage currents as a function of reverse voltage are shown in Fig. 3. In the legend of this figure, the first temperature refers to the substrate temperature of the LGBL while the second refers to the substrate temperature of the active layer. Moreover, the arrows point in the direction of increasing growth temperature.

The comparison presented in Fig. 3 reveals that the samples grown at a substrate temperature of 400°C result in devices with the lowest leakage current. Increasing the substrate temperature to 450°C caused a one order of magnitude increase in the current, while decreasing the substrate temperature to 300–350°C increased the current by more than two orders of magnitude in both cases. Moreover, the surface morphology of the sample grown at 450°C was very poor. Taken together, this data suggests that substrate temperatures in the range of 400°C are the most appropriate for growth of lattice mismatched material by MBE.

In prior studies that used LGBL's to accommodate lattice mismatch during the growth of InGaAs:GaAs [3], it was demonstrated that increasing the thickness of the buffer layer minimized the formation of threading dislocations. Moreover, a reduction in the substrate temperature during the growth of the buffer layer over that of the active layer also was found to improve device performance [3]. In order to study the effect of buffer layer thickness, devices from the second set of samples where the substrate temperature was 400°C and the buffer thickness was changed to 0.5 and 2.0 µm were characterized. The results of the room temperature reverse

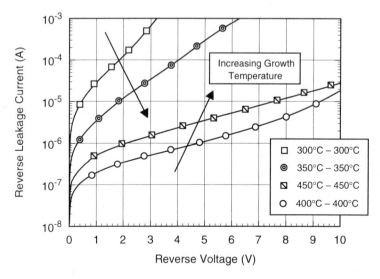

FIGURE 3. Reverse leakage current versus voltage for samples with 1µm thick LGBL's grown at temperatures ranging from 300°C to 450°C.

leakage current versus reverse bias are given in solid lines in Fig. 4. From this figure, it is evident that increasing and a decreasing the buffer layer thickness produced a substantial increase in the leakage current.

Although, it is routinely accepted that thicker buffers are more effective in reducing the threading dislocations that propagate through the active layer material, this data suggests that the step size is also an important factor. In this work, the buffer layers were divided into 50 equal steps resulting in a step size of 400 Å for the 2.0 µm sample and a step size of 200 Å for the 1 µm sample. Therefore, improved characteristics would be expected from a sample that had a 2 µm buffer with a step size comparable to that of the 1 µm sample. The thin, 0.5 µm buffer, also demonstrated inferior performance. This can most likely be attributed to the very fast gradient in the indium composition of this buffer layer.

The reverse leakage current of the sample that was grown at a reduced substrate temperature of 350°C is given by the dashed line in Fig. 4. At low biases, the leakage current of this sample is comparable to that of the sample grown at 400°C. As the bias increases, the current remains relatively constant, and is

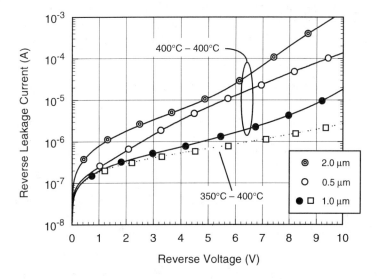

FIGURE 4. Reverse leakage current versus voltage for samples with LGBL thicknesses of 0.5, 1.0, and 2.0 µm. The curves labeled with circles represent characteristics typical of devices fabricated from material grown at a buffer temperature of 400°C and an active temperature of 400°C. In order to demonstrate the effect of a lower buffer temperature, an I-V curve (filled squares) is also shown for a sample grown with a 1 µm LGBL at a buffer temperature of 350°C and an active temperature of 400°C.

approximately one order of magnitude lower than that of the sample grown at 400°C at −10 V. The same dependence on buffer layer thickness was observed for the samples grown at buffer temperature 350°C. This is illustrated in Fig. 5 where the reverse currents for all the samples grown at a buffer temperature of 350°C are given by solid lines. Finally, when the substrate temperature during the growth of the buffer was reduced even further to 300°C while that of the active layer was maintained at 400°C, the device characteristics degraded beyond those grown at 400°C. This implies that, of the substrate temperatures studied, 350°C yielded the most promising results. In order to optimize further the device performance, additional experiments are needed to determine the substrate temperature required for the growth of the active layer.

PV CELL GROWTH AND CHARACTERIZATION

A more conventional PV cell structure was grown by MBE to evaluate the $In_{0.74}Ga_{0.26}As$ material for use as a PV converter in low temperature TPV systems. The structure consisted of a p^+-n diode as given in Table 3. A lattice matched $In_{0.74}Al_{0.26}As$ window was included to reduce the interface recombination velocity

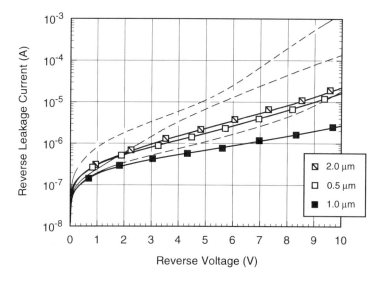

FIGURE 5. Reverse leakage current versus voltage for samples with LGBL thickness of 0.5, 1.0, and 2.0 μm grown at a buffer temperature of 350°C and an active temperature of 400°C. Typical I-V curves (dashed lines) for samples grown at 400°C (from Fig. 4) are included for reference.

TABLE 3. PV cell structure with a 2 μm thick linearly graded buffer that was grown at a buffer temperature of 350°C and an active temperature of 400°C.

Layer	Material	Doping (cm^{-3})	Thickness (nm)	Growth Temperature
Contact	In$_{0.74}$Ga$_{0.26}$As:Be	5x10^{18}	50	
Window	In$_{0.74}$Al$_{0.26}$As:Be	5x10^{18}	50	
Emitter	In$_{0.74}$Ga$_{0.26}$As:Be	6x10^{17}	200	400°C
Base	In$_{0.74}$Ga$_{0.26}$As:Si	5x10^{17}	2000	
n$^+$–Buffer	In$_{0.74}$Ga$_{0.26}$As:Si	5x10^{18}	50	
Linearly Graded Buffer	In$_x$Ga$_{1-x}$As:Si (0.53 < x < 0.74)	5x10^{18}	2000	350°C
Lattice Matched Buffer	In$_{0.74}$Ga$_{0.26}$As:Si	5x10^{18}	300	450°C
n$^+$–Substrate	InP:S	2–5x10^{18}		

at the front surface of the device. The buffer layer was 2 μm thick and the substrate temperature during the growth of the buffer and active layers was 350°C and 400°C, respectively. Devices ranging in size from 50 μm to 0.5 cm on a side were fabricated using the procedure described previously. The largest devices had a top ohmic grid with a metal coverage of approximately 12% to allow frontside illumination.

Typical forward dark current versus voltage curves are shown for various device sizes in Fig. 6. Moreover, values of saturation current density, J_0, are plotted versus device perimeter-to-area ratio in the insert of Fig. 6. The two most noticeable features of these plots is the shunt leakage that dominates the low bias characteristics of the large area devices and the increase in the saturation current density with device size. Such increases in the shunt leakage and saturation current have been attributed to point defects and dislocations present in the device active layer material [8]. However, because PV cells are typically operated at high current densities, the low bias leakage is not expected to pose any fundamental obstacles to excellent device performance. In fact, similar low bias leakage currents were observed in record efficiency GaAs solar cells grown by MBE [8]. The measured saturation current densities of these devices ranges from $1-7 \times 10^{-4}$ A/cm^2. Although these values are higher than those predicted from models that do not account for the dislocations [9], there is a good correlation with measurements made on material grown by OMVPE (most values are reported for $E_g = 0.6$ eV or higher, therefore this is based on an extrapolation of reported data to 0.55 eV) [10].

CONCLUSIONS

The results presented in this paper demonstrate that the use of LGBL's holds promise for the growth by MBE of high quality lattice mismatched InGaAs on InP substrates. It was found that the quality of the active layer material is very sensitive to the thickness of the buffer layer, the step size used in the linear grade, and the substrate temperature during the growth of the active and the buffer layers. Samples with a 1 µm thick buffer layer grown at a buffer temperature of 350°C and an active temperature of 400°C yielded devices with the best electrical performance. Moreover, analysis of double crystal x-ray diffraction results showed that the samples grown at a buffer layer temperature of 350°C tended to be more relaxed and had FWHM values comparable to samples grown at higher substrate temperatures. Optimization of the active layer substrate temperature is expected to further improve the $In_{0.74}Ga_{0.26}As$ material quality.

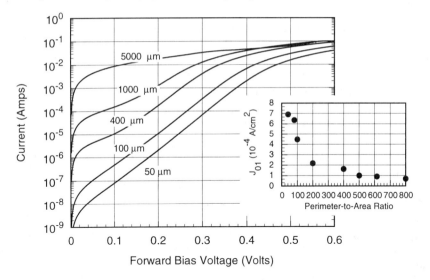

FIGURE 6. Forward current versus voltage for various sizes of p^+–n devices grown with a buffer thickness of 2 µm and a buffer and active temperature of 350 and 400, respectively. Each curve is labeled with the length of one side of the square device.

ACKNOWLEDGMENTS

This work was performed under a subcontract to Sensors Unlimited, Inc. funded by the Navy STTR program Phase I (N00014-94-C00262). The authors would like to thank M. Cohen and G. Olsen for the spectral response measurements and many insightful discussions regarding the growth of lattice mismatched InGaAs on InP. We would also like to thank M. Markey for the PV cell fabrication and characterization.

REFERENCES

1. Baldasaro, P. F., E. J. Brown, D. M. Depoy, B. C. Campbell, J. R. Parrington, "Experimental Assessment of Low Temperature Voltaic Energy Conversion," in *Proc. of 1st NREL Conference on Thermophotovoltaic Generation of Electricity*, pp. 29-43, 1994.
2. Nelson, R. E., "Thermophotovoltaic Emitter Development," in *Proc. of 1st NREL Conference on Thermophotovoltaic Generation of Electricity*, pp. 80-96, 1994.
3. Lord, S. M., B. Pezeshki, and J. S. Harris, Jr., "Investigation of High In Content InGaAs Quantum Wells Grown on GaAs by Molecular Beam Epitaxy," *Electronics Letters*, Vol. 28, No. 13, pp. 1193-1195, 1992.
4. Martinelli, R. U., T. J. Zamerowski, and P. Longeway, "2.6 μm InGaAs photodiodes," *Applied Physics Letters*, Vol. 53, No. 11, p. 989, 1988.
5. Linga, K. R., G. H. Olsen, V. S. Ban, A. M. Joshi, W. F. Kosonocky, "Dark Current Analysis and Characterization of InGaAs/InAsP Graded Photodiodes with x>0.53 for Response to Longer Wavelengths," *IEEE Journal of Lightwave Technology*, Vol. 10, No. 8, pp. 1050-1054, 1992.
6. Fisher-Colbrie, A., R. D. Jacowitz, and D. G. Ast, "Non-lattice Matched Growth of InGaAs on InP," *Journal of Crystal Growth*, Vol. 127, pp. 560-565, 1993.
7. Forrest, S. R., "Performance of InGaAsP Photodiodes with Dark Current Limited by Diffusion, Generation, Recombination, and Tunneling," *IEEE Journal of Quantum Electronics*, Vol. QE-17, No. 2, pp. 217-226, 1981.
8. Tobin, S., S. M.Vernon, C. Bajgar, S. Wajtczuk, M. R. Melloch, A. Keshavarzi, T. B. Stellwag, S. Venkatensan, M. S. Lundstrom, K. A. Emery, "Assessment of MOCVD- and MBE-Grown GaAs for High-Efficiency Solar Cell Applications," *IEEE Trans. on Electron Devices*, Vol. 37, No. 2, pp. 469-477, 1990.
9. Wojtczuk, S., E. Gagnon, L. Geoffroy, and T. Parodos, "InGaAs Thermophotovoltaic Cell Performance vs. Bandgap," in *Proc. of 1st NREL Conference on Thermophotovoltaic Generation of Electricity*, pp. 177-187, 1994.
10. Wilt, D. M., N. S. Fatemi, R. W. Hoffman, Jr., P. P. Jenkins, D. Scheiman, R. Lowe, and G. A. Landis, "InGaAs PV Device Development for TPV Power Systems," in *Proc. of 1st NREL Conference on Thermophotovoltaic Generation of Electricity*, pp. 210-220, 1994.

SESSION V:
TPV CELLS II

Polycrystalline-Thin-Film Thermophotovoltaic Cells

Neelkanth G. Dhere

Florida Solar Energy Center
300 State Rd 401, Cape Canaveral, FL 32920-4099

Abstract. Thermophotovoltaic (TPV) cells convert thermal energy to electricity. Modularity, portability, silent operation, absence of moving parts, reduced air pollution, rapid start-up, high power densities, potentially high conversion efficiencies, choice of a wide range of heat sources employing fossil fuels, biomass, and even solar radiation are key advantages of TPV cells in comparison with fuel cells, thermionic and thermoelectric convertors, and heat engines. The potential applications of TPV systems include: remote electricity supplies, transportation, co-generation, electric-grid independent appliances, and space, aerospace, and military power applications. The range of bandgaps for achieving high conversion efficiencies using low temperature (1000-2000 K) black-body or selective radiators is in the 0.5-0.75 eV range. Present high efficiency convertors are based on single crystalline materials such as $In_{1-x}Ga_xAs$, GaSb, and $Ga_{1-x}In_xSb$. Several polycrystalline thin films such as $Hg_{1-x}Cd_xTe$, $Sn_{1-x}Cd_{2x}Te_2$, and $Pb_{1-x}Cd_xTe$, etc have great potential for economic large-scale applications. A small fraction of the high concentration of charge carriers generated at high fluences effectively saturates the large density of defects in polycrystalline thin films. Photovoltaic conversion efficiencies of polycrystalline thin films and PV solar cells are comparable to single crystalline Si solar cells e.g., 17.1% for $CuIn_{1-x}Ga_xSe_2$ and 15.8% for CdTe. The best recombination-state density N_t is in the range of 10^{-15}-10^{-16} cm^{-3} acceptable for TPV applications. Higher efficiencies may be achieved because of the higher fluences, possibility of bandgap tailoring, and use of selective emitters such as rare earth oxides (erbia, holmia, yttria) and rare earth- yttrium aluminium garnets. As compared to higher bandgap semiconductors such as CdTe, it is easier to dope the lower bandgap semiconductors. TPV cell development can benefit from the more mature PV solar cell and opto-electronic (infrared detectors, lasers and optical communications) technologies. Low bandgaps and larger fluences employed in TPV cells result in very high current densities which make it difficult to collect the current effectively. Techniques for laser and mechanical scribing, integral interconnection, and multi-junction tandem structures which have been fairly well developed for thin-film PV solar cells could be further refined for enhancing the voltages from TPV modules.

Thin-film TPV cells may be deposited on metals or back-surface reflectors. Spectral control elements such as indium-tin oxide or tin oxide may be deposited directly on the TPV convertor. It would be possible to reduce the cost of TPV technologies based on single-crystal materials being developed at present to the range of US$ 2-5 per watt so as to be competitive in small to medium size commercial applications. However, a further cost reduction to the range of US ¢ 35-$ 1 per watt to reach the more competitive large-scale residential, consumer, and hybrid-electric car markets would be possible only with the polycrystalline-thin film TPV cells.

© 1996 American Institute of Physics

INTRODUCTION

Thermophotovoltaic (TPV) cells convert thermal energy to electricity. Modularity, portability, silent operation, absence of moving parts, reduced air pollution, rapid start-up, high power densities, potentially high conversion efficiencies, choice of a wide range of heat sources employing fossil fuels, biomass, and even solar radiation are key advantages of TPV cells in comparison with fuel cells, thermionic and thermoelectric convertors, and heat engines. The potential applications of TPV systems include: remote electricity supplies, transportation, co-generation, electric-grid independent appliances, and space, aerospace, and military power applications. The range of bandgaps for achieving high conversion efficiencies using low temperature (1000-2000 K) black-body or selective radiators is in the 0.5-0.75 eV range. Present high efficiency convertors are based on single crystalline materials such as $In_{1-x}Ga_xAs$, GaSb, $Ga_{1-x}In_xSb$, and $In_{1-x}Ga_xAs_{1-y}Sb_y$. Several polycrystalline thin films such as $Hg_{1-x}Cd_xTe$, $Sn_{1-x}Cd_{2x}Te_2$, and $Pb_{1-x}Cd_xTe$, etc., have great potential for economic large scale applications in cheaper, civilian, terrestrial applications. A wide variety of deposition techniques such as sputtering, close-space sublimation, vacuum evaporation, electrodeposition, screen printing, chemical bath deposition, etc., have been developed for preparation of polycrystalline thin films and PV solar cells. Photovoltaic conversion efficiencies comparable to single crystalline Si solar cells have been achieved e.g., 17.1% for $CuIn_{1-x}Ga_xSe_2$ and 15.8% for CdTe.

TPV cell development can benefit from the more mature PV solar cell and opto-electronic (infrared detectors, lasers and optical communications) technologies. This paper reviews applicable polycrystalline-thin film PV cell technologies and presents ideas towards the utilization of polycrystalline thin films for TPV modules.

EFFECT OF HIGH FLUENCES

A blackbody emitter operating at a temperature of 2000 K which is approximately one-third that of the sun can provide light flux density to the cell of 90 W cm^{-2}. This flux density is ~1000 times that of the one-sun air-mass one intensity of ~0.1 W cm^{-2}. It is instructive to study the effect of high fluences on the efficiency and limitations of TPV cells. Many concepts developed for high concentration PV cells are directly applicable to the functioning of the TPV cells operating at high fluences [1].

Diffusion Length

The midgap recombination states affect the minority-carrier diffusion length and the intrinsic junction rectification. The diffusion length L is given by

$$L = \sqrt{D\tau}$$

where D is the minority carrier diffusion coefficient and τ is the minority carrier lifetime. The diffusion coefficient D is related to the carrier mobility μ by

$$D = \frac{\mu kT}{q}$$

and the carrier lifetime τ is given by

$$\tau = \frac{1}{\sigma v N_t}$$

where σ is the recombination-state capture cross-section, v is the minority carrier velocity, and N_t is the density of mid-gap recombination states which can be calculated from the expression [1]

$$N_t = \frac{D}{\sigma v L^2}$$

For a direct-bandgap semiconductor with an intragrain mobility μ of 1000 cm^2 V^{-1} s^{-1}, an atomic-size recombination state cross-section of 10^{-14} cm^{-2}, and assuming optical absorption length of 1 μm and the diffusion length of 2 μm yields an allowable recombination-state density of 6.5×10^{15} cm^{-3} [1]. For a crystal atomic density of 4×10^{22} cm^{-3}, this would represent one defect every 6×10^6 atoms.

Intrinsic Junction Rectification

The allowable mid-bandgap density of states N_t to maintain intrinsic junction rectification can be calculated as follows. The junction dark current is given by

$$J = J_{01} e^{(\frac{qV}{kT})} + J_{02} e^{(\frac{qV}{2kT})}$$

where J_{01} is the intrinsic-saturated-reverse-current density and J_{02} the extrinsic-saturated-reverse-current density due to the defect states. The intrinsic current term varies more rapidly with voltage. Hence it will dominate at high voltages. The extrinsic defect density below the value at which the two currents are equal will maintain intrinsic junction rectification [1], i.e.

$$J_{01} e^{(\frac{qV}{kT})} = \frac{J_{sc}}{2}$$

then

$$e^{(\frac{qV}{2kT})} = \sqrt{(\frac{J_{sc}}{2J_{01}})}$$

i.e.

$$J_{02} e^{(\frac{qV}{2kT})} = \frac{J_{sc}}{2} = J_{02} \sqrt{(\frac{J_{sc}}{2J_{01}})}$$

hence

$$J_{02} = \sqrt{(\frac{J_{sc}J_{01}}{2})}$$

We have

$$J_{01} = q\sqrt{\frac{D}{\tau}}\left[\frac{n_i^2}{N_D}\right]$$

also

$$J_{02} = \frac{qw\sigma v n_i N_t}{2}$$

where n_i is the intrinsic carrier density. Solving for the recombination state density N_t, we have [1]

$$N_t \sigma v = \left(\frac{2J_{sc}\sqrt{D}}{qw^2 N_D}\right)^{2/3}$$

Since we have [1]

$$qN_D w^2 = 2\epsilon\phi_B$$

hence we have

$$N_t \sigma v = \left(\frac{J_{sc}\sqrt{D}}{\epsilon\phi_B}\right)^{2/3}$$

where ϵ is the dielectric constant and ϕ_B is the built-in-voltage of the pn junction. Using the values of the dielectric constant ϵ of $12 \times 9 \times 10^{-14}$ F cm^{-1} and the built-in-voltage ϕ_B of 0.45 V, for the above semiconductor, and a short circuit current density of 3 A cm^{-2}, a recombination-state density below ~3 x 10^{16} cm^3 will be required to maintain an intrinsic junction rectification.

Thus for a TPV cell the diffusion length and intrinsic junction rectification criteria set limits to the recombination-state density of 6.5x10^{15} cm^{-3} and 3x10^{16} cm^{-3} respectively. Until recently these values could be met only by the single crystal materials. However, the recent progress in the polycrystalline-thin-film photovoltaic cell technology has made it possible to consider these cells for TPV applications. The best values of the product recombination state density times the depletion region width N_t w values are few times 10^{-11} A cm^{-2}, while the depletion region width w is a few thousand Å [2]. Hence the best recombination-state density N_t is in the acceptable range of 10^{-15}-10^{-16} cm^{-3}. More importantly, the low recombination-state density is in the region of interest for effective collection of photogenerated carriers. A small fraction of the high concentration of charge carriers generated at high fluences would effectively saturate the larger density of defects in polycrystalline thin films.

Efficiency Improvement at Higher Fluences

Prior to turning attention to the polycrystalline-thin-film photovoltaic cell technology, it would be interesting to consider the effect of high fluences on the cell efficiency. As the intensity of radiation (with hv > E_g) incident on the cell is increased, there is a proportionate increase in the short-circuit-current densities. There is also an increase in the open circuit voltage. This increases the photovoltaic conversion efficiency. The correlation between the open circuit voltage V_{oc} and concentration can be derived by considering the expression for the open circuit voltage [1]

$$V_{oc}(C) = \frac{akT}{q} \ln\left(\frac{J_{sc}(C)}{J_{0a}}\right)$$

where a is the diode quality factor. Then [1]

$$V_{oc}(c) = \frac{akT}{q} \left[\ln\left(\frac{J_{sc}(1)}{J_{oa}}\right) + \ln C\right]$$

$$= V_{oc}(1) + \frac{akT}{q} \ln C$$

Interestingly the increase in the open circuit voltage and consequently the device efficiency increases more rapidly in the case of cells with an initially poor diode quality factor. Of course, to begin with, the value of the open circuit voltage is low at low intensities and then increases rapidly and approaches the intrinsic limiting open circuit voltage. Overall it can be seen that the PV conversion efficiencies will be higher at higher fluences available in TPV applications.

BANDGAP DEPENDENCE OF SEMICONDUCTOR DOPING

Wide bandgap semiconductors such as CdTe have a tendency to self-compensate shallow levels of deliberately incorporated extrinsic donor or acceptor impurities through spontaneous generation of native point defects, such as vacancies, antisite defects, host interstitials, or their complexes. The self-compensation tries to minimize the free energy of the sample by bringing the Fermi level E_F to the intrinsic level E_i. Deliberate extrinsic doping moves the Fermi level E_F close to the conduction or the valence band and creates a metastable phase. The metastable phase could be brought to the equilibrium state by minimizing the total free energy of the system by generating a collection of native defects with appropriate ionization levels in the bandgap and letting the free carriers from the Fermi sea into these levels. Self-compensation has been avoided by several methods such as ion implantation by introducing dopants at concentrations far above the equilibrium thermodynamic limits.

The range of bandgaps for achieving high conversion efficiencies of TPV devices using low temperature (1000-2000 K) black-body or selective radiators is in the 0.5-0.75 eV range. The lower values of bandgap make it easier to dope the semiconductors to the appropriate concentrations.

MATERIALS FOR TPV CELLS

Figure 1 shows a chart of the bandgap and lattice constants of II-VI and IV-VI compound semiconductors some of which would be of interest for TPV cell fabrication. Existence of ternary- or pseudoternary solid solutions over the entire ranges of compositions have been denoted by lines joining the binary compounds. The ranges of direct or indirect bandgap ternaries are represented by solid or dashed lines respectively. It can be seen that there are continuous ranges of direct bandgaps for the ternary and pseudoternary compounds $Hg_{1-x}Cd_xTe$, $Hg_{1-x}Zn_xTe$, and $Pb_{1-x}Zn_xS$ covering the region 0.5-0.75 eV range. Other ternary, pseudoternary, and pseudoquaternary compounds which show direct bandgaps in most of or all of the 0.5-0.75 eV range are $Pb_{1-x}Cd_xTe$, $Pb_{1-x}Zn_xTe$, $Sn_{1-x}Cd_{2x}Te_2$, $Pb_{1-x}Cd_xSe$, $Pb_{1-x}Zn_xSe$, and $Pb_{1-x}Cd_xS$. The list does not show other III-V ternaries such as $Cu_{1-x}Fe_xSe_2$ which have bandgaps in the region of interest. The important point to be made is that several material combinations present a range of bandgaps which would allow bandgap engineering for achieving the best performance. In most TPV applications, the main requirement is the output-power density [3]. For the TPV convertors, the optimum bandgap for achieving the maximum output-

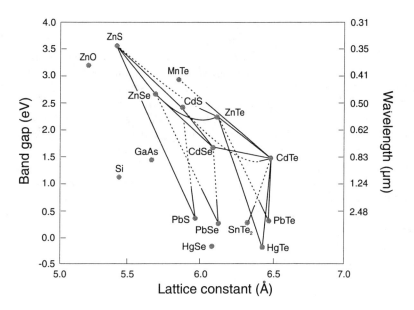

FIGURE 1. Chart of the bandgap and lattice constants of II-VI compound semiconductors.

power density is approximately 0.2 eV lower than that for achieving the maximum conversion efficiency [3]. Thus a compromise may be essential between achieving either the maximum power output and the maximum conversion efficiency depending on each application. It can be seen that there is great variety of materials which could be of interest for such an optimization. The materials of interest cited in the following include only one material from each family viz. $Hg_{1-x}Cd_xTe$, $Sn_{1-x}Cd_{2x}Te_2$, and $Pb_{1-x}Cd_xTe$, even though it may be understood that any of all the other materials mentioned above could be useful.

POLYCRYSTALLINE-THIN-FILM SOLAR CELLS

Photovoltaic modules based on polycrystalline thin films have been targeted as likely candidates for low-cost energy production by the U.S. Department of Energy (DOE) and a segment of the PV industry. $CuIn_{1-x}Ga_xSe_2$ and CdTe have emerged as strong potential candidates for low-cost, large-scale manufacture of PV solar cells. They offer low material cost, potential scalability and automation of the fabrication processes, and efficiencies competitive with the presently predominant crystalline-Si technology.

Of the several processes employed for the preparation of $CuIn_{1-x}Ga_xSe_2$ thin films, vacuum coevaporation, selenization of sputtered metallic precursors, and rapid isothermal processing have resulted in device efficiencies over 10%, the present best efficiency being 17.1% [4-6]. CdTe has long been recognized as an important PV material for the fabrication of low-cost, high efficiency solar cells because of its near-ideal bandgap of 1.5 eV and high optical absorption coefficient. PV conversion efficiencies over 10% have been achieved by most of the over fifteen processes employed for the deposition of CdTe thin films, the best efficiency being 15.8% by close-space sublimation (CSS) on borosilicate glass substrates coated with high quality SnO_2:F window layers [4,7]. $CuIn_{1-x}Ga_xSe_2$ thin-film solar cells which are fabricated in the substrate configuration usually consist of ~1 µm Mo layer on low-cost sodalime glass by sputtering or e-beam evaporation, followed by $CuIn_{1-x}Ga_xSe_2$ absorber, a thin heterojunction partner CdS layer by solution growth, transparent conducting ZnO:Al window by sputtering or metalorganic chemical vapor deposition (MOCVD), and Ni\Al front grid vacuum evaporated through a mechanical mask. CdTe solar cells fabricated in the superstrate structure usually consist of 7059 borosilicate or sodalime glass coated with SnO_2:F transparent conducting window, heterojunction partner CdS layer grown by chemical bath deposition (CBD), CdTe absorber layer, and back contact of doped graphite and silver paste.

Siemens Solar Inc (SSI), International Solar Energy Technology (ISET), and Energy Photovoltaics (EPV) all in USA have been fabricating large-area thin-film $CuIn_{1-x}Ga_xSe_2$ modules with the best power outputs of 43.1 W from 3832 cm^2 (SSI) and 10.4 W from 938 cm^2 (SSI), 4.8 W from 846 cm^2 (ISET), and 5.7 W from 791 cm^2 (EPV) [4,5]. The key strength of emerging $CuIn_{1-x}Ga_xSe_2$ thin-film technology is their stability verified at NREL for over six years. Golden Photon Industries (GPI) and Solar Cells Inc (SCI) in the US, Matsushita Battery in Japan, and BP Solar in UK are actively developing CdTe module fabrication with sizes ranging from 1-8 ft^2. Absorbers are deposited by spray (GPI), elemental vapor deposition (SCI), screen printing (Matsushita), and electrodeposition (BP Solar) [4,5]. The reported large-area module power outputs are 27.5 W from 3528 cm^2 (GPI), 60.3 W from 7200 cm^2 (SCI), 10.0 W from 1200 cm^2 (Matsushita), and 38.2 W from 4540 cm^2 (BP Solar). Both GPI and SCI have announced plans to construct large multimegawatt plants.

COMPARISON WITH SINGLE-CRYSTAL CELLS

It is useful to compare the basic parameters of the best polycrystalline-thin-film and single-crystal PV solar cells with the ideal cell which converts all the photons with energy above the absorber bandgap to electricity. Fig. 2 shows the comparison of the best active-area current densities of single crystal Si and GaAs and polycrystalline-thin-film CdTe, $CuIn_{1-x}Ga_xSe_2$, and $CuInSe_2$ solar cells with the ideal current densities at various values of the bandgap [8]. Percentages of the

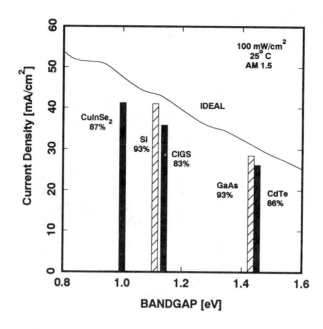

FIGURE 2. Comparison of the best active-area current densities of single-crystal and polycrystalline-thin-film solar cells with the ideal-cell current densities (from ref. 8).

ideal values achieved for each case have also been shown. The current densities are lower for the polycrystalline-thin-film cells because of their less efficient red response resulting from smaller diffusion lengths, lower blue collection limited by absorption in the CdS heterojunction partner layer, and less efficient antireflection coatings.

Comparison of the open-circuit voltages for the single-crystal and polycrystalline-thin-film solar cells with the calculated average open-circuit voltages of ideal cells shows that the open circuit voltages of polycrystalline-thin-film cells are ~15% lower than those of single-crystal cells. This is primarily because the forward current in polycrystalline cells is at least an order of magnitude larger. Diode quality factors under illumination for the best CdTe and $CuIn_{1-x}Ga_xSe_2$ cells are ~2 and 1.3 respectively [8]. The larger diode quality factors result in lower fill factors for polycrystalline cells. The conversion efficiencies of polycrystalline cells are approximately two-thirds of those of single crystal cells (Fig. 3) [8]. The comparison shows that even though the values of the various parameters are lower for the polycrystalline cells, the overall values are respectable in view of the fabrication economics. The difference between the efficiencies of polycrystalline and single-crystal cells has been narrowing steadily over the last few years and the trend is expected to continue.

Several polycrystalline thin films such as $Hg_{1-x}Cd_xTe$, $Sn_{1-x}Cd_{2x}Te_2$, and

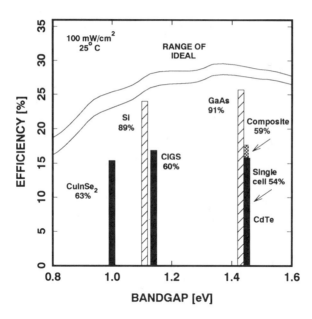

FIGURE 3. Comparison of the best active-area current densities of single-crystal and polycrystalline-thin-film solar cells with the ideal-cell current densities (from ref. 8).

$Pb_{1-x}Cd_xTe$, etc., have great potential for economic large scale applications in cheaper, civilian, terrestrial TPV applications. Greatly enhanced carrier mobilities can be achieved because the high dielectric permittivities of group II-VI and IV-VI binary and pseudo-ternary compound semiconductors heavily screen the field of ions and thus reduce the perturbation of potential considerably.

The major losses of the present polycrystalline-thin-film CdTe solar cells are absorption in CdS heterojunction partner layer accounting for ~2.5 mA cm^{-2} in the current density at maximum power and junction quality factor and series resistance losses which account for 180 mV and 40 mV respectively in the voltage at maximum power [9]. A good junction formation is essential for achieving high efficiencies. High efficiency and excellent diode factor were observed in CdS\CdTe cells grown by atomic layer epitaxy (ALE), even though a mixed ALE $CdS_{1-x}Te_x$ layer limited the low wavelength response [10]. In the present CdTe solar cells, compromise is made between improvement of the junction quality factor and limiting optical absorption near the surface. Thus the interface layer thickness is chosen to be large enough to result in an acceptable value of the junction quality factor and thin enough to limit the optical absorption in the interface layer. Part of these constraints as well as the losses in the CdS heterojunction partner layer would be eliminated in a TPV cell where the absorption in the heterojunction partner and the interface layers would be negligible for the IR radiation and at the same time the junction quality could be improved considerably.

Ohmic contacts to p-type absorber can be achieved either by contacting with a metal having a value of the work function higher than that of the semiconductor or by highly doping the semiconductor. In the case of CdTe it is neither possible to find a suitable metal having a larger work function nor is it possible to dope the semiconductor sufficiently heavily to achieve a thin depletion layer which could be easily tunneled by charge carriers. Stable and fairly efficient back contacts to CdTe thin-film solar cells have been achieved with HgTe and Cu doped graphite. It is easier to heavily dope the lower bandgap p-type $Hg_{1-x}Cd_xTe$, $Sn_{1-x}Cd_{2x}Te_2$, and $Pb_{1-x}Cd_xTe$, etc absorbers. Even for the best CdTe solar cells, the series resistance is ~0.5 Ω cm^2 while the series resistance of best $CuIn_{1-x}Ga_xSe_2$ cells is as low as 0.25 Ω cm^2. It would be easier to obtain lower series resistance values by heavily doping the lower bandgap p-type $Hg_{1-x}Cd_xTe$, $Sn_{1-x}Cd_{2x}Te_2$, and $Pb_{1-x}Cd_xTe$ absorbers. Of course $Hg_{1-x}Cd_xTe$ has disadvantage of more stringent health and safety procedures. Hence it may be preferable to employ $Sn_{1-x}Cd_{2x}Te_2$, or $Pb_{1-x}Cd_xTe$, etc.

Higher efficiencies may be achieved because of the higher fluences, possibility of bandgap tailoring, and use of selective emitters such as rare earth oxides (erbia, holmia, yttria) and rare earth- yttrium aluminium garnets. TPV cell development can benefit from the more mature PV solar cell and opto-electronic (infrared detectors, lasers and optical communications) technologies. Low bandgaps and larger fluences employed in TPV cells result in very high current densities which would require specially developed front contact grids. Techniques for laser and mechanical scribing, integral interconnection, and multi-junction tandem structures which have been fairly well developed for thin-film PV solar cells could be further refined for enhancing the voltages from TPV modules. Thin-film TPV cells may be deposited on metals or back-surface reflectors. Spectral control elements such as indium-tin oxide, tin oxide, or cadmium stannate may be deposited directly on the TPV convertor.

COST OF TPV CELLS

The most costly items in a TPV system are the emitter and the cells. Majority of the present TPV systems are utilized in small-scale aerospace and military applications. The total power output of the systems deployed at present is small. Hence the cost of TPV systems and the cells is not a major issue. It would be necessary to reduce the cost to the range of US$ 2-5 per watt so as to be competitive in small to medium size commercial applications [11]. The technologies being developed at present may be able to reach this goal. However, the cost will have to be reduced to the range of US ¢ 35-$ 1 per watt to reach the large-scale applications in the residential and consumer markets as well as the much larger application in hybrid-electric car market [11]. These markets are very competitive and the consumers are cost conscious. For example, it has been estimated that the cost of a 450 W TPV generator for a self-powered gas-fired residential warm air furnace would have to be limited to US$ 75 equivalent to the cell cost of ~US ¢ 17 per watt [12]. It may not be possible to achieve these goals with the TPV cells based on single crystal materials. Such a cost reduction would be possible with the polycrystalline-thin film TPV cells.

THIN-FILM PHOTOVOLTAICS AND MATERIALS LABORATORY

A well-equipped thin-film photovoltaics and materials laboratory has been established at the Florida Solar Energy Center. Both CdTe and $CuIn_{1-x}Ga_xSe_2$ polycrystalline thin-film solar cells are being prepared routinely using processes amenable to large-scale manufacture. It would be possible to undertake research and development work on TPV cells based on polycrystalline thin films of $Hg_{1-x}Cd_xTe$, $Sn_{1-x}Cd_{2x}Te_2$, and $Pb_{1-x}Cd_xTe$, etc. for economic large scale applications in cheaper, civilian, terrestrial appliances.

CONCLUSIONS

$Hg_{1-x}Cd_xTe$, $Sn_{1-x}Cd_{2x}Te_2$, and $Pb_{1-x}Cd_xTe$, etc polycrystalline-thin films have great potential for economic applications in cheaper, civilian, terrestrial TPV applications. The PV conversion efficiencies of the present polycrystalline-cell are approximately two-thirds of the single-crystal cell efficiencies. The overall values are respectable in view of the fabrication economics and further improvements are expected in the future. The best values of the product recombination-state density N_t times the depletion region width $N_t w$ values are few times 10^{-11} A cm^{-2}, while the depletion region width w is a few thousand Å. Hence the best recombination-state density N_t is in the acceptable range of 10^{-15}-10^{-16} cm^{-3}. More importantly, the low recombination-state density is in the region of interest for effective collection of photogenerated carriers.

The range of bandgaps between 0.5-0.75 eV is useful in TPV cells. For the TPV convertors, the optimum bandgap for achieving the maximum output-power density which often is the main requirement, is approximately 0.2 eV lower than that for achieving the maximum conversion efficiency. Several group II-VI compounds could be of interest for such an optimization. Direct bandgaps over the entire range are available from the group II-VI compounds $Hg_{1-x}Cd_xTe$, $Hg_{1-x}Zn_xTe$, and $Pb_{1-x}Zn_xS$ while other compounds such as $Pb_{1-x}Cd_xTe$, $Pb_{1-x}Zn_xTe$, $Sn_{1-x}Cd_{2x}Te_2$, $Pb_{1-x}Cd_xSe$, $Pb_{1-x}Zn_xSe$, and $Pb_{1-x}Cd_xS$ cover direct bandgaps in most or all of the range of interest.

As compared to higher bandgap semiconductors such as CdTe, it is easier to dope the semiconductors having bandgaps 0.5-0.75 eV used in TPV cells. The higher doping makes the application of ohmic contacts easier reducing the series resistance. The other advantages of polycrystalline-thin film TPV cells based on $Hg_{1-x}Cd_xTe$, $Sn_{1-x}Cd_{2x}Te_2$, and $Pb_{1-x}Cd_xTe$, etc. are enhanced carrier mobilities, negligible IR losses, and possibilities of bandgap tailoring, direct deposition of spectral control elements, deposition of back-surface reflectors, laser and mechanical scribing, integral interconnection, and multi-junction tandem structures.

It would be possible to reduce the cost of TPV technologies based on single-crystal materials being developed at present to the range of US$ 2-5 per watt so as to be competitive in small to medium size commercial applications. However, a further cost reduction to the range of US ¢ 35-$ 1 per watt to reach the more competitive large-scale residential, consumer, and hybrid-electric car markets would be possible only with the polycrystalline-thin film TPV cells.

ACKNOWLEDGEMENTS

This work was supported by National Renewable Energy Laboratory Contract # XG-2-11036-5. The author is thankful to Dr. Tim Coutts of NREL for suggesting the topic, for useful discussions, and for providing the figure 1 and to Dr. James R. Sites for useful discussions, and for providing the figures 2 and 3.

REFERENCES

1. Fraas, L. M., in Current Topics in Photovoltaics, T. J. Coutts and J. D. Meakin (eds.), Academic Press, 1985, pp. 168-221.
2. Mauk, P. H., Tavakolian, H., and Sites, J. R., "Interpretation of Thin-Film Polycrystalline Solar Cell capacitance", IEEE Trans. Electron Devices, **37**, 422-427 (1990).
3. Iles, P. A., Chu, C., and Linder, E. "The influence of bandgap on TPV convertor efficiency", AIP Proc. 2nd NREL Conference on Thermophotovoltaic Generation of Electricity, Colorado Springs, CO, July 17-19, 1995.
4. Ullal, H., Zweibel, K., and von Roedern, B. G., "Thin-Film CdTe and $CuInSe_2$ Photovoltaic Technologies" Proc. ISES Solar World Congress, Budapest, Hungary, Aug. 23-27, 1993, pp. 187-193.
5. Dhere, N. G., Kuttath, S., Lynn, K. W., Birkmire, R. W., and Shafarman, W. N., "Polycrystalline $CuIn_{1-x}Ga_xSe_2$ Thin Film PV Solar Cells Prepared by Two-stage Selenization Process Using Se Vapor" Proc. IEEE First World Conf. Photovoltaic Energy Conversion, Waikoloa, Hawaii, 1994, pp. 190-193.
6. Contreras, M. A., Gabor, A. M., Tennant, A. L., Asher, S., Tuttle, J. R., and Noufi, R., "16.4% Total-area Conversion Efficiency Thin-Film Polycrystalline $MgF_2/ZnO/CdS/ Cu(In,Ga)Se_2/Mo$ Solar Cell" *Prog. in Photovoltaics: Res. &*

Appl. **2**, 287-292 (1994).

7. Ferekides, C., Britt, J., Killian, L., "High Efficiency CdTe Solar Cells by Close Spaced Sublimation", Proc. 23rd IEEE Photovoltaic Specialists' Conference, Louisville, KY, 1993, pp. 389-393.

8. Eisgruber, I. L. and Sites, J. R., "Status of Polycrystalline Thin Film Solar Cells", AIP Conf. Proc. 12th NREL Photovoltaic Program Review, Denver, CO, 1993, pp. 407-413.

9. Sites, J. R. and Liu, X., "Six-year Efficiency Gains for CdTe and $CuIn_{1-x}Ga_xSe_2$ Solar Cells: What Has Changed?", Proc. IEEE First World Conf. Photovoltaic Energy Conversion, Waikoloa, Hawaii, 1994, pp. 119-122.

10. L. Skarp, E. Antilla, A. Rautiainen, and T. Suntola, "ALE-CdS/CdTe-PV-Cells", Int. J. Solar Energy, **12**, 137-142 (1992).

11. Ostrowski, L. J., Pernisz, U. C., and Fraas, L. M., "Thermophotovoltaic Energy Conversion: Technology and Market Potential", AIP Proc. 2nd NREL Conference on Thermophotovoltaic Generation of Electricity, Colorado Springs, CO, July 17-19, 1995.

12. Nelson, R. E., "Grid-independent residential power systems", AIP Proc. 2nd NREL Conference on Thermophotovoltaic Generation of Electricity, Colorado Springs, CO, July 17-19, 1995.

Characteristics of GaSb and GaInSb Layers Grown by Metalorganic Vapor Phase Epitaxy

H. Ehsani, I. Bhat, C. Hitchcock, J. Borrego and R. Gutmann

Center for Integrated Electronics and Electronics Manufacturing
Department of Electrical, Computer and Systems Engineering
Rensselaer Polytechnic Institute, Troy, NY 12180-3590

Abstract. GaInSb and GaSb layers have been grown on GaSb and GaAs substrates using metalorganic vapor phase epitaxy (MOVPE) with trimethylgallium, trimethylindium and trimethylantimony as the sources. As grown layers were p-type with the carrier concentration in the mid-10^{16} cm^{-3} range. N-type layers were grown using diethyltellurium as the Te source. Incorporation of Te in high concentration showed compensation and secondary ion mass spectrometry (SIMS) result showed that only 2.5% of Te are active when 2×10^{19} cm^{-3} of Te was incorporated. The carrier concentration measured in n-type samples increased as the temperature is lowered. This is explained by the presence of second band close to the conduction band minima. Silane, which is a common n-type dopant in GaAs and other III-V systems, is shown to behave like p-type in GaInSb-. P-n junction structures have been grown on GaSb substrates to fabricate TPV cells.

INTRODUCTION

There is renewed interest in the use of photovoltaic cells for converting the energy of radiant heat sources, other than the sun, into useful electrical energy. In this application, the so-called thermophotovoltaic (TPV) cells use the same principle of operation as photovoltaic cells (PV), except that the radiant heat source is other than the sun. Since these heat sources are at a much lower temperature than the sun (generally in the range 900°C to 1500°C), different materials with much lower band gaps are required (1). Ternary compounds such as InGaAs or GaInSb are ideally suited for this purpose since their band gaps can be varied to match the radiant spectrum. However, epitaxial growth of these is difficult since suitable lattice matched substrates are not available. Hence, various dislocation reduction techniques should be used to grow high quality layers. Epitaxial layers of InGaAs for TPV applications have been reported earlier and pilot production of TPV cells based on InGaAs materials are underway (1, 2). However, P- and As-based materials are expensive in the long run because one has to use large amount of dangerous gases such as arsine or phosphine during growth. The substrates used are also generally very expensive. Therefore, an alternative to InGaAs materials should be explored in order to reduce the cost of TPV cells.

In this work, we concentrate on GaInSb materials with energy gaps in the range from 0.5 eV to 0.75 eV grown on GaSb substrates. GaInSb is preferable to GaInAs because one need not have to use arsine or phosphine gases to grow the epitaxial layers. Moreover, the substrates, namely GaSb, should be less expensive in the long run compared to InP because bulk growth of GaSb is carried out at a lower temperature and does not require high pressure furnaces. There is strong evidence that the interface state density in a properly oxidized GaSb can be less than 10^{11} per cm^2 so that GaInSb based TPV cell will have potentially higher efficiency. GaSb and $Ga_xIn_{1-x}Sb$ materials have received considerable interests for applications as infrared detectors, in optoelectronics and high speed devices (3-8). Unlike other commonly known III–V semiconductors, such as GaAs and InP, the growth of Sb-based materials is difficult because vapor pressure of Sb over GaSb is very low. Thus, only a narrow range of V/III ratios and growth temperatures can be used to grow a good quality layer. Beyond this narrow range, either Sb droplets or Ga droplets are seen on the growth surface. Another difficulty is that the surface oxide can not be thermally removed by heating the wafer to high temperature prior to growth because the substrate will decompose.

There is limited number of work carried out on the growth of GaSb and GaInSb materials using the MOVPE technique (9-11). Also, few doping studies on GaInSb layers are available. This paper addresses the growth and extrinsic doping of high quality $Ga_xIn_{1-x}Sb$ layers grown on GaSb substrate. The effect of growth temperature, V/III ratio, and etching procedures on the quality of the layers was studied. P–type and n–type doping of $Ga_xIn_{1-x}Sb$ layers for TPV fabrication was also investigated.

EXPERIMENTAL PROCEDURE

The growth of $Ga_xIn_{1-x}Sb$ was carried out in a low pressure (100 Torr) horizontal MOVPE system with rf heating. Precursors for the growth of $Ga_xIn_{1-x}Sb$ were trimethylgallium (TMGa), trimethylindium (TMIn), and trimethylantimony (TMSb), each held in a temperature controlled bath at -10°C, at 20°C and at 5°C, respectively. Both GaAs and GaSb wafers were used as substrates. Since GaSb wafers are not available in semi-insulating form, GaAs wafers were used for making Hall Effect measurements. After cleaning the wafers in organic solvents, GaAs was etched in Caro's etch (solution containing $H_2SO_4:H_2O_2:H_2O$, 5:1:1 by volume) for two minutes and GaSb was etched in 1% Br-methanol solution for 30 seconds. Other etches such as Caro's etch and H_2O_2+acetic acid etch were also used for GaSb.

Hall effect measurements were used for electrical characterization of the layers grown on GaAs substrates. Crystalline quality and the lattice parameter of the layers were determined by the double crystal x–ray diffraction. Fourier transform infrared (FTIR) spectrophotometer was used to measure the transmission spectra of layers grown on GaAs substrates from which the thickness and the approximate bandgap could be determined. The thickness of the layers grown on GaSb wafers was determined by measuring the change in the weight before and after the growth, and was generally found to be the same as that grown on GaAs. The FTIR spectra on GaSb wafers do not give thickness fringes since the refractive indices of GaInSb and GaSb are very close.

RESULTS AND DISCUSSIONS

GaSb Layers

GaSb was first grown on GaSb and GaAs substrates in order to optimize the growth conditions for binary layer growth. The growth temperature was varied from 560°C to 630°C and the ratio of TMSb/TMGa partial pressure was varied from 1.1 to 4. The best layers were obtained at a substrate temperature of 600°C, with the TMSb/TMGa ratio between 1 and 2.5. The typical mole fraction of TMSb and TMGa were 2.2×10^{-4} and 1.3×10^{-4} respectively. The growth rate at lower temperature was slow and the morphology was poor. Figure 1 shows the growth rate of GaSb on GaSb as a function of the growth temperature. At higher temperature (=630C), the surface morphology was also poor, probably caused by the degradation of GaSb surface during heat up. The surface morphology of the GaSb layers on GaAs substrates were rough, except for the layers grown at 560°C and below.

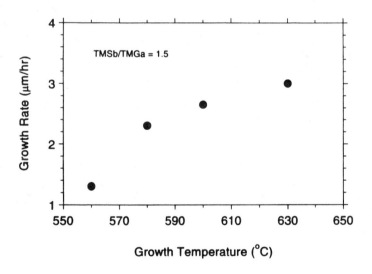

FIGURE 1. Growth rate of GaSb films versus growth temperature.

GaInSb Layers

The growth of $Ga_xIn_{1-x}Sb$ was initiated by first growing a thin (about 1000Å) layer of GaSb. In some cases, step grading of the composition was carried out to accommodate the lattice mismatch between GaInSb top layer and the substrate. All the growth runs for GaInSb were carried out at 600°C and the ratio of TMSb/(TMGa+TMIn) partial pressure were kept between 1.2 and 2. Several etchants such as dilute Br_2–methanol, Caro's etch, and $H_2O_2+CH_3COOH$ were tried to etch GaSb wafers, but Br-methanol solution was found to be the best. The surface morphology of all the layers were mirror-like, but layers grown on Br-methanol etched wafers had better crystal quality as measured by the double crystal x–ray diffraction. Figure 2 represents the surface morphology of a 11 μm thick $Ga_{0.8}In_{0.2}Sb$ layer grown on GaSb substrate. The surface morphology is very smooth, except for the presence of a few hillocks. Since these hillocks have the same size and crystalline orientation, we believe that they originated at the interface. Better cleaning procedure or dust free surrounding may eliminate these hillocks.

The double crystal x-ray diffraction spectra of a 2 μm thick $Ga_{0.8}In_{0.2}Sb$ grown on GaSb substrate is shown in Fig. 3. The full width half maximum (FWHM) is about 450 arc-secs and is typical of layers grown on lattice mismatched substrates (11). Hall measurements of layers grown on GaAs substrates showed that all the undoped materials (both GaSb and GaInSb) are p-type with the carrier concentrations in the mid 10^{16} cm^{-3} range at room temperature. The residual p-type doping is probably caused by the native defects and not by the carbon contamination since undoped bulk grown GaSb is also found to be p-type with a hole concentration of the order of 10^{16} cm^{-3}. In bulk GaSb, the p-type conduction is attributed to the native defect, Ga on Sb site (8). The room temperature mobilities were in the range 500-600 cm^2/Vs for most of the undoped layers. Similar results were observed by others when GaSb layers were grown on GaAs substrates (9-11). Since the crystalline quality and the surface morphology of the layers grown on GaAs are poor, the value of mobility measured in these layers do not necessarily represents those of the layers grown on GaSb substrates.

Extrinsic Doping

Since as grown materials are p-type, extrinsic doping is required to get n-type layers. Moreover, for TPV cell fabrication, one should be able to control the p-type concentration in the range 10^{16} to mid-10^{18} cm^{-3}. Therefore, extrinsic p-type doping is also investigated. N-type doping was achieved using tellurium and p-type doping was obtained by silicon.

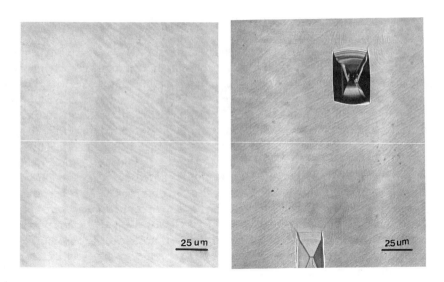

FIGURE 2. Surface morphology of a 11μm thick $Ga_{0.8}In_{0.2}Sb$ TPV structure grown on GaSb substrate. Scattered hillocks were observed especially near the edges.

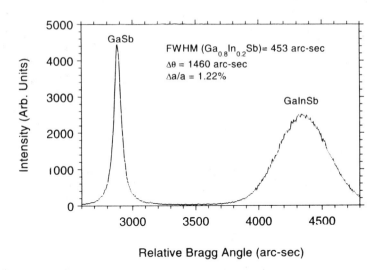

FIGURE 3. Double crystal X-ray diffraction spectra of a 2μm thick $Ga_{0.8}In_{0.2}Sb$ layer grown on GaSb substrate.

Doping with Tellurium

We have investigated Te from diethyltellurium (DETe) as the n-type dopant in $Ga_xIn_{1-x}Sb$. Figure 4 shows the carrier concentration at room temperature and at 77K as a function of the DETe flow rate. Two points should be noted here. First, the carrier concentration measured decreases as the DETe flow rate is increased. Second, the carrier concentration at 77K is higher than that at room temperature. Decrease of carrier concentration with the increase in DETe mole-fraction (MF) indicates compensation of Te-donors when incorporated in high concentration. This fact is supported by the fact that the Hall mobility decreased as the DETe mole fraction is increased (see Fig. 5). SIMS measurement was carried out on a layer doped with 2 1×10^{-7} MF of DETe and the spectrum is shown in Fig. 6. The Te concentration in this layer is about 2×10^{19} cm^{-3}, whereas the carrier concentration measured is about 5.3×10^{17} cm^{-3}, indicating that only 2.5% of Te is electrically active. The layer should be grown with lower DETe mole fraction than was used here. The system is being modified to deliver DETe in an effuser mode so that the lower MF of DETe can be delivered to the reactor.

The increase in the carrier concentration of n-type GaInSb layer with the decrease in temperature cannot be explained by the simple impurity ionization model. This type of behavior is caused by the presence of a second conduction band minimum (L-minimum) very close to the minima at the center of the Brillouin zone (Γ). Carriers can be excited to the L band as the temperature is increased. Figure 7 shows the measured Hall coefficient as a function of temperature for a Te-doped GaInSb sample. Also shown is a first order theoretical curves of the Hall coefficient assuming two conduction band minima separated by an energy $\Delta E_{L\Gamma}$ of 80 meV. The curves have been calculated for mobility ratio of $\mu_\Gamma/\mu_L=10$ and density of states $N_L/N_\Gamma=11.4$ which correspond to values of GaSb (13). Notice that the data can be better fitted when the energy separation between the L and the Γ band minima is 130 eV. In InSb the calculated separation between the L and the Γ band minima is 0.78 eV and in GaSb is 0.08 eV. The separation of 0.13 eV corresponds to an In concentration of approximately 7%. X-ray measurements show the In concentration to be 15%. It should be mentioned that the theoretical curves drawn depend upon the mobility ratio and the density of states assumed.

Doping with silicon

Si is widely used for n-type doping in GaAs and GaInAs. It is an amphoteric dopant and its behavior depends on the growth methods and the growth conditions. In InSb, Si is known to behave as p-type impurity (14). Behavior of silicon in GaInSb has not been studied before. Since the control of p-type doping level is necessary to fabricate TPV cells, we carried out extrinsic p-type doping of GaInSb using silicon as the dopant. Silicon was supplied by 600 ppm silane in hydrogen. The GaInSb layers were grown at 600°C on both GaSb and GaAs substrates. Hall

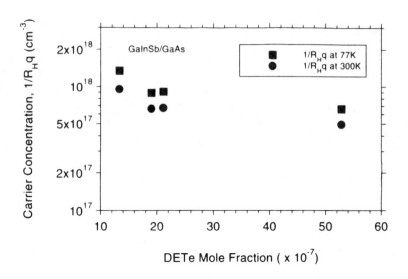

FIGURE 4. Measured carrier concentration versus DETe mole-fraction for GaInSb layers grown on GaAs substrates. Note that the concentration falls with the DETe flow and that the value at 77K is higher than that at 300K (see text for details).

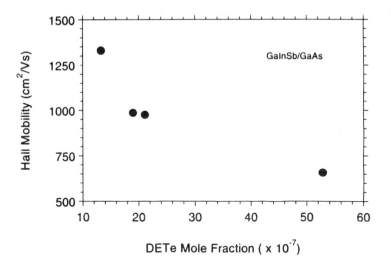

FIGURE 5. Hall mobility of n type GaInSb/GaAs layers at 300K versus DETe mole-fraction. The electron mobility and the carrier concentration reduced with increasing DETe flow indicating increased defects and high compensation.

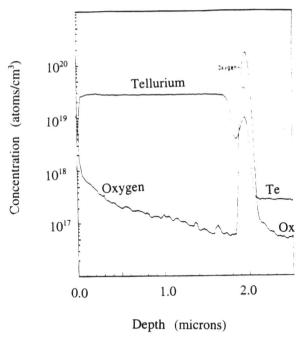

FIGURE 6. SIMS profile of a Te-doped GaInSb layer (MF of DETe=21 x 10^{-7}). Only 2.5% of the 3×10^{19} cm^{-3} Te atoms are electrically active.

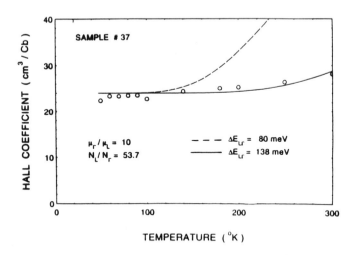

FIGURE 7. Simulated Hall coefficient of an n type GaInSb layer as a function of temperature assuming two-band conduction. Also shown are the experimental points.

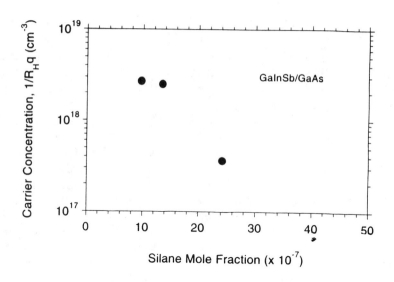

FIGURE 8. Hole concentration of GaInSb/GaAs layers versus silane mole fraction. Silicon behaves like an acceptor in GaInSb.

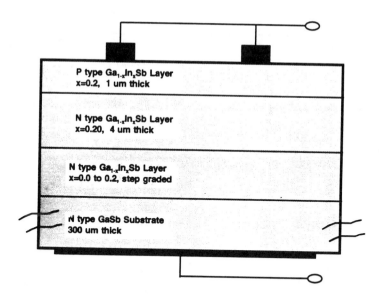

FIGURE 9. Scematic of a GaInSb TPV structure grown on GaSb substrate.

measurements showed that Si is p-type in $Ga_xIn_{1-x}Sb$. Figure 8 shows the measured carrier concentration as the silane mole fraction is increased. P-type carrier concentration as high as 2×10^{18} cm^{-3} was achieved with silane doping. Similar to that of Te, the doping concentration decreases as the silane flow increased, indicating compensation.

Device Characteristics

A TPV device structure was grown on GaSb substrate using DETe for n-type doping and silane for p-type. The structure is shown schematically in Fig. 9. Circular, small area p-on-n devices with radii of 700 microns were fabricated by mesa etching. The I-V curve of a typical $Ga_{0.8}In_{0.2}Sb$ p-on-n diode is shown in Fig. 10. The device exhibited a diode ideality factor of approximately 1.2 and a saturation current density of about 2×10^{-5} A/cm^2. The shunt resistance for this device was too low to be measured by our testing apparatus; however, we could place a lower bound on the shunt resistance of 6000 ohms, corresponding to an areal shunt resistance of 100 ohms-cm^2, if the leakage is assumed to occur uniformly across the junction, or a linear shunt resistance of 2700 ohms-cm if the leakage is assumed to occur only at the exposed junction edges. Fabrication and testing of large area TPV cells are underway.

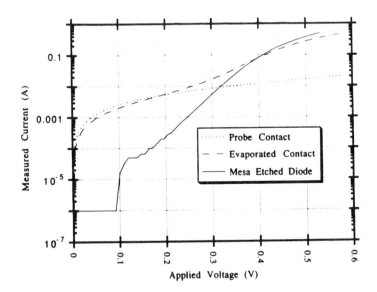

FIGURE 10. I-V characteristics of a small area p-n junction diode fabricated on the structure shown in figure 9.

CONCLUSIONS

GaInSb has been investigated as possible material system for TPV application. This material was grown on GaSb and GaAs substrates using OMVPE. As grown layers are p-type with a carrier concentration in the mid-10^{16}cm^{-3} range. Extrinsically doped p-type and n-type layers have been grown using Te and Si as the dopants, respectively. Unlike other III-V system such as GaAs, silicon was found to behave as p-type in GaInSb. The measured carrier concentration in n-type layers increased as the temperature of measurement is decreased. This phenomena was attributed to the presence of second band very close to the conduction band minima in GaInSb. A first order simulation assuming two-band conduction adequately explains the behavior of n-type GaInSb layers. Device structures have been grown on GaSb substrates and the results show that the materials are of sufficiently high quality, suitable for TPV devices.

ACKNOWLEDGMENT

We would like to thank J. Barthel for technical assistance and Dr. G. Charache for making the necessary arrangements to send samples for SIMS measurements.

REFERENCES

1. Wojtczuk, S., Gagnon, E., Geoffroy, L., and Parodos, T., "InGaAs Thermophotovoltaic Cell Performance vs. Bandgap", *Proceedings of First NREL Conf. on TPV Generation of Electricity, AIP Conf. Proc.*, Vol. 321, 1984, pp. 177-187.
2. Sharps, R. et al., *Proc. of First NREL Conf. on TPV Generation of Electricity, AIP Conf. Proc.*, Vol. 321, 1994, p. 1994.
3. Basu, S., Basu, N., Bartman, P., *Mater. Sci. Engg*, B9, 47 (1991).
4. Itani, Y., Asahi, H., Kaneko, T., Okuno, Y., and Gonda, S., *J. Appl. Phys.* 73, 1161-1167 (1993).
5. Iyer, S., Hegde, A. Abdul-Fadi, Bajaj, K., and Mitchel, W., *Phys. Rev. B* 47, 1329-1339 (1993).
6. Zaza, L.J.G., Montojo, M.T., Castano, J.L., and Piqueras, J., *J. Electrochem. Soc.* 136, 1480-1484 (1989).
7. Karouta, F., Marbeuf, A., Joullie, A., and Fan, J.H., *J. Crystal Growth* 79, 445-450 (1986).
8. Milnes, A.G. and Polyakov, A.Y., *Solid State Electron.* 36, 803-818 (1993).
9. Haywood, S.K., Mason, N.J., and Walker, P.J., *J. Cryst. Growth* 93, 56-61 (1988).
10. Chidley, E.T.R., Haywood, S.K., Mallard, R.E., Mason, N.J., Nicholas, R.J., Walker, P.J., and Warburton, R.J., *J.Cryst. Growth* 93, 70-78 (1988).
11. Bougnot, G., Delannoy, F., Pascal, F., Grosse, P., Giani, A., Kaoukab, J., Bougnot, J., Fourcade, R., Walker, P.J., Mason, N.J., and Lambert, B., *J. Cryst. Growth* 90, 502-508 (1991).
12. Su, Y.K., Juang, F.S., and Wu, T.S., *J.Appl. Phys.* 70, 1421-1424 (1991).
13. Modelung, O., *Physics of III-V Compounds*, J. Wiley & Sons, Inc., New York, 1964, pp. 130-131.
14. Thompson, P.E., Davis, J.L., and Simons, D.S., "Low Temperature p- and n-Type Doping of InSb Grown on GaAs Using Molecular Beam Epitaxy", *Proceedings of the Materials Research Society Symposium*, Vol. 216, 1991, pp. 221-226.

Recombination Lifetime in Ordered and Disordered InGaAs

R.K. Ahrenkiel, S.P. Ahrenkiel, and D.J. Arent

National Renewable Energy Laboratory
1617 Cole Boulevard
Golden, Colorado 80401

Abstract. The ternary semiconductor $In_xGa_{1-x}As$ has been a key component of current thermo photovoltaic energy converters. These studies indicate that the bandgap of the epitaxial films, that are lattice-matched to InP, show varying degress of ordering of the metal sublattice depending upon growth temperature. The bandgap of partially ordered films is lowered by as much as 75 meV. The transport of carriers in ordered films is dominated by domain trapping.

INTRODUCTION

The ternary semiconductor $In_xGa_{1-x}As$ has been a key component of current photovoltaic energy converters. The recombination lifetimes are intimately related to photovoltaic efficiency. Both the short-circuit current and open-circuit voltage are functions of the minority-carrier (low injection) and recombination (high injection) lifetimes. Thus, maximizing the lifetime is a crucial component of device development.

Recent work has described the recombination process in epitaxial thin films that are lattice-matched to InP and have the composition $In_{0.53}Ga_{0.47}As$ (1) using the radio-frequency photoconductive decay (RFPCD) technique. For this composition, lifetimes as large as 12 µs were measured in undoped $In_{0.53}Ga_{0.47}As$ films. The lifetime decreased with increasing doping levels as radiative recombination became dominant. In addition, this work also measured recombination rates in lattice-mismatched, small bandgap compositions. In the latter, recombination at dislocations became dominant.

Atomic ordering in ternary semiconductors has been a subject of great interest in recent years. Ordering in the ternary system InGaP that is lattice-matched to GaAs has been thoroughly studied. In that system, ordering on the cation sublattice lowers the bandgap energies up to 150 meV from those of disordered crystals as

© 1996 American Institute of Physics

predicted by theory (2). The ordering in InGaAs is qualitatively similar to that in InGaP. Transmission electron diffraction (TED) data (3) have shown that ordering in these materials occurs on {111} planes (CuPt-type ordering), and only two of the four possible ordered "variants" are observed. Partially ordered materials typically displays inhomogeneity on the μm or even sub-μm scale. This inhomogeneity is observed as intensity fluctuations in dark-field TEM images, and is associated with the formation of microdomains. The strength of the ordering and the geometry of the ordered microdomains can vary considerably with growth conditions and substrate miscut (4).

One consequence of the ordering phenomena is that carrier transport is affected by the local domains. One impact of the ordered domains is that charge separation of electrons and holes occurs when type II heterointerfaces are formed (5). The consequence of this effect is to produce spatially indirect recombination as observed by photoluminescence decay studies (6,7). One effect of the ordering is to produce extremely long minority-carrier decay times because of the local domain trapping. A related effect in ordered InGaAs is that extremely high electron mobilities have been observed in the 2-dimensional electron flow associated with smaller bandgap domains (8). Here we will describe the effects of ordering in lattice-matched $In_{0.53}Ga_{0.47}As$ on InP on the photoconductive decay using the RFPCD technique.

GROWTH AND CHARACTERIZATION

Undoped double heterostructures were grown on (100) Fe doped InP cut 2° toward the nearest (110) by atmospheric pressure organometallic vapor phase epitaxy using trimethylgallium, trimethylindium, phosphine (PH_3), arsine (AsH_3), diethylzinc for p-type dopant, and H_2 carrier gas in a vertical geometry. Samples were grown at a growth rate of 8 μm/hr and a V/III ratio of 54:1, and consisted of an InP buffer, 1.5 μm of InGaAs, and 0.075 μm InP cap. The substrate temperature during the InGaAs growth was varied from 550°C to 650°C in order to induce various amounts of ordering in the samples. Composition was determined by double crystal rocking curve x-ray diffraction where the relative lattice mismatch between $In_{0.53}Ga_{0.47}As$ and InP substrate was found to be less than 10^{-3} (250 arcsec). Owing to previous reports on the unreliability of using photoluminescence peak energies to determine the band gap energy of $In_{0.53}Ga_{0.47}As$ due to the influence of excitation intensity on the peak position (9), the band gap energy was deduced from infrared absorption measurements. The band gap energies were determined by fitting the absorption behavior to the form $\alpha(h\nu) = A(h\nu - E_g)^{1/2}$ cm^{-1}, as theoretically predicted for a direct gap semiconductor (10). The band gap energy was corrected for variation in composition, including strain (11), such that all reported values are for lattice matched $In_{0.53}Ga_{0.47}As$.

TED experiments were performed on $In_{0.53}Ga_{0.47}As$ samples on InP using a Philips CM-30 transmission electron microscope (TEM). The samples were prepared in (110) cross section by gluing to silicon blocks and dimpling from both sides. The samples were then cooled with $l-N_2$ and milled for 1-2 hr with 4 keV Ar ions at 0.5 mA.

Figure 1 shows the observed variation in 300 K band gap energy as a function of growth temperature. The observed lowering of E_g is approximately 10% of the normal band gap energy (0.751 eV), but only 20% of the expected shift for a perfectly ordered material (12). Thus, we can expect that the lower band gap material will be inhomogeneous; e.g., it will possess a distribution of domains with varying geometry or amounts of order, similar to the behavior and structure observed in partially ordered $GaInP_2$ (13).

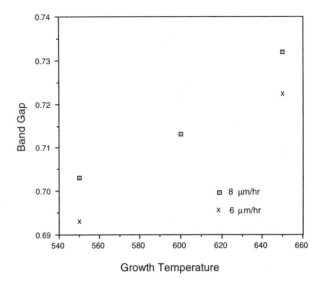

Figure 1. Bandgap of $In_{0.47}Ga_{0.53}As$ versus growth temperature. The two data points at 550° C and 650° C correspond to different growth rates. The larger bandgap in each case corresponds to 8 µm/hr. and the smaller bandgap corresponds to 6 µm/hr.

The diffraction from these samples showed variations in ordering strength and perfection. Figure 2a shows a [110] TED pattern from an $In_{0.53}Ga_{0.47}As$ with a low band gap of only E_g=0.644 eV as determined by absorption experiments. The presence of two strongly ordered variants is revealed by the two sets of 1/2{111} and equivalent superstructure spots that are not observed in diffraction patterns from random alloys. The diffraction was acquired from a selected area roughly 0.3 µm in

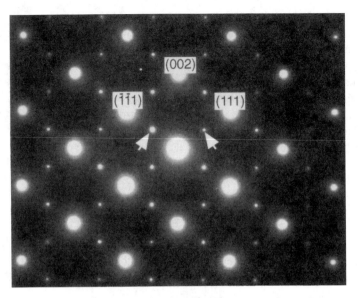

Figure 2a. [110] diffraction pattern from CuPt ordered InGaAs. The sample shows strong ordering on two sets of {111} planes.

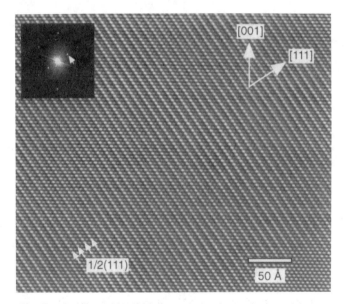

Figure 2b. Fourier-filtered [110] high-resolution lattice image of CuPt ordered InGaAs showing strong (111) ordering within a large domain. The inset is the fast-Fourier transform of the original image.

diameter. High-resolution lattice images were obtained to observe the distribution of ordering on an atomic scale. The domains were quite large and uniform, and we did not observe significant overlap of the two ordered variants. Fig. 2b is a [110] lattice image taken from a large ordered domain. The objective aperture allowed contributions from low-index reflections with d d{200}. The image was Fourier-filtered to reveal the underlying lattice. The cubic lattice constant of the alloy at this composition is 5.88 Å. In addition to the periodicity of the zincblende lattice, the ordering produces a 1/2(111) periodicity that appears as alternating light and dark (111) planes. Some random material is seen at the edge of the image.

A common feature observed in TED experiments is diffuse streaking along the growth direction, which has been attributed to imperfections of the ordered domains (14,15). Ordering appears to become less uniform within the domains of material showing strong streaking, and the domains in these samples display a complicated arrangement of tilted boundaries between ordered/ordered and ordered/random material (16). We find some indications of streaking in nearly all of our InGaAs samples, although the streaking is nearly minimal for the sample in Fig. 2. Future work may reveal connections between the domain microstructure and the RFPCD signal.

TRANSPORT THEORY

The RFPCD technique measures the photo-induced or excess conductivity following excitation from a pulsed light source. The excess conductivity of a semiconductor after pulsed excitation is given by:

$$\Delta\sigma = q(\mu_n \Delta n(t) + \mu_p \Delta p(t)) \qquad 1)$$

Here μ_n (μ_p) are the electron (hole) mobilities and the excess electron and hole carrier concentrations are given by n(t) and p(t), respectively. Because of transient charge neutrality, the latter are usually equal and the expression can be written in terms of the excess carrier concentration $\rho(t)$.

$$\Delta\sigma = q(\mu_n + \mu_p)\rho(t) \qquad 2)$$

In the simplest case, a single recombination mechanism controls the excess conductivity and $\rho(t)$ can be written in terms of a minority-carrier lifetime τ.

$$\Delta\sigma = q(\mu_n + \mu_p)\rho_0 \exp(-t/\tau) \qquad 3)$$

When recombination is Shockley-Read-Hall limited, the conductivity decay is more complicated as τ is a function of the injection level $\delta(t)/N$ where N is the majority-carrier density. That situation will be discussed later in this section.

Domain Trapping Effects

The term trapping is used to apply to capture of a free carrier at a site without subsequent recombination. That site can be a point defect or a smaller bandgap domain as will be described here. These domains are formed by ordering and were described in the previous section. Trapping can be described in the transport equation by a capture rate and an emission rate. If one defines a domain capture probability by R_c and an emission probability by R_e, a pair of coupled differential equations can be written to describe the density of excess carriers that are free and are occupying traps as follows:

$$\frac{d\rho}{dt} = -\frac{\rho}{\tau_r} - R_c \rho + R_e \rho_t$$

$$\frac{d\rho_t}{dt} = R_c \rho - R_e \rho_t \quad\quad 4)$$

Here ρ is the excess electron or hole density and ρ_t is the density of the same carrier in traps. The solution of this coupled pair is easily solved numerically. Here the Matlab version 4.2c software easily provided a solution to these equations for $\rho(t)$ and $\rho_t(t)$. This solution is applicable to a single type of domain as characterized by capture and emission rates, R_c and R_e. It is applied in the next section to data a case where that situation seemed to be applicable. The solution for multiple types of domains is more complicated and will be described in future work.

By detailed balance, one can show that the capture and emission rates in equilibrium are equal and that the emission rate can be written as:

$$R_e = A \exp(-\Delta E/KT) \quad\quad 5)$$

where ΔE is the depth of the potential well and A is a weak function of temperature. Also, each particular ordered region may have a different value of ΔE depending upon the geometry of the region.

Band Diagram for Order-Disordered Domains

Owing to the similarity of the effect of the ordering on the band gap properties of InGaAs and InGaP, one may expect the interface between ordered and disordered regions of the InGaAs to display a Type II offset (5,6). A schematic of a typical Type II heterointerface is shown in Fig. 3. Here we see that the conduction band minima lie in the ordered regions and the valence band maxima lie in disordered

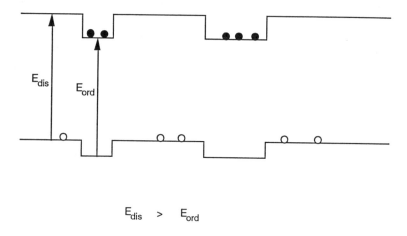

Figure 3. Schematic of possible band alignment or ordered and disordered regions assuming a type II heterointerface.

regions. Consequently, charge separation of electrons and holes is produced by the band structure. Here, the electrons are trapped in the ordered domains and the holes in the disordered domains. Consequently, recombination is retarded and thermal excitation is required to get electrons and holes in the same potential well. Thus, very long effective recombination lifetimes are expected in such structures. Our data will show that to be the case for these ordered structures.

RADIO-FREQUENCY PHOTOCONDUCTIVE DECAY MEASUREMENTS

We applied the radio-frequency photoconductive decay technique (RFPCD) of reference (1) to measure the recombination lifetime of electrons and holes in the disordered and ordered thin films described above. RFPCD is basically a contactless technique using thin films that are inductively coupled to a high frequency (~430 MHz) bridge circuit. The bridge senses unbalanced photo-eddy-currents in the films that are induced by a pulsed YAG laser (1.064 μm) or by doubled YAG (532 nm) excitation with a pulse width of 3.0 ns FWHM. The minimum lifetime resolution here is about 5 ns and the sample size could be as small a 1 mm by 1 mm by 1.0 μm thick and still produce usable signals. The measurements here were made at room temperature (~290 K). The laser power per pulse could be varied from less than 0.1 mJ to approximately 400 mJ using a pair of rotating polarizers as variable attenuators. The laser beam diameter is about 5 mm. By varying the laser energy by orders of magnitude, the various domains could be filled with charge. The energy density was used as a tool to sort out the effect of various domains on carrier kinetics and recombination.

As a background monitor of the effects of ordering, a completely disordered film (Sample K752) was measured by RFPCD over a wide range of incident energy densities. These data are shown in Fig. 4 for the extremes of incident energy. Curve A was measured with a pulse energy of 60 mJ and Curve B was measured at a pulse energy of 60 µJ. This represented the extremes of pulse energies used in this set of measurements. One sees that there is no significant change in the shape of the photoconductive decay curve. There are two lifetimes seen here that can be ex-

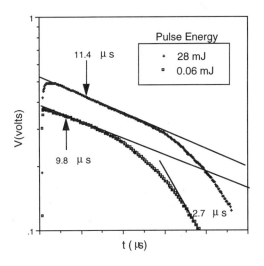

Figure 4. RFPCD decay of completely disordered sample at two injection levels, 28 mJ pulse energy and 0.06 mJ pulse energy.

plained by a saturated Shockley-Read-Hall recombination center (17). The initial decay represents the sum of majority and minority-carrier lifetimes whereas the final decay is the minority-carrier lifetime. Excess carrier decay of this type is commonly observed in silicon and in AlGaAs which are Shockley-Read-Hall limited materials. In the data of Fig. 4, the high-injection lifetime is about 11 µs and the low-injection lifetime is about 3 µs.

Fig. 5 shows the decay of sample K751 that was grown at 550° C as described in the previous section. TED measurements showed that this sample is highly ordered. The excitation energy used for the measurement of Fig. 5 is 60 mJ, the same

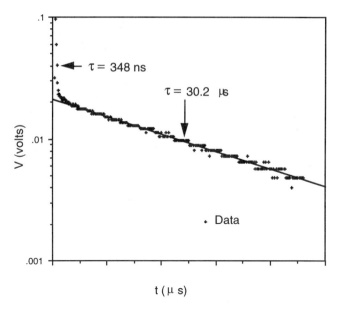

Figure 5. RFPCD decay of highly ordered sample (E_g = 0.703 eV) using 60 µJ excitation pulse.

at for Curve A of Fig. 4. The RFPCD data indicate a fast, initial decay of 112 ns and a very long final decay of 30.3 µs. These data are extremely different from that of Fig. 4, Curve A. The decay curve of Fig. 5 has been modeled using Eq. 4. An excellent fit was obtained and reasonable recombination lifetime, capture, and emission rates were produced. In such cases, modeling shows that the initial decay is produced by trap capture.

The initial response of the sample K751 to the YAG pulse is shown in Fig. 6 expanding the measurement to 2 µs full scale. Curve A shows the fast decay of Fig. 5 in detail with the noted decay time attributed to domain capture of excess carriers. Curve B shows the same sample using a pulse energy of 60 mJ. The RFPCD signal rises somewhat slowly (about 100 ns compared to 4 ns system response) and then saturates. This is indicative of well filling as the RFPCD system is still linear over this signal range. This is followed by a fast decay (358 ns) and a slower decay (1.06 µs). Finally, not shown here is the asymptotic decay of 30 µs that is shown in Fig. 5. These data are qualitatively indicative of well filling and redistribution of charge among the potential wells. Finally at very long times, carriers are emitted from the deepest level with the 30 µs decay time measured earlier.

The emission rate from the various wells will be a function of domain morphology and will vary with different ordering structures. As an example, Fig. 4 shows

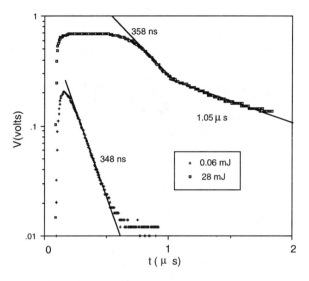

Figure 6. RFPCD decay of the sample of Fig. 5 using a high-injection (28 mJ) excitation pulse energy. Also shown is magnified presentation of the initial decay at low injection (60 µJ) that was shown in Fig. 5.

the RFPCD of sample K753 that was also grown at 550° C and highly ordered. The curve of Fig. 4 was measured by RFPCD at the lowest pulse energy of 60 µJ and showed the usual fast and slow component decay. The slow component in this case is 192 µs and is indicative of a relatively deep potential well. The primary difference between samples K751 and K753 is substrate orientation. Sample K751 is aligned 2 degrees from (110) and K753 is aligned 4 degrees from (110). As ordering is very sensitive to substrate orientation, large differences in the domain structure are expected.

Applications to Thermophotovoltaic Technology

Although ordered InGaAs has not been a component of TPV device design, some results found here could potentially impact the technology. Although the experimental bandgap has only been lowered by about 65-75 meV, band theory indicates that bandgap reduction as much as 300 meV may be obtained for perfectly order InGaAs. That would allow lattice-matched devices to be fabricated in the 450 to 600 meV bandgap range. Lattice-matching would produce much larger recombination lifetimes due to the absence of dislocation losses. This effect in term would boost efficiencies appreciably.

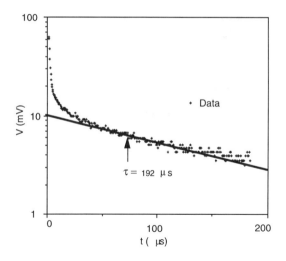

Figure 7. RFPCD decay of a highly ordered film (E_g = 0.693 eV) using low-injection (60 µJ) excitation energy.

CONCLUSIONS

The effects of ordering in ternary InGaAs on carrier were measured and found to be significant. Ordering effects may find applications in future TPV devices.

ACKNOWLEDGMENTS

The authors would like to thank C. Kramer, K. Bertness and J. M. Olson for assistance with the growth of the samples. This work was partially funded by the U.S. Department of Energy, under contract DE-AC36-83CH10093.

REFERENCES

1. Ahrenkiel, R.K., Wangetsteen, T., Al-Jassim, M.M., Wanlass, M., and Coutts, T., *AIP Conference Proceedings* **321**, 1994, p. 412.
2. Wei, S. H. and Zunger, A., *Appl. Phys. Lett.* **56**, 662 (1990).
3. Ueda, O., Hoshino, M., Takechi, M., Ozeki, M., Kato, T., and Matsumoto, T., *Jour. Appl. Phys.* **68**, 4268 (1990).
4. Arent, D. J., Bertness, K. A., Bode, M., Kurtz, Sarah R., and Olson, J. M., *Appl. Phys. Lett.*

62, 1806 (1993).
5. Samulelson, L., Pistol, M. E., and Nilsson, S., *Phys. Rev. B.* **33**, 8776 (1986).
6. Fouquet, J. E., Robbins, V. M., Rosner, J., and Blum, O., *Appl. Phys. Lett.* **57**, 1566 (1990).
7. Delong, M. C., Ohlsen, W. D., Vioh, I., Taylor, P. C., and Olson, J. M., *Jour. Appl. Phys.* **70**, 2780 (1991).
8. Ueda, O. and Nakata, Y., *Proceedings of the 16th International Conference on Defects in Semiconductors*, 1991.
9. Bassignana, I.C., Miner, C. J. and Puetz, N., *Jour. Appl. Phys.* **65(11)**, 4299, 1989.
10. Pankove, J., *Optical Processes in Semiconductors*, New York: Dover Publications, Inc., 1957.
11. Wang, T. Y. and Stringfellow, G. B., *Jour. Appl. Phys.* **67**, 344, 1990.
12. Wei, S.-H and Zunger, A., *Appl. Phys. Lett.* **56**, 662, 1990.
13. Horner, G. S., Mascarenhas, A., Alonso, R. G., Froyen, S., Bertness, K.A., Olson, J. M., *Phys. Rev. B,* **49**, 1727, 1994.
14. Gomyo, A., Suzuki, T., Kobayashi, K., Kawat, S., Hino, I., and Yuasa, T., *Appl. Phys. Lett.* **50**, 673, 1987.
15. Otsuka, N., Ihm, Y. E., and Hirotsu, Y.*J., Crystal Growth* **95**, 43, 1989.
16. Seong, T.-Y., Norman, A. G., Booker, G. R., and Cullis, A. G., *Appl. Phys. Lett.* **75**, 7852, 1994.
17. Ahrenkiel, R. K., Keyes, B. M., and Dunlavy, D. J., *Jour. Appl. Phys.* **70**, 225, 1991.

The Influence of Bandgap on TPV Converter Efficiency

P.A. Iles, C. Chu and E. Linder

Applied Solar Energy Corporation
City of Industry, CA 91745-1002, USA

Abstract. The paper reviews the effect of the energy bandgap of TPV converters on their conversion efficiency. Several sources are analyzed, monochromatic, narrow band (selective emitters) and broad band (blackbodies). We find that at low bandgaps, conversion efficiency falls off. The conclusion drawn should help in estimates in all cases of practical efficiencies to be used in TPV system analysis.

BACKGROUND

TPV applications have increased the interest in the performance of PV converters made from low bandgap semiconductors. The converters are to be used in the near infrared region, over a wavelength range extending from 1μm to 3μm. Solar PV efficiencies have been optimized, mostly for converters using higher bandgap materials. These higher bandgap efficiencies are usually quoted as goals for TPV converters (eg 45-50% for the best UNSW silicon cells) (1).

Work on non-solar cells (1-3) has confirmed that for narrow bandwidth illumination (such as lasers or LED's), measured conversion efficiencies are higher then those obtained under sunlight. The reason is the reduced losses resulting from illuminating photons with energies close to the bandgap of the semiconductor material used in the cell. The analysis of non-solar cells (which included estimates for selective emitters) also showed that as the bandgap of the semiconductor was reduced (to convert longer wavelength light), the conversion efficiency for near-monochromatic sources was also reduced.

Here we examine the dependence of conversion efficiency from sources with different bandwidths near-monochromatic (lasers), narrowband (selective emitters) and broadband (blockbodies). We conclude that similar fall-off in efficiency with bandgap is found for all three types of source. The estimates provide practical targets for TPV converters.

EFFICIENCY VERSUS BANDGAP ESTIMATES

We use a well-proven format, based on equation (1), which avoids the need to know the detailed physical parameters for the semiconductors being analyzed.

$$\text{Efficiency} = \frac{J_{sc} \, V_{oc} \, CFF}{\text{Input Power Density}} \quad (1)$$

Where J_{sc} is the generated/collected current density
V_{oc} is the open-circuit voltage
CFF is the ratio of the maximum power to $I_{sc} \times V_{oc}$

This approach can also include the best-achieved values for these three PV parameters, to provide a pragmatic model which has been found useful for optimizing cell performance (by design and fabrication methods), or for optimizing cascade cell combinations.

We can trace the qualitative effects of bandgap on efficiency by considering the effects on the three PV parameters, J_{sc}, V_{oc}, and CFF.

Short-Circuit Current (J_{sc})

J_{sc} is determined by the spectral response (SR), the amperes/watt generated by the cells. In turn, SR depends on the quotient of the external quantum efficiency for given illuminating wavelengths, divided by the photon energy for these wavelengths. The bandgap of the converter is involved only in the location of the cut-off wavelength where the cell SR ceases. J_{sc} increases linearly with illumination intensity, but the changes in J_{sc} do not affect efficiency directly. For the same reason, we can postulate idealized illumination intensity (i.e. emissivity = 1), and we can assume maximum generation/collection efficiency for the cell without serious impact on the efficiency versus bandgap estimates. The electrical power density from the cell increases with intensity, with linear current increase and logarithmic voltage increase. Below we will discuss the importance of both PV conversion efficiency and available power density for given source conditions.

Open Circuit Voltage (V_{oc})

Open circuit voltage is given by equation (2)

$$V_{oc} = n \cdot kT \ln \left[\frac{J_{sc}}{J_o} \right] \quad (2)$$

Where n is the diode ideality factor, taken as unity here
   kT is the Boltzmann factor, = 0.026V at 300°K
   J_{sc} is the short circuit current
   Jo is the saturation current density

In practice, Jo is the sum of several current contributions from recombination around the PN junction. Here we assume the ideal (lowest) Jo-values for n=1, and in (3) we show an empirical expression which agrees well with other Jo values derived in cell modeling (3, 4a, 5, 6)

$$J_o = 1.5 \times 10^5 \exp\left[\frac{-E_g}{kT} \right] \quad (3)$$

Where E_g is the energy bandgap (eV).

By combining (2) and (3) and putting $J_{sc} = 1.5 A/cm^2$ we find that $V_{oc} = E_g - 0.3$, and this proportionality leads to the major impact of bandgap logarithmically with J_{sc} (and the illuminating intensity), and this slow variation has been included in the estimates given below.

Curve Fill Factor (CFF)

There is an intrinsic CFF value (CFFo) which depends on V_{oc}. The empirical dependence is given by reference 4(b) as

$$CFFo = \frac{v - \ln(v + 0.72)}{v + 1} \quad (4)$$

Where $v = \frac{V_{oc}}{nkT}$

CFFo varies logarithmically with V_{oc}, and this slow dependence has been included in efficiency estimates using equation (1). We used (4) with n=1, and (2) and (3) to estimate values for V_{oc}, CFFo and V_{oc} CFFo as functions of bandgap. The results are shown in Figure 1.

Figure 1: Voc and CFF as a Function of Bandgap

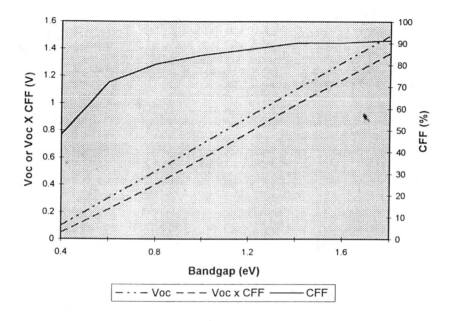

CFF also depends on parasitic losses in the cell, both series and shunt resistances. At the high intensity levels typical of TPV operation shunt losses are reduced, even when some degree of lattice-mismatch is incurred in growth of the cell. For Theoretical estimates we have neglected the resistance losses, and use only CFFo values. In a few cases, where experimental results are included to show the approach to idealized conditions we have used measured CFF values.

APPLICATION TO THREE TYPES OF SOURCE

We will investigate the dependence of PV converter efficiency on bandgap, using three different illumination sources. For each source type, we will give the assumptions, and will plot the estimates of conversion efficiency versus bandgap.

Monochromatic Sources

Even for direct-gap semiconductors like GaAs, the absorption for wavelengths near the cut-off value corresponding to the bandgap decreases gradually, not abruptly. For monochromatic sources we allowed for the fall-off in SR near the bandgap wavelength of the semiconductor. The ideal spectral response (A/W) at wavelength λ is given by (5)

$$SR = \frac{\lambda}{hc} = \frac{\lambda}{1.24} \quad \text{where } \lambda \text{ is in } \mu m. \tag{5}$$

By (5) SR is an increasing function of λ. However because the absorption of the longer wavelengths decreases near the cut-off edge, and because the charge carriers generated by the longer wavelengths must travel further to be collected at the PN junction, the measured SR falls off near the bandage. Using experimental SR values, we found that the optimum conversion efficiency was obtained when the illuminating peak wavelength was about 0.1µm to 0.2µm below the cut-off wavelength. In Figures 2, 3 we show the normalized SR (the ratio of measured SR to ideal SR) values for a space design GaAs cell and for a good quality UNSW Si cell (1). We also plot the estimated conversion efficiencies for various laser wavelengths for these cells, and the fall-off in efficiency follows the fall-off in normalized SR. The peak efficiencies are 0.1µm to 0.2µm below the cut-off wavelength.

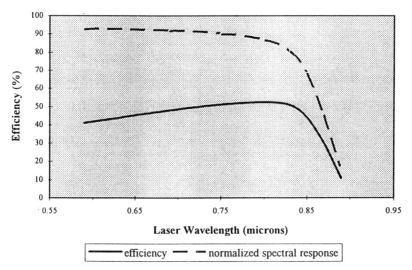

Figure 2: Practical GaAs Cell Efficiency vs. 1 W/cm² Laser Wavelength

For GaAs cells, efficiency values over 50% are shown, in line with reported values (2, 3). For Si cells; peak efficiency values around 45% were obtained. Although the input power density used for these figures was $1 W/cm^2$ at each wavelength (typical of current laser output), the efficiency values that Green reported in (1) at several wavelengths, and lower power input ($\sim 0.05\ W/cm^2$) fall close to the plotted line in Figure 3. This shows the relative insensitivity of efficiency on input power, and supports the use of idealized emitter conditions below.

Figure 3: Practical Silicon Cell Efficiency vs. 1 W/cm^2 Laser Wavelength

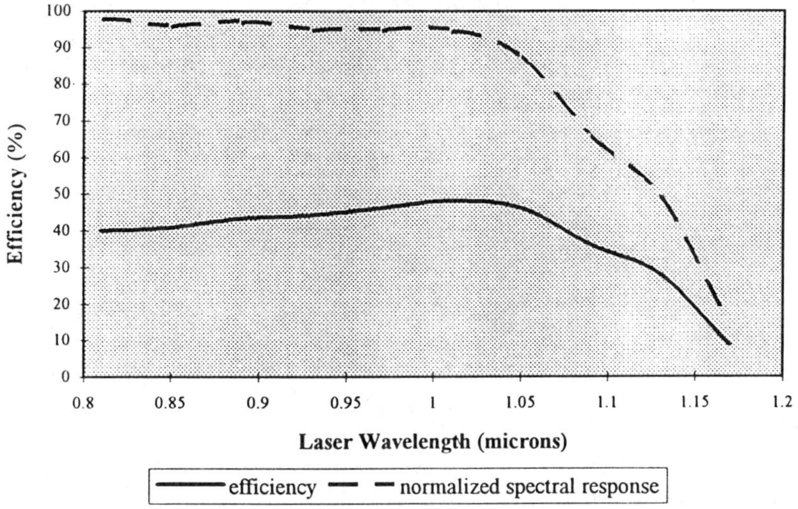

The next figure (Figure 4) shows the dependence of conversion efficiency on the bandgap of the converter material, for laser illumination.

Based on Figures 2 and 3, the converter cut-off wavelength in Figure 4 was set 0.1µm above the laser wavelength. The ideal SR value for each laser wavelength was used, and J_{sc} estimated as the product of SR x $1 W/cm^2$ input at that wavelength. V_{oc} values were estimated for each bandgap using (2) and (3), and the appropriate J_{sc} value. The CFFo value corresponding to each V_{oc} value and the V_{oc}, CFFo products were obtained from Figure 1.

Figure 4: Ideal Efficiency vs. PV Converter Bandgap (1 W/cm² Laser)

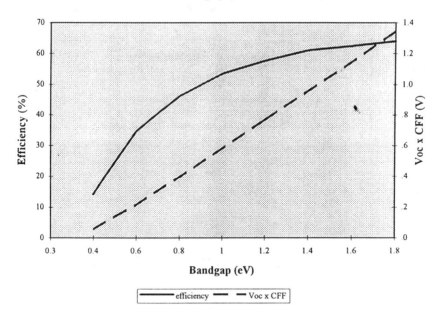

In Figure 4, the fall-off in efficiency at low bandgaps is caused mostly by the reduced V_{oc} CFFo product. The J_{sc} values decreased from 2.4A/cm² at low Eg values to 0.475 A/cm² at larger bandgaps. These J_{sc} variations are the result of the larger number of charge carriers generated for longer wavelengths since the current density is the constant power input divided by the photon energy which depends on the reciprocal of the wavelength. At higher bandgaps, the increase in V_{oc} CFFo was offset by the lower J_{sc} values, resulting in a flattening of the efficiency values. In the next two figures we will include the monochromatic efficiency versus bandgap curve as a background, to show that the curve in Figure 4 forms an envelope for the low bandgap efficiencies for selective emitters and blackbody sources.

Narrowband Sources (Selective Emitters)

We analyzed four selective emitters containing ytterbium, erbium, holmium and neodymium. For each emitter we selected the temperature which gave optimum emission efficiency. For these narrowband sources, we adopted a rule similar to that used for monochromatic sources, selecting the bandgap of the converter material to place the cut-off wavelength around 0.2μm beyond the long wavelength edge of the selective emitter output. Idealized emitter conditions were assumed - 100% selectivity, with no out-of-band emission - and the peak intensity was taken as the appropriate blackbody output contained in the envelope of the emitted pulse. The emitter pulse shapes were taken from references 7-9.

Figure 5 shows calculated efficiencies for the bandgaps assigned to give best output for each emitter.

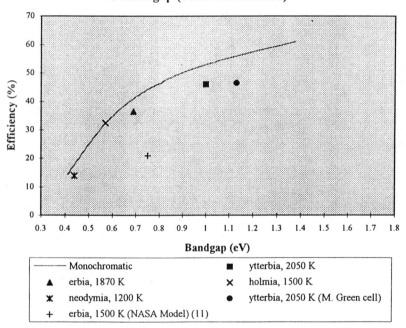

Figure 5: TPV Converter Efficiency vs. Bandgap (Selective Emitters)

—	Monochromatic	■	ytterbia, 2050 K
▲	erbia, 1870 K	✕	holmia, 1500 K
✻	neodymia, 1200 K	●	ytterbia, 2050 K (M. Green cell)
+	erbia, 1500 K (NASA Model) (11)		

The efficiency values lie close to the envelope from Figure 4, for monochromatic sources. Figure 5 includes estimates for Nelson's ytterbia emitter using the experimental results reported by Martin Green's UNSW group (1), and also NASA - Lewis results for an erbia emitter heated to 1500°K.

We also considered mixed emitters, containing erbia and holmia, heated to 1500K and 1775K. We selected the converter bandgap best suited for holmia (0.565ev). The efficiency for the mixed oxide emitter was 29.5% (1500K) and 31.3% (1775K), lower than the efficiencies for the separate emitters. However the power output was higher (almost double) and this may be an important factor in selecting the best emitter for TPV use.

We also analyzed the possibility of using series-connected converters matched to each emitter wavelength in the mixture, and found that both the power output and efficiency were lower. Even if the emitter outputs were designed to generate equal currents (by varying the composition ratio of holmia and erbia), the resultant efficiency and power output did not justify the added complexity.

Blackbody Sources

We extended the analysis of the power density versus bandgap for blackbodies at several temperatures (12). The efficiency was calculated by setting the input power density as the sum of the absorbable power for the blackbody source up to the bandedge cut-off wavelength, and 10% of the BB radiation above the cut-off wavelength. Most estimates show that this is a practical assumption, when using selective optical filters, possibly supported by reflection from the back surface of the convertor.

For TPV conversion, it is assumed that the 90% of the non-absorbed radiation can be returned to the BB source, thereby increasing the BB emission efficiency. Incidentally, applying the same definition of efficiency for Si and GaAs space solar cells with 20% efficiency, versus their efficiency values to 25% and 30% respectively. However the reflected radiation does not increase the emission efficiency of the sun.

We considered three BB temperatures (1275K, 1500K, and 2050K), and the estimated variation of efficiency versus bandgap are shown in Figure 6.

Figure 6: PV Conversion Efficiency vs. Bandgap from Blackbody Sources

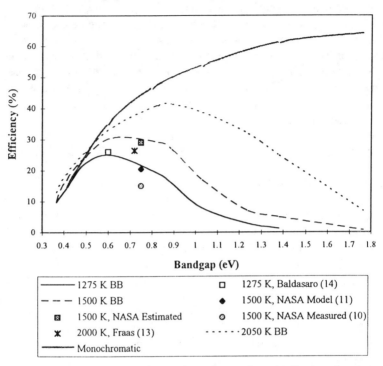

We have also shown some experimental results, and again for low bandgaps the efficiency fall-off is similar to the ideal cases considered above. However the experimental power densities are well below the ideal estimates.

The fall-off in efficiency at higher bandgaps is the result of decreasing J_{sc} and an increasing V_{oc} CFFo product. In the definition of efficiency, the input power density increases at lower bandgaps, and as a result, the peak of the output power density is about 0.2 to 0.25 eV below the peak of the efficiency values. We will discuss this further in the next section.

COMMENTS AND CONCLUSIONS

Using idealized assumptions to show the dependence of efficiency on bandgap, we showed that for lower bandgaps typical of the temperatures used for TPV conversion, the efficiency fall-off at lower bandgap is similar for all three source types considered here.

In most TPV operations, the output power density is the main requirement. As we saw for the BB sources, it may be necessary to select the bandgap which gives higher power density even if the efficiency is slightly lower. Also, it may be necessary to use converters which do not have the optimum bandgap, because the material properties may be better developed. Si cells have been intensively developed for PV uses, and can give higher output than other materials which may have preferable bandgaps. GaSb cells (14) are also a good example where converter performance is better than nearby medium bandgap materials which have preferable bandgaps. The analysis showed that the major practical requirements are to increase the source emission efficiency, and also to take advantage of the 30% increased J_{sc} available from an AR coating. Also, the material properties of the converter must be improved, to achieve the maximum V_{oc} (and CFFo)

The analysis showed that for the three BB temperatures, (1275K, 1500K, 2050K) the maximum efficiency was obtained for converters which absorbed only 15%, 15.5% and 23.5% of the BB output. For all temperatures, 25% of the BB output is contained up to the wavelength giving peak BB output. This shows that BB sources are poorly used, and confirms that high beyond-bandgap reflectance is necessary. The rapid fall-off in efficiency at low bandgaps also suggests that attempts to use cascade cells to convert BB radiation beyond the BB peak may not have much advantage. For selective emitters, the peak emission wavelength allows the use of higher bandgap converters, and this leads to the interest in using highly optimized Si cell with high temperature selective emitters. Also as pointed out in reference 10, selective emitters can provide optimized emission with lower heat input from the thermal sources.

Overall, when converters with low bandgaps must be used (for low temperature sources), lower efficiencies and power output must be expected.

The limiting efficiency values in this study should be useful to indicate practical PV conversion efficiencies for use in estimating the overall energy efficiency of TPV system.

REFERENCES

1. MA Green et.al "45% Efficient Silicon Photovoltaic Cell under Monochromatic Light", IEEE Electron Device Letters, Vol. 13, No. 6 June 1992, pp. 317-8.
2. P.A. Iles "Non-Solar Photovoltaic Cells", Con. Rec. 21st IEEE Photovoltaic Spec. Conf., May 1990, pp. 420-425.

3. L.C. Olsen et.al "High Efficiency Monochromatic GaAs Solar Cells", Conf. Rec. of 22nd IEEE Photovoltaic Speciality Conference, 1991, pp. 419-424.
4. (a) M.A. Green "SOLAR CELLS", Operating Principles, Technology and System Applications, (1982), p. 88.
   (b) M.A. Green "Accuracy of Analytical Expressions for Solar Cell Fill Factors" Solar Cells Vol. 8, (1983), pp. 3-16.
5. M.W. Wanlass et.al "Practical Considerations in Tandem Cell Modeling" Solar Cells Vol 27 (1989) pp. 191-204.
6. M.E. Nell and A.M. Barnett, IEEE Trans. Electron Dev. Ed-34 (1987), p. 257.
7. R.E. Nelson "Thermophotovoltaic Emitter Development" Publication 1.* pp. 80-96.
8. P.L. Adair and M.F. Rose "Composite Emitters for TPV Systems" Publication 1, pp. 245-262.
9. R.A. Lowe et.al "Radiative Performance of Rare Earth Garnet Thin Film Selective Emitters" Publication 1, pp. 291-297.
   Note: this reference contains details of several of D.L. Chubb's papers on selective emitters.
10. D.M. Witt et.al "InGaAs PV Device Development for TPV Power Systems", Publication 1, pp. 210-220.
11. R.K. Jain et.al "Modeling of Low Bandgap Solar Cells for Thermophovoltaic Applications" Publication 1, pp. 202-209.
12. (a) P.A. Iles, "Photovoltaic Principles Used in Thermophotovoltaic Generation" Publication 1, pp. 67-79.
    (b) S. Wojtczuk et.al "InGaAs Thermophotovoltaic Cell Performance vs Bandgap" Publication 1, pp. 177-187.
13. L.M. Fraas et.al "Electric Power Production Using New GaSb Photovoltaic Cells with Extended Infrared Response", Publication 1, pp. 44-53.
14. P.A. Baldasoro et.al "Experimental Assessment of Low Temperature Thermal Voltaic Energy Conversion", Publication 1, pp. 29-43.

* Publication 1 is:
AIP Conference Proceedings 321, First NREL Conference on Thermophotovoltaic Generation of Electricity, June 1994.

Development of p-on-n GaInAs TPV Devices

P. R. Sharps and M. L. Timmons

Research Triangle Institute
Research Triangle Park, NC 27709

Abstract. We report on the development of p-on-n $Ga_xIn_{(1-x)}As$ thermophotovoltaic devices. For $Ga_{0.47}In_{0.53}As$ cells, lattice-matched to InP and having a bandgap of 0.73 eV, the n-on-p configuration gives better results than the p-on-n configuration. However, for $Ga_{0.32}In_{0.68}As$ cells, grown on InP and having a bandgap of 0.58 eV, the p-on-n polarity has superior performance for cells with similar step-graded buffer layers. The improvement in the p-on-n devices is due to reduced dark currents and increased open circuit voltages (V_{oc}). Optimized back surface field layers produce these effects. Because of the absorption of long wavelength light in the base region of low bandgap materials, a high quality back surface acts as a minority carrier mirror and reduces recombination in buffer layers. We have been able, so far, to get more effective back surface field layers with the p-on-n configuration. While the n-on-p polarity may offer the advantage of lower cell emitter sheet resistivity, the p-on-n device offers lower free carrier absorption of long wavelength radiation, better spectral control of incoming radiation, and improved large scale manufacturing ability. Results for several different p-on-n cell structures for the 0.73 eV and 0.58 eV $Ga_xIn_{(1-x)}As$ compositions will be reported.

INTRODUCTION

Considerable effort has gone into the development of n-on-p $Ga_xIn_{(1-x)}As$ photovoltaic and thermophotovoltaic (TPV) devices [1-3]. The n-on-p polarity has been chosen for two reasons: (1) Higher electron mobility has meant longer minority carrier diffusion lengths in p-type base layers when compared to n-type layers, and (2) the higher electron mobility gives lower emitter sheet resistivities for n-type layers. The lower sheet resistivities are particularly important for TPV devices, which are designed to run at high currents. Excessive resistive losses lead to a reduction in fill factor and a degradation of device performance.

However, the work done at Varian Corporation in the late 1980's [4,5] and also the more recent work done by the Solar Research Corporation [6] have indicated that p-on-n GaAs solar cells are completely suitable for high current, concentrator solar cells. The Varian cells were designed to operate at a 1,000-sun concentration, while the Solar Research Corporation cells have operated at 469 suns. The p-on-n polarity cell has also been chosen for the large scale manufacturing of GaAs photovoltaic devices [4,7]. There are a number of reasons for this, historical as well as technical. The improved control of doping in an n-type base layer compared to a p-type base layer, especially in large organometallic vapor phase epitaxy (OMVPE) reactors, has been one of the technical reasons that the p-on-n polarity was chosen. For TPV applications, the p-on-n polarity has an additional benefit. Because of lower free carrier absorption in n-type layers, as compared to p-type layers [8], the p-on-n polarity will absorb less sub-bandgap radiation through free-carrier absorption. This may facilitate spectral control of long wavelength radiation. For all of these reasons, the development of p-on-n $Ga_xIn_{1-x}As$ TPV devices is worth considering.

In the present paper we report results for both p-on-n and n-on-p 0.73 eV $Ga_{0.47}In_{0.53}As$ cells lattice-matched to InP and also for p-on-n 0.58 eV $Ga_{0.32}In_{0.68}As$ cells, grown on InP substrates with step-graded buffer layers.

EXPERIMENTAL

$Ga_xIn_{1-x}As$ devices with bandgaps of 0.73 eV and 0.58 eV are grown using OMVPE. The OMVPE system operates at atmospheric pressure in a pressure-balanced, vent-run configuration. All gas switching is computer controlled. The reactor tube is vertical, with substrates placed horizontally on an RF heated graphite susceptor. The exhaust gases are burned and then filtered. The growth temperature is 640 °C. Trimethylgallium, ethyldimethylindium, trimethylaluminum, 100% arsine, 100% phosphine, diethylzinc, and dilute hydrogen selenide are the Ga, In, Al, As, P, Zn, and Se sources, respectively. Zinc is the p-type dopant, and Se is the n-type dopant.

Conventional InP substrate preparation techniques are used, as are conventional solar cell processing techniques to fabricate devices from wafers. Contacts to p-InP substrates are layers of Au/Zn/Au, while contacts to n-InP substrates are layers of Au-Ge/Ni/Au. Both contacts are alloyed to ensure low resistivity. For front surface contacts to n-$Ga_xIn_{1-x}As$, layers of Cr/Au are used, which are subsequently alloyed. Front surface contacts to p-$Ga_xIn_{1-x}As$ are Ti/Au. Standard etches are used for the removal of the cap/front contact layer and for the mesa etch.

For 1 cm^2 devices, the grid pattern for the front metal contact has ten percent obscuration, and a central busbar with another ten percent obscuration. For 0.25 cm^2 devices, the total busbar and grid obscuration is fourteen percent.

RESULTS AND DISCUSSION

Lattice-Matched Devices

The structures for the n-on-p and p-on-n 0.73 eV $Ga_{0.47}In_{0.53}As$ cells are shown in Figure 1 and Figure 2, respectively. The devices are double heterostructures, with InP layers acting both as a back surface field (BSF) and as a window.

The results for the two different polarity devices are shown in Table 1. Only one 4 cm^2 wafer of each polarity was grown, and while the data shown in Table 1 are for the best cell from each wafer, they are fairly typical of all the cells on each wafer. The devices were measured under AM0 illumination. Pertinent cell parameters include open circuit voltage (V_{oc}), short circuit current (J_{sc}), and fill factor. The n-on-p cell outperforms the p-on-n cells in terms of voltage and current, but the fill factors are similar for the currents measured in the devices. Spectral response measurements on p-on-n cells indicate a poorer red response, and this accounts for both the lower current and voltage in the p-on-n cell. Further optimization of the lattice-matched p-on-n cell is believed possible, to increase both the current and the voltage. A single p-on-n wafer does not give a good statistical sampling.

Lattice-Mismatched Devices

Several different structures have been considered for the lattice-mismatched devices. A schematic of a typical structure is shown in Fig. 3 for p-on-n devices. Two different methods of optimizing the step-graded buffer layers have been developed, and several different back surface field layers have also been used.

Different methods of growing the step-graded buffer layers give different surface morphologies. Fig. 4 shows scanning electron micrographs of the surfaces of devices grown with two different step-graded buffer layers, Buffer A and Buffer B. Differences between the two buffers can be seen with the eye. Buffer B shows much less cross hatching. At 2000x, devices grown with Buffer B have a "smooth" appearance with very little surface topography. At 0.58 eV, Buffer B gives slightly improved cell performance. The improvement in the buffer structure may help to improve device performance for greater latter mismatch, and hence $Ga_xIn_{(1-x)}As$ devices with smaller bandgaps can be developed using Buffer B. We are continuing work in this area.

Results for several different p-on-n lattice-mismatched cells are contained in Table 2. Spectral response measurements confirmed that all of these cells have a bandgap of 0.58 eV. Cell 6-2547-6 is the best p-on-n cell that we have seen to date, and the I-V curve for the cell is shown in Fig. 5. The dark I-V curve for 6-2547-6 is shown in Fig. 6, and

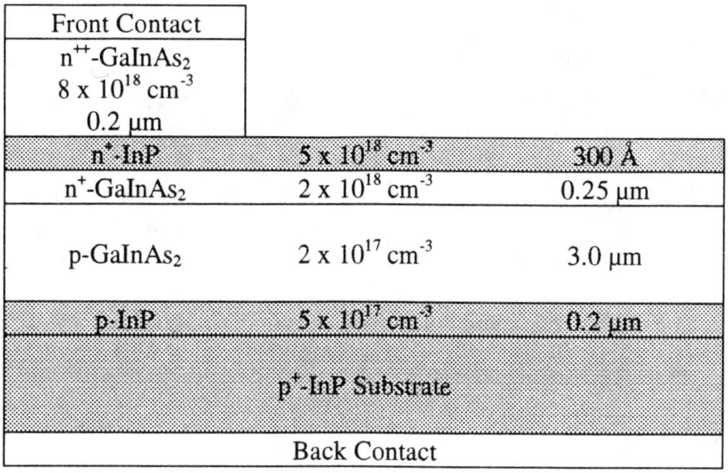

Front Contact		
n^{++}-GaInAs$_2$ 8×10^{18} cm^{-3} 0.2 μm		
n^+-InP	5×10^{18} cm^{-3}	300 Å
n^+-GaInAs$_2$	2×10^{18} cm^{-3}	0.25 μm
p-GaInAs$_2$	2×10^{17} cm^{-3}	3.0 μm
p-InP	5×10^{17} cm^{-3}	0.2 μm
p^+-InP Substrate		
Back Contact		

Figure 1. Schematic of the n-on-p lattice-matched Ga$_{0.47}$In$_{0.53}$As device.

Front Contact		
p^{++}-GaInAs$_2$ 1×10^{19} cm^{-3} 0.2 μm		
p^+-InP	5×10^{18} cm^{-3}	300 Å
p^+-GaInAs$_2$	5×10^{18} cm^{-3}	0.3 μm
n-GaInAs$_2$	2×10^{17} cm^{-3}	3.0 μm
n-InP	5×10^{17} cm^{-3}	0.2 μm
n^+-InP Substrate		
Back Contact		

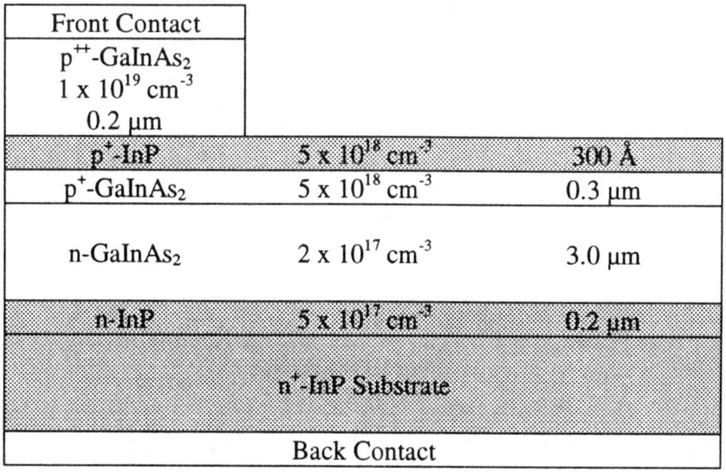

Figure 2. Schematic of the p-on-n lattice-matched Ga$_{0.47}$In$_{0.53}$As device.

Table 1. Cell Results for Lattice-Matched Devices. All cells are 1 cm^2 in area, and are measured under AM0 illumination.

Cell #	6-2454-4	6-2458-3
Cell polarity	n-on-p	p-on-n
V_{oc}, mV	382	340
J_{sc}, mA	37.2	31.7
fill factor	~70%	~70%

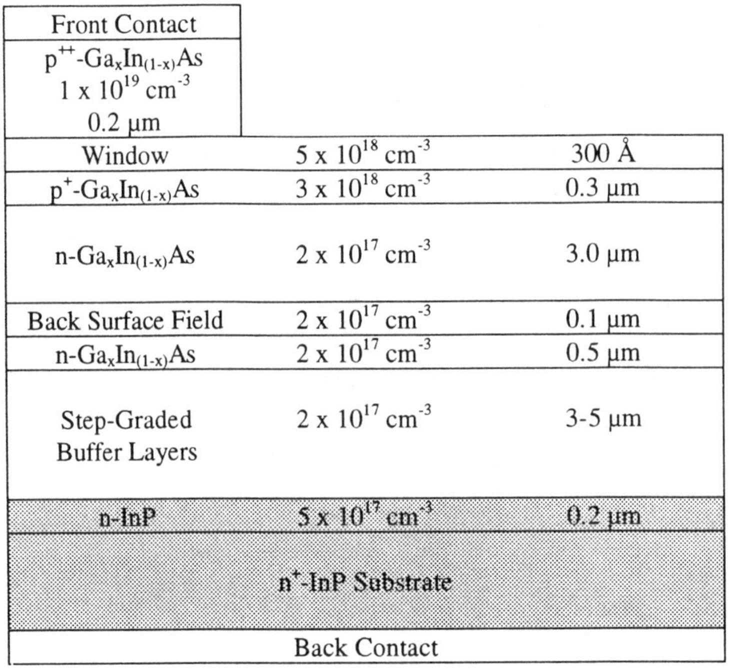

Front Contact		
p^{++}-$Ga_xIn_{(1-x)}As$ 1×10^{19} cm^{-3} 0.2 µm		
Window	5×10^{18} cm^{-3}	300 Å
p^+-$Ga_xIn_{(1-x)}As$	3×10^{18} cm^{-3}	0.3 µm
n-$Ga_xIn_{(1-x)}As$	2×10^{17} cm^{-3}	3.0 µm
Back Surface Field	2×10^{17} cm^{-3}	0.1 µm
n-$Ga_xIn_{(1-x)}As$	2×10^{17} cm^{-3}	0.5 µm
Step-Graded Buffer Layers	2×10^{17} cm^{-3}	3-5 µm
n-InP	5×10^{17} cm^{-3}	0.2 µm
n^+-InP Substrate		
Back Contact		

Figure 3. Schematic of the p-on-n lattice-mismatched $Ga_{0.32}In_{0.68}As$ device.

Table 2. Results for several p-on-n lattice-mismatched devices with varying buffer structures and back surface field layers. All cells are 0.25 cm^2 in area, have a bandgap of 0.58 eV, and are measured under AM0 illumination.

Cell #	6-2527-11	6-2528-6	6-2530-1	6-2531-3	6-2547-6
Cell structure	Buffer B BSF 1	Buffer B BSF 2	Buffer A BSF 1	Buffer A BSF 2	Buffer B BSF 3
V_{oc}, mV	266	271	271	254	287
J_{sc}, mA/cm^2	50.8	55.0	44.9	49.2	54.4
fill factor	66.5%	70.2%	64.5%	65.3%	69.5%

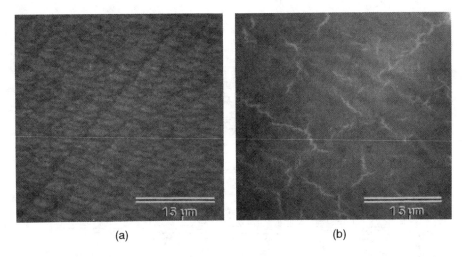

Figure 4. Scanning electron micrographs of the surface of two different p-on-n devices grown with two different buffer structures. a) Cell 6-2531-3, grown with Buffer A, at 2,000x. b) Cell 6-2527-7, grown with Buffer B, at 2,000x.

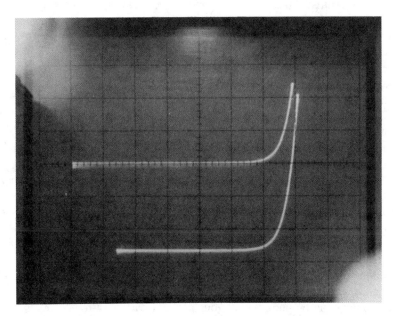

Figure 5. I-V curve for 6-2547-6, the best p-on-n cell grown to date, under AM0 illumination. The cell area is 0.25 cm^2 in area. The x-axis divisions are 0.1 Volts, and the y-axis divisions are 5 mAmps.

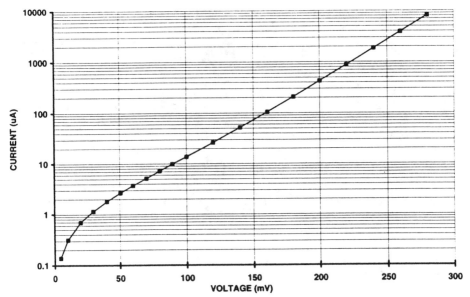

Figure 6. Dark I-V curve for cell 6-2547-6, the best 0.25 cm² p-on-n device to date. The diode saturation current I_o is 2×10^{-6} A/cm², and the diode ideality factor n is 1.18.

analysis of the curve gives a diode saturation current I_o of 2×10^{-6} A/cm², and a diode ideality factor n of 1.18.

We also obtained good device uniformity over 4-cm² substrates. Fig. 7 shows results for four 1 cm² p-on-n cells processed from wafer 6-2540. The cell structure is similar to 6-2527-11. For the four devices on 6-2540, the voltages are within 3.5%, the currents are within 3.4%, and the fill factors are all about 61%. The difference between cell 6-2527-11 and the four cells from 6-2540 can be explained by the differences between the grid design for the 0.25 cm² cell and the grid design for the 1 cm² cell.

By way of comparison, n-on-p devices that we have grown and processed do not achieve the same level of performance as the p-on-n devices that we have been describing. To date, the best 1 cm² n-on-p device (6-2524-3) has a V_{oc} 0f 201 mV, a J_{sc} of 41 mA/cm², and a fill factor of ~65%, all measured under AM0 illumination. The obvious difference in the two polarities is that the p-on-n has a much lower dark current. We do not yet completely understand the source of the dark currents in the n-on-p devices and why they do not achieve the same level of performance as the p-on-n devices.

As mentioned in the introduction, one concern with p-on-n as compared to n-on-p devices for TPV applications is high sheet resistivity in the emitter. Because TPV devices are high current devices, the high sheet resistance will limit device performance. To further consider the effects of high currents on the types of p-on-n devices that we have

	1	2
Cell #	246	255
V_{oc}, mV	45.5	47.1
J_{sc}, mA/cm^2	~61%	~61%
fill factor	3	4
	250	251
	45.8	46.5
	61.5%	61.1%

Figure 7. Summary of device results for wafer 6-2540, with a structure similar to cell 6-2527-11. All cells are 1 cm^2 in area, and are measured under AM0 illumination.

been developing, cell 6-2527-11 was measured under high current conditions. The resultant I-V curve is shown in Fig. 8. The J_{sc} for the cell is 2.0 Amps/cm2, the V_{oc} is 350 mV, and the fill factor is 51.7%. The high current measurement was made using equipment designed for measurement of higher bandgap, lower current devices; so, some of the resistance seen in Fig. 8 is due to the measuring system itself. The measured fill factor of 51.7% is actually a lower limit on the fill factor of the device. The p-on-n cells that have been grown have not yet been optimized

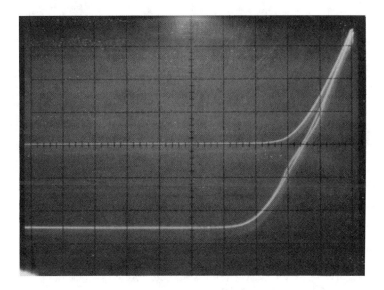

Figure 8. I-V curve for cell 6-2527-11 under high current conditions. The x-axis divisions are 0.1 Volts, and the y-axis divisions are 200 mAmps.

for high current applications, and further development of p-on-n $Ga_xIn_{(1-x)}As$ devices for TPV applications is necessary.

Further examination of the Varian work [4,5] on concentrator cells provides considerable encouragement in regards to optimization of the p-on-n cells for TPV applications. The Varian cells were 0.212 cm^2 in area, and generated over 3 Amps at their intended operating point, for a current density of over 14 A/cm^2. Sheet series resistance problems were avoided by increasing the emitter thickness to 0.7 µm. The devices had excellent performance at 942 sun concentration. Increasing the emitter doping level does not necessarily improve device performance, since at a certain level the minority carrier diffusion length decreases, reducing the J_{sc}. The Varian work indicates that further development of the p-on-n $Ga_xIn_{(1-x)}As$ TPV cell should be pursued.

CONCLUSIONS

We have demonstrated improved performance of the p-on-n, compared to the n-on-p, $Ga_xIn_{(1-x)}As$ TPV device, under AM0 illumination. The considerable improvement in V_{oc} in the p-on-n cell is attributed to a reduction in the dark current. Further work is necessary for the optimization of the p-type emitter thickness and doping for the devices to be suitable for high current applications.

REFERENCES

1. T. J. Coutts and J. P. Benner, The First NREL Conference on Thermophotovoltaic Generation of Electricity, *AIP Conference Proc. 321*, (1994).
2. M. W. Wanlass, J. S. Ward, K. A. Emery, and T. J. Coutts, "GaxIn1-xAs Thermophotovoltaic Convertors", *24th IEEE PVSC*, 1994, pp. 1685-1691.
3. M. W. Wanlass, J. S. Ward, K. A. Emery, T. A. Gessert, C. R. Osterwald, and T. J. Coutts, "High-Performance Concentrator Tandem Solar Cells, Based on IR-Sensitive Bottom Cells," *Solar Cells*, **30**, 363 (1991).
4. M. Ladle Ristow, J. C. Chen, M. S. Kuryla, and H. F. MacMillan, "A High Yield Manufacturing Demonstration for High-Efficiency GaAs Concentrator Solar Cells," *22nd IEEE PVSC*, 1991, pp.128-132.
5. M. S. Kuryla, M. Ladle Ristow, L. D. Partain, and J. E. Bigger, "22.3% Efficient 12 Cell 1000 Sun GaAs Concentrator Module," *22nd IEEE PVSC*, 1991, pp. 506- 511.
6. J. B. Lasich, A. Cleeve, N. Kaila, G. Ganakas, M. Timmons, R. Venkatasubramanian, T. Colpitts, and J. Hills, "Close Packed Cell Arrays for Dish Concentrators," *24th IEEE PVSC*, 1994, pp. 1938-1941.
7. G. C. Datum and S. A. Billets, "Gallium Arsenide Solar Arrays - A Mature Technology," *22nd IEEE PVSC*, 1991, pp. 1422-1428.
8. R. H. Bube, *Electron Properties of Crystalline Solids*, Academic Press, 1974, pp. 403-412.

SESSION VI:
EMITTER DESIGN AND TESTING

A Fluidized Bed Selective Emitter System Driven By a Non-premixed Burner

U. Ortabasi*, K. O. Lund[+] and K. Seshadri[+]

UNITED INNOVATIONS , Encinitas, CA 92024

[+]*Center for Energy and Combustion Research, Dept. of Applied Mechanics and Engineering Sciences, University of California, La Jolla, CA 92093 - 0411*

Abstract. One of the key priorities in the development of Thermophotovoltaic power technology is a highly efficient heat-source/emitter system that is robust and stable. This paper describes a tightly coupled burner/selective emitter combination that integrates two novel concepts that are now under development: A fluidized bed emitter that consists of hollow, submillimeter spheres as the sources of radiant energy and a non-premixed, self regulating burner. The rational behind the proposed system is to combine the unique intrinsic features of both concepts to provide the TPV community with an enabling technology.

The fluidized bed provides excellent heat transfer, temperature uniformity, high radiant power density, reduced substrate and combustion background, robustness, thermal shock resistance, minimal contamination and long operational life.

The paper discusses a fluidized bed system that consists of selectively emitting, hollow Ho-YAG spheres with 500 micron diameter and 10 - 100 micron shell thickness operating at 1500K. Key issues related to heat transfer and radiation transport in the fluidized bed are analyzed. The collective emitter efficiency and power density of a fluidized bed are discussed.

The non-premixed burner achieves very high temperatures, has a low emission in toxic byproducts, provides self regulating stability, eliminates flashback hazards and is operable with hydrogen.

The paper concludes with a description of a complete fluidized bed TPV system including an elliptic/parabolic transfer optics and a photovoltaic cavity converter that boosts the flux density received by the photovoltaic cells.

INTRODUCTION

After a relatively inactive period of almost a decade, TPV is now emerging as a promising heat-to-electric energy conversion technology with a broad spectrum of potential applications ranging from a few watts to megawatt systems. TPV is suitable for use with a multitude of conventional and renewable primary energy sources like fossil fuels, nuclear, radioisotopes, solar, bio-fuels and hydrogen. When used with hydrogen, the emission of toxic by-products can be several orders lower than the best emission levels obtainable with present day combustion systems.

In principle there are no thermodynamic barriers that would prohibit TPV system efficiencies in excess of 40%. However, material limitations at the component level and "mismatch" of the components at the system level have been the stumbling blocks that slowed down the progress over the last 30 years. A critical evaluation of the TPV key issues, from combustion stage to the power conditioned electrical output, is given in a recent paper by D.C. White and H.C. Hottel (1). As pointed out in that paper and in agreement with many others in the field, TPV technology requires a total system approach. To have a high performance TPV device, all system subcomponents will have to be optimized and carefully tuned to each other to provide an efficient energy flow throughout the system. Recent progress in the area of low bad-gap solar cells and the advent of new, selective materials emitting in the respective frequency bands, now together provide a new degree of freedom that will allow to achieve unprecedented system performances.

The overall block diagram of the fluidized bed TPV system under study is shown in Fig 1. The present paper describes the first two principal stages of such a system: The Non-Premixed Burner (NPB) and the Fluidized-Bed Radiant Emitter (FBRE). Fig. 2 is a schematic illustration that shows NPB and FBRE thermally coupled via a heat transfer system that uses helium as the working gas. Although the present study uses hydrogen as the reactant, other gaseous- and liquid fuels like natural gas, propane, diesel oil, kerosene etc., can also be used as the source of thermal energy. Similarly, the particular selective emitter beads chosen for this study can also be replaced by other selective- or black body radiator beads that are sufficiently transparent to the useful part of the spectrum converted by the cells. Thus the fluidized bed concept presented here has general applicability. The paper is mostly concentrated on the second stage i.e. the fluidized bed radiant emitter. The non-premixed burner is briefly described below. Detailed description of the mechanisms and the structure of non-premixed combustion burners are given in references (2) and (3).

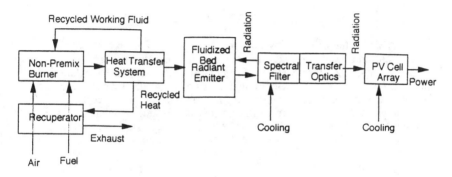

FIGURE 1. Principal Components of the Fluidized Bed TPV System

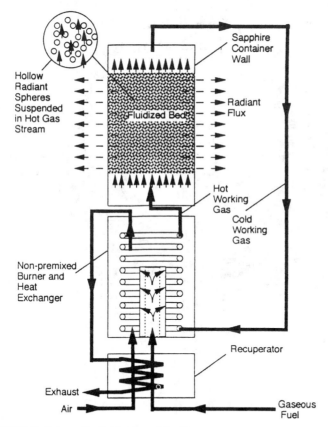

FIGURE 2. Schematic Illustration of the Non-premixed Burner/ Fluidized Bed Radiant Emitter System

NON-PREMIXED BURNER

Most of the present-day burners employ premixed combustion, i.e. the reactant and the oxidizer are mixed prior to the combustion process. In NPB systems, also referred to as Counter-Flow Burners, the reactant and the oxidizer are mixed in the combustion zone where the flame is created. Fig. 3 shows the basic principles of such a system. The burner consists of a porous cylinder through which gaseous fuel is introduced radially into the reaction zone. The counter flowing, oxidizing gas is introduced perpendicular to the axis of the cylinder. In non-premixed combustion the flame will always be stabilized where the flux of the fuel and the flux of the oxidizer are in stochiometric proportions. Thus, complete combustion is always obtained.

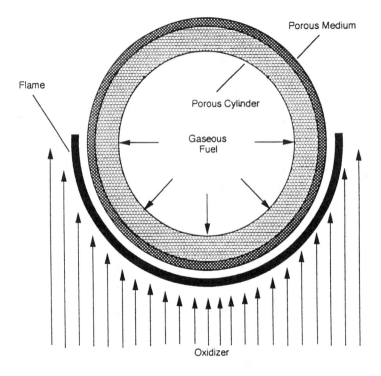

FIGURE 3. Principles of Non-Premixed or Counter-Flow Burner

The temperature profile of the flame and the flame stand-off distance depend on the characteristic residence time of the reactants in the flame. Strain rate, which is defined as the gradient of the flow-velocity normal to the flame surface is a measure of the characteristic residence time. In non-premixed combustion the strain rate depends on the flow velocity of the reactants which can be accurately controlled. Thus, the maximum flame temperature and the concentrations of various species formed in the reaction zone can be accurately controlled by changing the flow rate of reactants. NPB systems allow the designer to control the flow rate of reactants in such a way as to achieve maximum combustion efficiency and lowest emission of toxic by-products such as CO, HC and NO_x.

NPB's have a number of other advantages over the premixed systems. In premixed systems the burning velocity of the flame depends on the composition and the initial temperature of the reactants. The burning velocity of the flame increases with increasing pre-heat temperature of the reactants which in turn increases the possibility of flashback. Therefore in practical systems the incoming reactants cannot be heated beyond a certain value. Since the burning velocity of the pre-mixed flame is strongly influenced by the composition of the reactant streams, the range of composition and temperature of the reactant streams over which a premixed flame can be stabilized is somewhat limited. By contrast, in non-premixed systems flames can be stabilized over a wide range of composition, preheat temperature and flow velocity of the reactant streams, because the possibility of flashback is completely eliminated. Thus the NPB flames can be operated over wider limits than the conventional pre-mixed combustor. The use of hydrogen gas as a NPB fuel is completely safe because of the absence of flashback hazard. In contrast, hydrogen cannot be used safely in pre-mixed systems, because of the greater danger of flashback.

FLUIDIZED BED RADIANT EMITTER

Around the turn of the century it was recognized that very fine particles and fibers of refractory materials could be brought rapidly and efficiently to incandescence when heated in a gas flame. Shortly thereafter the selective emissivity of the well known "Gluehstrumpf" was discovered (4). Filaments of the selectively absorbing Ceroxide, coated with a thin layer of Thoroxide were woven to form a so called "Mantel" which then became known as the Welsbach Mantel. This porous structure, when heated in the colorless flame of a Bunsen burner, emitted intense radiation in the visible. Gas driven Coleman lanterns are the commercial outgrowth of this discovery.

Recently renewed interest in TPV sparked many imaginative concepts to improve the radiative and mechanical properties of the selective emitters. R. E. Nelson of Tecogen (5) reported on new, fibrous rare earth emitter structures with improved mechanical strength and better view factor when illuminating photoconverter arrays. Theoretical and experimental work on thin film selective emitters were carried out by Donald L. Chubb, R. A. Lowe and co-workers at NASA, LeRC and Kent State University in Ohio, respectively(6). J. B. Milstein, et. al. (7) of the University of Massachusetts and New Materials, Inc. investigated ytterbium oxide - aluminum oxide systems in porous form. P. L. Adair and M. F. Rose (8) of Space Power Institute in Alabama have developed composite rare earth oxide filaments with non-selective "binders" and demonstrated improved structural reliability over Welsbach type mantels. It is beyond doubt that these novel approaches have advanced the selective emitter technology significantly. The authors strongly believe however, that in spite of these incremental progresses, significant work is required to improve the selective radiant emitters to bring about a robust, cost effective and efficient TPV device that is competitive. Below we describe a fluidized bed selective emitter system (see Fig. 4) that promises significant improvements in the following key areas:

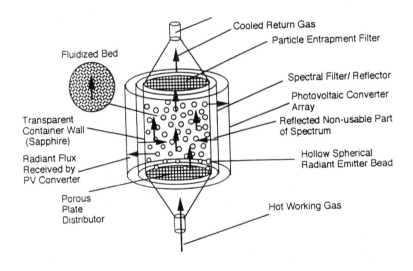

FIGURE 4. Schematic Illustration of a Fluidized Bed Radiant Emitter PV Converter

- Increased Power Density and Long Term Stability
- Uniform Exitance and Large View Factor
- Improved Conforming of Emitter and Heat Source
- Uniform Heat Transfer and Efficient Thermal Coupling
- Circumvention of Fragility and Stress Build-up Problems
- Minimization of Parasitic Off-band Radiation
- Fast Thermal Response Time without Thermal Shock
- Multi-Emitter Composite Source by Simple Physical Mixing
- Isolation of Emitter from Reactants and Combustion Process
- Flexible, Multi-fuel Capability including Liquid Fuels

The Fluidized State

Fluidized Bed Technology is used routinely to day in the chemical-, petroleum-, and power generation industries. It is a well established method to control and accelerate reactions in multiphase flows where heat transfer, surface contact and reaction uniformity are important. An excellent treatise of the theoretical aspects of fluidization is given in a recent book by D. Gidaspow (9).

The basic principle of the concept is simple. Fig. 5a. shows an experiment where particles such as sand are poured into a tube provided with a porous plate distributor at the bottom. Gas is then forced upward through the particle bed. This flow causes a pressure drop across the bed. When this pressure drop is sufficient to support the weight of the particles, the bed is said to be at minimum fluidization. The gas velocity at this point is called minimum fluidization velocity. The porosity ε at minimum fluidization is determined from the height of the bed. Figures 5b. and 5c. show bubbling and slugging of the bed as the flow rate of the gas is gradually increased. Below, for the sake of simplicity, we will only consider the minimum fluidization state, although bubbling or slugging may turnout to be more beneficial from a heat transfer point of view. The one dimensional steady gas-solid flow in the absence of acceleration is described in terms of the momentum balance equation:

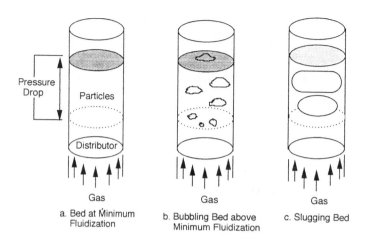

FIGURE 5. Fluidization Regimes at Increasing Gas Velocities

$$\text{Buoyancy} = \text{Drag} - \text{SWL} + \text{GWF} + \text{SPG}, \tag{1}$$

where, SWL = solids wall friction, GWF = gas wall friction, and SPG = solids pressure gradient. In the absence of gas wall friction and solid stress transmitted by the particles, Eqn. (1) can be written as Buoyant Force = Drag, or

$$(1-\varepsilon)\,[\rho_s - \rho_g]\,g = (\beta_A/\varepsilon)\,[v_g - v_s], \tag{2}$$

where, ε = porosity of the bed (gas volume fraction), ρ_s = solid density, ρ_g = gas density, g = gravitational acceleration, β_A = friction coefficient, v_g =

gas flow velocity and v_s = solid flow velocity. β_A can be obtained from Ergun equation (9) to be:

$$\beta_A = 150 \{ [(1-\varepsilon)^2 \mu_g / [\varepsilon (d_p \phi_s)^2]\} + \qquad (3)$$
$$1.75 \{ [\rho_g |v_g - v_s|(1-\varepsilon)] / (d_p \phi_s)\}.$$

By definition, at minimum fluidization velocity v_s is zero. Thus Eqns. (2) and (3) give a relation between ε_{mf} and $U_{mf} = \varepsilon_{mf} v_{gmf}$, where ε_{mf}, U_{mf} and v_{gmf} are porosity, superficial gas velocity (volumetric flow/tube area) and the actual gas velocity at the point of minimum fluidization, respectively. d_p is particle diameter and ϕ_s is the sphericity of the particle. In the following uniform particle size and perfect sphericity i.e. $\phi_s = 1$ will be assumed.

For small particles or for Reynolds numbers, i.e.[($d_p U_{mf} \rho_f)/ \mu$] < 20) substitution of Eqn. (3) into Eqn. (2) yields the following expression for the U_{mf}:

$$U_{mf} = (d_p^2 \Delta\rho\, g\, \varepsilon_{mf}^3) / (150\, \mu\, [1-\varepsilon]). \qquad (4)$$

For spherical particles of uniform size distribution we expect the porosity at minimum fluidization velocity to be close to that of a bed packed in a cubic mode, with $\varepsilon_{fm} = 1 - \pi/6 = .476$. The quantities $\Delta\rho$ and μ are the solid-gas density difference and the viscosity of the gas, respectively.

A Sample Case:

Operational Temperature : 1500 K

Emitter Material: Ho, ρ_s = 8.75 g/cm^3

Type of Spheres: Hollow, uniform size, sphericity = 1

Sphere Dimensions: Outer Radius = .5 mm, wall thickness = .1mm

Effective Sphere Density: <ρ> = 4.27 g/cm^3

Gas : Air; μ(1500K) = 540 g/cm sec, ρ_g(1500K) = .235 mg/cm^3

Porosity: $\varepsilon_{fm} = .476$ (cubic mode packed bed)

Superficial Velocity: U_{fm} (From Eqn. 4) = 26.6 cm/sec (~ 1 km/hr)

Actual Gas Velocity: $v_g = U_{fm}/\varepsilon_{fm}$ = 55.9 cm/sec (~ 2 km/hr)

The calculated gas velocities are quite low (comparable to a very light breeze). The respective pressure drop in a 100 cm tall radiant bed will be only in the order of 0.2 atmospheres. These results indicate that the energy required to circulate the working gas would be minimal and no technical difficulties in this area are expected.

Heat Transfer and Radiative Properties of the Fluidized Bed

It is assumed that the particles making up the fluidized bed are hollow spheres made of 25% Ho-YAG material described in reference (10). The diameter and the shell thickness are 0.5 mm and 0.01 mm, respectively. Thin flat samples of Ho-YAG mounted on a platinum foil have been investigated by the authors of references (6), (10) and (11). Although some of the materials related observations they made apply to the present concept, radiation transfer in a fluidized bed involves significantly different mechanisms than in the bulk. Some of these differences are mentioned here briefly.

Heat Transfer

Thin selective emitter studies reported (10) involve emitter thicknesses ranging from 0.2 to 2mm. The samples are mounted on a platinum foil substrate that thermally couples the heat source with the emitter material. A temperature gradient in the order of 200 to 300K exist between the substrate and surface of the sample being viewed. The magnitude of the gradient increases with increasing temperature.

The hollow spheres considered in this study have shell thicknesses ranging from 10 to 100 microns i.e. they are thinner by a factor of 20 than their flat counterparts. The spheres are fully immersed in the hot working gas so that a uniform temperature distribution for each bead is achieved. Temperature gradients across the fluidized bed are possible, but they can be minimized by controlling the state of fluidization i.e. the level of agitation, heat transfer and mixing. Thus in the fluidized bed the temperature gradient related losses can be considered negligible.

Emitter Efficiency

In the study reported by Lowe et. al.(10) the flat emitter samples are mounted on a substrate that contributes to the off-band radiation flux. This parasitic background lowers the efficiency η_E of the emitter as can be seen from Eqn. 5:

$$\eta_E = Q_b / (Q_l + Q_b + Q_u) \tag{5}$$

where, Q_b, Q_l and Q_u represent the radiated power within the band, below the band and above the band (6). This adverse effect of the substrate background on the efficiency becomes worse with decreasing emitter thickness as reported by Chubb and co-workers (6). In the case of filamentary rare earth emitters the ceramic support structure and the hot combustion products contribute to the off-band radiation contamination. These trends do not effect the fluidized bed performance for the following reasons:

- The hollow beads in the fluidized bed are basically suspended in the working gas and are self supporting. Thus, there is no background contribution associated with a structure that is attached to the emitter. As the fluidized emitter bed is radiation-wise isolated from the combustion chamber, the back ground emissions from hot combustion products can be totally blocked from reaching the PV cells. Given the fact that the emissivity of the heat transfer gas can be assumed negligible, the only other off-band radiation load that exist in the fluidized bed is the transparent container wall that holds the emitter particles. This source of parasitic radiation is mitigated by the fact that the total emitter surface in the fluidized bed is several orders larger than the surface area of the container wall. For example, if a container tube that is 100 cm high and 10 cm in diameter is filled with spheres of 500 μm diameter and fluidized in the cubic mode (porosity, $\varepsilon_{fm} = .476$), a surface area enhancement of about 280 is achieved. This emitter surface-to-wall surface ratio can be improved by going to smaller spheres and/or larger container diameters. It is important to note however, that the actual emitter efficiency η_E does not improve linearly with the increasing surface area in the bed. The effective extinction coefficient of the fluidized bed as a result of absorption, scattering and diffraction by the spheres diminishes the exitance i.e. the flux leaving the container envelope per unit area. The complex phenomena that control the actual exitance from a given fluidized bed is highlighted under the section on power density.

- Another factor that also negatively effects the efficiency of the thin, flat emitters is the surface heat waste due to convection and conduction losses. These loss mechanisms, particularly the convective heat removal from the emitting surface, seem to be quite significant as manifested by the large temperature gradients measured between the substrate side (heater side) and the viewed side of the emitter sample (10).

Power Density

One of the key advantages of the small particle emitter bed concept is the significantly increased emissive surface area as compared to monolithic type emitters, occupying the same amount of volume. To a large extend this is also true for other large emitters in form of a shell, like a hollow tube with a thin wall. The relative enhancement of power density in a fluidized bed becomes obvious, if we assume that the small particles are largely transparent to their own band emission (e.g. hollow spheres with extremely thin walls) and also absorb, scatter and diffract independently. This can easily be seen, if we assume that both emitters are radiating at the same temperature and are surrounded by the same blackbody. Given these conditions, the ratio of the power densities generated by the respective emitters would be close to the ratio of their surface areas .Thus the exitance generated by the fluidized bed emitter would be several orders higher than the power density radiated from the surface of the monolithic emitter.

Although the fluidized bed consisting of small hollow spheres is a good way to increase the flux density, the thought experiment above is only indicative of the principle and it oversimplifies the real nature of radiation transport in such a system. Below we highlight some key parameters and issues that limit the power densities obtainable from fluidized bed radiant emitters.

Radiation Transport in Fluidized Bed

Radiation transport in a packed bed of transparent- and semi-transparent particles with emissive properties has been studied in the past (12). For a general review of various analytical methods that has been developed to predict the radiative properties of porous media, the reader is referred to a paper by M. Kaviany and B. P. Singh (13). To our knowledge there are no analytical models available that describe the radiation transport in a fluidized bed consisting of selectively emitting hollow spheres with very thin walls. Such a model is now being developed by the authors. Below we discuss some of the particular issues associated with such a system.

Dependent versus Independent Radiation Properties. To formulate the integro-differential equations that describe the radiation transport in a porous medium the knowledge of the radiation properties of the medium is required. These are the averaged values for scattering coefficient $<\sigma_S>$, absorption coefficient $<\sigma_A>$ and the particle scattering phase function $<\Phi>$. The scattering and absorption processes are called *dependent*, if the scattering and absorbing characteristics of a particle in a medium are influenced by neighboring particles and are called *independent*, if the presence of neighboring particles has no effect on the absorption and scattering by a single particle. The assumption of independent scattering greatly simplifies the task of obtaining the radiative properties of the medium. Independent theory is best applicable to systems with larger porosity and predicts higher transmission for the porous bed. Below we explore to what extent the fluidized bed under consideration lends itself to be treated by the independent theory.

Criteria for the independent Theory. Limits on the validity of independent transport are a minimum value of porosity and a minimum value of the ratio C/λ, where C is the average inter-particle spacing based on rhombohedral packing. This particular arrangement gives the maximum particle concentration for a given inter-particle spacing C. This assumption in turn leads to the relations between inter-particle clearance C, the porosity ε and the wavelength λ involved. These relations are:

$$C/d = [.9047/(1-\varepsilon)^{1/3}] - 1 \qquad (6)$$

$$C/\lambda = (\alpha_R/\pi)[.9047/(1-\varepsilon)^{1/3}] - 1 \qquad (7)$$

where, $\alpha_R = (\pi d/\lambda)$, d = particle diameter and λ = wavelength of the photons interacting with the particles.

An earlier work by Hottel et. al. has identified the limits of independent radiation transport as $C/\lambda \geq 0.4$ and $C/d \geq 0.4$ (i.e., $\varepsilon \geq 0.73$).

Independent Radiation Transport and State of Minimum Fluidization. For the Ho-YAG system under consideration we have d = 500 microns and $\lambda \cong 2$ microns @ 1477K. This value is about the center of the emission band reported in reference (9). Using Eqn. 6. and the criteria by Hottel et. al. for independent radiation transport and assuming rhombohedral porosity i.e. $\varepsilon = 0.73$ we find that the average particle spacing C must be equal or larger than 200 microns. Using this value we calculate $C/\lambda = 100$ and $C/d = 0.4$. Thus, both criteria are satisfied and the system at minimum fluidization can be modeled according to independent radiation transport theory. According to reference (13), the average spectral scattering coefficient $<\sigma_{\lambda S}>$ for a fluidized bed, consisting of uniformly distributed spheres and $\varepsilon \geq .73$ can be written as:

$$<\sigma_{\lambda S}> = [\,3\,(1-\varepsilon)\,A_{\lambda S}\,]/[\,4\pi R^3\,] \quad (8)$$

with,

$$A_{\lambda S} = \eta_{\lambda S}\,\pi\,R^2$$

$$\eta_{\lambda S} = [\,\int_{4\pi} I_{\lambda S}\,r^2\,d\Omega\,]/[\,\pi\,R^2\,I_{\lambda i}\,],$$

where, R = physical radius of the spherical particle, $A_{\lambda S}$ = spectral scattering cross-section of the particle, $\eta_{\lambda S}$ = spectral scattering efficiency, $I_{\lambda S}$ = local scattered spectral intensity and the quantity $[\,\pi\,R^2\,I_{\lambda i}\,]$ = spectral power arriving at the sphere. Similar treatments can be given to the respective averaged absorption coefficient and the particle phase function.

•*Need for Experimental Data.* There are two distinct phenomena that limit the validity of the independent theory in the case of the spherical hollow sphere:

1.) As previously stated, at the onset of minimum fluidization the bed is likely arranged in a cubic mode. Thus the bed porosity is 0.476. This value is substantially less than the required porosity of 0.73 that is necessary to fulfill the criteria for independent radiation transport. By increasing the working gas velocity beyond minimum fluidization, the porosity of 0.73 can be reached. It is conceivable, however that at this point bubbling or slugging may be induced. Thus the concept of average particle distance loses its meaning and the independent radiation transport theory loses validity.

2.) The hollow sphere structure of the radiant emitter particles have different transmission properties than solid transparent spheres studied in the past. As seen in Fig. 6 the photons passing through the sphere undergo multiple reflections in the hollow region of the sphere. They also may be trapped within the shell as a result of total internal reflections. This violates the independent theory requirements as many absorptions and scatterings are happening within the minimum elemental volume that is required to warrant the independent radiation transport.

Both phenomena described under 1.) and 2.) are complex and difficult to model analytically. Relatively simple, straight forward experiments however could be easily performed to measure the collective optical properties of the

information would help the designer to maximize the flux densities that are obtainable from fluidized bed emitter systems.

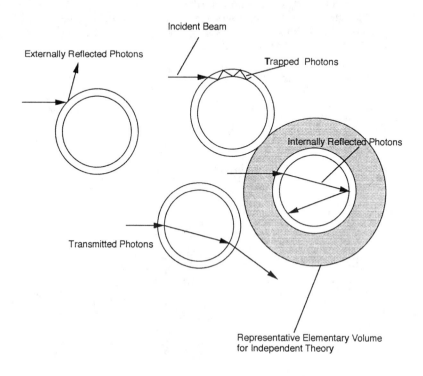

FIGURE 6. Complex Photon Paths in the Fluidized Bed

FLUIDIZED BED TPV SYSTEM CONCEPT

Fig. 7 is a schematic illustration of the fluidized bed TPV system under consideration (the heat recovery is not shown). Helium gas (working gas) is heated by a non-premixed burner via a heat exchanger and (working gas) is heated by a non-premixed burner via a heat exchanger and is forced through the distributor plate at a predetermined flow rate. Rising gas fluidizes and heats the hollow, spherical emitter particles simultaneously. At the required design temperature intense radiant flux emanates from the transparent sapphire envelope which also structurally contains the fluidized bed. A spectral filter reflects back the non-usable part of the irradiant flux. A symmetrical and highly reflective transfer optics concentrate the radiation onto the two photovoltaic cavity converters positioned at the opposite ends of the housing that supports and integrates the emitter and the transfer optics.

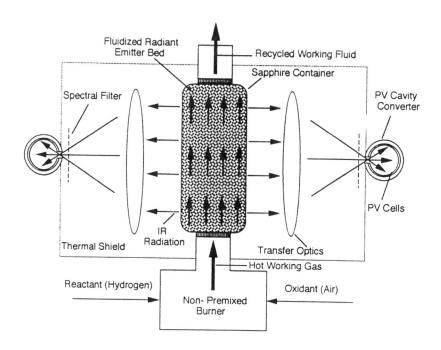

FIGURE 7. Schematic Illustration of Non-Premixed Burner / Fluidized Bed TPV System

Fig. 8 is a more detailed illustration of the optical design. The transfer optics section consists of an elliptical first stage (a quasi collimator) and a compound parabolic concentrator (CPC) second stage. The center of the fluidized bed coincides with the foci of the overlapping ellipses on either side. The design requirement is that all rays entering the aperture of the second stage are within the acceptance angle of the CPC and are received by the cavity.

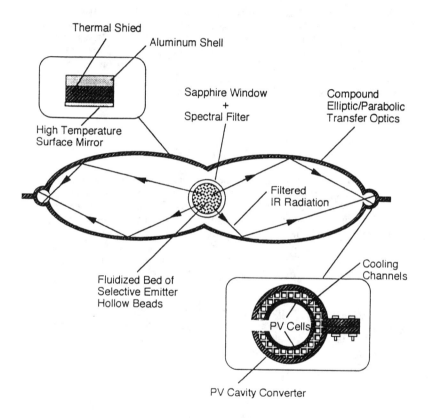

FIGURE 8. Cross-sectional Top View of Fluidized Bed and Optics

The Photovoltaic Cavity Converter (PVCC) has been described by Ortabasi (14). Here we highlight only the most relevant features.

Inside the cavity the photovoltaic cells are mounted on the highly reflecting interior surface coated for example with gold. To minimize the recombination losses in the bulk, the cell thickness is kept to a minimum and good thermal contact is established between the cell and the cavity wall for efficient cooling. Once the concentrated photon flux enters the cavity the escape probability is determined by the ratio A_c / A_{ap} where, A_c and A_{ap} are the interior surface area of the cavity and the aperture area of the cavity, respectively. Depending on the concentration ratio of the second stage (CPC), the escape probability can be made very small so that the photons entering the cavity are basically trapped inside the cavity. A system like this allows the use of very extensive grid pattern on the cells assuming that the grid itself is highly reflective. If the escape probability for the photons can be made small enough, even the anti-reflective coating on the cells can be omitted as all photons trapped are ultimately absorbed in the active areas of the cell array after a finite number of reflections. A high conversion efficiency is expected from PVCC because of the inherent features outlined above and in ref. (14).

CONCLUSIONS

A fluidized bed radiant emitter, thermally driven by a non-premixed burner will advance the state-of-the-art TPV technology in many ways. Key advantages of such a system are increased efficiency and power density, reduced parasitic thermal load, circumvention of thermal shock, and elimination of material fragility problems. The system has multi-fuel capability and can be used also with a variety of radiant emitters suitable for TPV. If hydrogen is used for the non-premixed combustion the environmental impact of TPV electric power generation would be close to zero.

The project is at an advanced stage of conceptual design. There is a need for experimental work that requires a prototype fluidized bed. Resulting data, when combined with a modified version of independent transport theory, will allow the optimization of the fluidized bed radiant power output.

REFERENCES

1. White, C. W. and Hottel, H. C., " Important Factors in Determining the Efficiency of TPV Systems," presented at the First NREL Conference on, Thermophotovoltaic Generation of Electricity, Copper Mountain, CO, July 17 -19, 1994.

2. Puri, I. K., and Seshadri, K., Combustion and Flame, **10**, 137 - 150.

3. Trees, D., et. al., Combustion Science & Technology, (Submitted April, 1994).

4. Pohl, R. W., Optik und Atom Physik, 10th Edition, Springer Verlag, 1958, ch. 16, p. 290.

5. Nelson, R. E., " Thermophotovoltaic Emitter Development," presented at the First NREL Conference on Thermophotovoltaic Generation of Electricity, Copper Mountain, CO, July 17 -19, 1994.

6. Chubb, D. L., et. al., " Emittance Theory for Thin Film Selective Emitter ," presented at the First NREL Conference on Thermophotovoltaic Generation of Electricity," Copper Mountain, CO, July 17 -19, 1994.

7. Milstein, J. B., et. al., Some Characteristics of a Novel Direct Thermal to Optical Energy Converter Medium "presented at the First NREL Conference on Thermophotovoltaic Generation of Electricity," Copper Mountain, CO, July 17 -19, 1994.

8. Adair, P. L. and Rose, M. F., "Composite Emitters for TPV Systems," presented at the First NREL Conference on Thermophotovoltaic Generation of Electricity, Copper Mountain, CO, July 17 -19, 1994.

9. Gidaspow, D., Multiphase Flow and Fluidization, New York: Academic Press, 1994.

10. Lowe, R. A., et. al., " Radiative Performance of Rare Earth Garnet Thin Film Selective Emitters," presented at the First NREL Conference on Thermophotovoltaic Generation of Electricity, Copper Mountain, CO, July 17 -19, 1994.

11. Good, B. S., et. al., " Temperature-Dependent Efficiency Calculations for a Thin-film Selective Emitter," presented at the First NREL Conference on Thermophotovoltaic Generation of Electricity, Copper Mountain, CO, July 17 -19, 1994.

12. Singh, B. P. and Kaviany, M., Int. J. Heat & Mass Transfer, **34**, pp. 2869 - 2881 (1991).

13. Kaviany, M. and Singh, B. P., Advances in Heat Transfer, **23**, pp. 133 - 186 (1993).

14. Ortabasi, U., Space Technology, **13**, p. 513 - 523 (1993).

SiC IR Emitter Design for Thermophotovoltaic Generators

Lewis M. Fraas*, Luke Ferguson*,
Larry G. McCoy[†] and Udo C. Pernisz[†]

*JX Crystals Inc., 1105 12th Ave NW, Suite A2, Issaquah, WA 98027
[†]Dow Corning Corp., Midland, MI 48686-0994

Abstract. An improved ceramic spine disc burner / emitter for use in a thermophotovoltaic (TPV) generator is described. A columnar infrared (IR) emitter consisting of a stack of silicon carbide (SiC) spine discs provides for both high conductance for the combustion gases and efficient heat transfer from the hot combustion gases to the emitter. Herein, we describe the design, fabrication, and testing of this SiC burner as well as the characterization of the IR spectrum it emits. We note that when the SiC column is surrounded with fused silica heat shields, these heat shields suppress the emitted power beyond 4 microns. Thus, a TPV generator using GaSb photovoltaic cells covered by simple dielectric filters can convert over 30% of the emitted IR radiation to DC electric power.

Introduction

Silicon carbide is an excellent IR emitter material. It has a high emissivity, good thermal conductivity, and good thermal shock resistance. Since its high temperature limit is somewhere between 1800 K and 2000 K with a corresponding peak wavelength for blackbody IR emission ranging between 1.45 and 1.6 microns, it is an excellent material for use with low bandgap thermophotovoltaic cells. For example, GaSb cells respond out to a wavelength of 1.7 microns.

In our early experiments, a simple emitter / burner using off-the-shelf SiC tubing was designed and operated with a heat exchanger and a blower. The burner design consists of a SiC tube with combustion occurring at its top end with hot combustion gases being blown downward to the open SiC tube bottom end. This SiC emitter tube is contained within a quartz tube and the exhaust gases then blow upward outside the emitter tube being confined by the outer quartz tube. The SiC tube is thus heated on a inside-down pass and then again on an outside-up pass. When operated without filter recycling, this burner runs with very low CO and hydrocarbon emissions. This burner also runs quietly as expected.

When this simple emitter was then surrounded with IR filters, the temperature for the same flow conditions rose. However, detailed measurements indicated that both the emitter temperature and the chemical to radiation efficiencies need to be improved. We believe that this can be done with custom SiC parts designed to promote more surface contact with the hot gases and more turbulent flow.

Herein, we describe an improved ceramic spine disc burner / emitter possessing both high conductance for the combustion gases and efficient heat transfer from the hot combustion gases to the emitter. We describe the design, fabrication, and testing of this SiC burner as well as the characterization of the IR spectrum it emits. We begin by describing the design concepts. We then proceed to the results of measurement and calculation, and finally, we present our conclusions.

Design Concepts

There are two important concepts associated with the SiC burner design described here. The first concept is the use of SiC spine discs to promote more hot gas contact area and turbulence and to tailor the emitter column uniformity. The second concept is the use of fused silica heat shields to tailor the IR emission spectrum. In the following, we describe these two concepts in successive order.

Figures 1 and 2 show the SiC spine disc burner. Figure 1 shows a vertical cross section through the entire thermophotovoltaic generator cylinder. Combustion air and fuel are supplied at the top of the cylinder. Exhaust gases also exit at the top. The supply air meanders through a stainless steel heat exchanger where it is preheated by the counterflow exhaust gas. The temperatures of both the supply air and the exhaust gas at the lower end of this heat exchanger are close to 800°C but below 900°C in order to avoid extensive oxidation of the stainless steel heat exchanger plates. The preheated air then passes downward through a second ceramic spine disc heat exchanger where it is heated still more. Fuel is supplied from the top directly through both heat exchangers. The fuel and preheated air are then mixed at the lower end of the ceramic heat exchanger and thence enter a combustion chamber. As chemical energy is added through combustion, the combustion gases are heated to well over 1400°C. These hot gases then meander downward through a SiC spine disc IR emitter section, heating the emitter to at least 1400°C. The hot gases then return upward outside the emitter spine disc heating them in a second upward pass. Thence, the exhaust gases are cooled from approximately 1400°C to below 900°C in the upward pass, percolating through the ceramic heat exchanger into the stainless steel heat exchanger.

The exhaust gases on the upward pass by the emitter spine discs are confined by a cylindrical tube which can be either a transparent fused silica tube or a SiC emitter tube. In any case, the emitter is surrounded by at least one fused silica tube. Thermophotovoltaic cell circuits surround the IR emitter and receive a fraction of the emitted IR and convert it to DC electric power. These circuits can be either liquid cooled or air cooled through cooling fins as shown. Infrared filters are bonded directly to the cells.

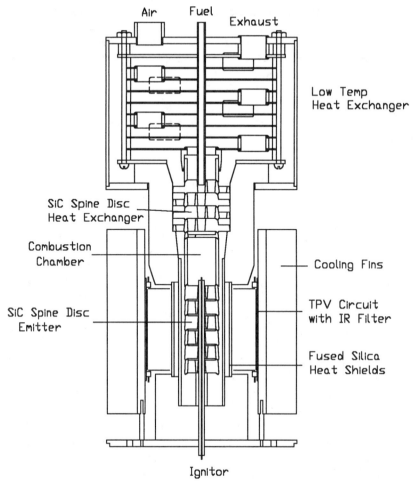

FIGURE 1. Vertical cross section through TPV generator cylinder.

The ceramic burner / emitter / regenerator sections are shown in more detail in figure 2. Note that the emitter section consist of identical SiC spine discs. In the present embodiment, the SiC emitter spine discs contain seven holes with one central hole surrounded by a hexagonal array of six holes. The emitter is made up by stacking or nesting several discs together. The center hole allows the insertion of an ignitor into the combustion chamber and also serves to center the discs. When each disc is rotated 30 degrees relative to adjacent top and bottom discs as is indicated in the A-B and C-D sections, the hot gas is forced to percolate back and forth through the hexagonal hole array as it traverses the column efficiently transferring energy to the emitter.

Figure 2 also shows the spine disc heat exchanger. Note that these heat exchanger discs have an additional array of six holes on a larger diameter. This second outer array of holes forces the exhaust gases to percolate on the up-pass thereby transferring energy to the supply air passing downward through the inner array of holes in these discs. Fuel is fed through the center hole.

In this generator, GaSb cells are used along with a SiC emitter operating in the 1700 K to 2000 K temperature range. GaSb cells are sensitive to IR energy for wavelengths less than approximately 1.7 microns and SiC is a near blackbody emitter. As a blackbody, SiC emits energy at all wavelengths. However, when a SiC emitter is surrounded by fused silica, it is convenient to divide the spectrum into three parts, i.e. wavelengths less than 1.7 microns, wavelengths between 1.7 and 4 microns, and wavelengths longer than 4 microns.

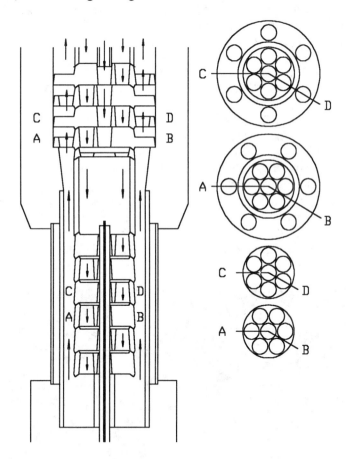

FIGURE 2. Vertical cross section through spine disc emitter and heat exchanger sections with disc orientation detail shown at right.

Since fused silica absorbs wavelengths longer than 4 microns and reradiates the absorbed energy in both directions, a single fused silica tube surrounding a SiC emitter will act as a heat shield returning half of the absorbed energy beyond 4 microns back to the emitter. More generally for N silica heat shields, the energy radiated beyond 4 microns can be reduced by $1/(N+1)$. In other words, the combination of a SiC emitter surrounded by N fused silica heat shields acts as a selective emitter with a high emissivity for wavelengths less than 4 microns but with a low emissivity for wavelengths longer than 4 microns. For the case of two silica tubes as shown in figure 1, one confining the exhaust gases and a second serving as a convection barrier, the effective emissivity is over 75% for wavelengths less than 4 microns and less than 33% for longer wavelengths. A simple dielectric filter can then be used. It can be designed to transmit for wavelengths less than 1.7 microns and to reflect wavelengths between 1.7 and 4.0 microns. The IR filter can transmit again beyond 4.0 microns where the fused silica heat shields have suppressed the emitted energy.

Results of Measurement and Calculation

A prototype high-density SiC emitter assembly using pre-ceramic polymer technology was fabricated by Dow Corning. Parts were first formed by isopressing and machining, and were subsequently fired and sintered at high temperature. Initial trial runs at JX Crystals now show higher temperatures and improved fuel-to-IR radiation efficiencies.

Figure 3 shows a photograph of SiC emitter spine discs and a two element spine disc column. Although figures 2 and 3 show a 30 degree relative rotation from disc to disc, the relative rotation can be varied from this amount through the column in order to optimize the IR radiation intensity uniformity along the length of the TPV receivers.

FIGURE 3. Photographs of the SiC spine discs fabricated by Dow Corning. Eight of these discs were stacked to form the TPV emitter column described here.

TABLE 1. Burner Uniformity Experimental Results

Spine Disc Orientation*	°C (top)	°C (bottom)
/\/\/\/\	1100	1000
////\/\/	1050	1100
/\\\/\/\	1100	1100

* Eight stacked spine discs, "/ /" or "\ \" imply the same orientation for adjoining discs, "/ \" or "\ /" imply 30 degree offset orientation for adjoining discs.

After assembling a SiC spine disc burner / emitter column, we first conducted experiments with different spine disc rotations in order to optimize temperature uniformity from top to bottom of the column as measured with a pyrometer. Table 1 shows the results of our experiments on uniformity for a spine disc burner. As can be seen from this table, depending on the relative orientation of the spine discs, the column can be made to run hot at the top, hot at the bottom, or uniform in temperature from top to bottom.

We also measured the spectral output from our spine disc emitter with two fused silica heat shields. Measurements were made with a small spectrometer and a cooled InSb detector over the range from 1 to 5 microns. These spectral measurements confirmed the concept that the fused silica heat shields effectively suppress the IR power emitted beyond 4 microns.

A dielectric filter can be designed using alternating high and low index materials. The IR transmission curve for a 17 layer dielectric filter design is shown in figure 4. This filter transmits energy below 1.7 microns to the GaSb cells and reflects the energy between 1.7 microns and 4 microns very efficiently back to the emitter.

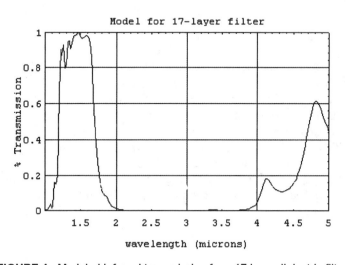

FIGURE 4. Modeled infrared transmission for a 17 layer dielectric filter.

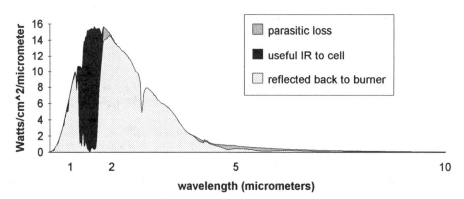

FIGURE 5. Energy partition for 1800 K emitter spectrum

The filter of figure 4 works very well with a selective SiC / fused silica emitter operating at 1800 K as is shown in figure 5. The envelope curve in figure 5 derives from our emitter spectral measurements. In figure 5, the black region represents the energy transmitted to the cell, and the light region represents the energy reflected back to the emitter effectively allowing a higher emitter temperature. Only the gray region represents lost energy. Quantitative calculation shows that the black region represents 7.9 W/cm^2 and the gray region represents 1.9 W/cm^2. Of the 7.9 W/cm^2 received by the cells, 3.1 W/cm^2 should be converted to electricity. The TPV receiver efficiency should therefore be 3.1/(7.9+1.9) = 32%.

Conclusions

The essential components for infrared energy control in our efficient thermophotovoltaic generator are a SiC blackbody emitter, fused silica heat shields, a dielectric filter, and low bandgap GaSb photovoltaic cells. The SiC blackbody emitter radiates energy over all wavelengths. Fused silica heat shields then absorb energy in wavelengths longer than 4 microns and return a large fraction of this energy back to the emitter. Infrared energy in wavelengths shorter than 4 microns is transmitted through the fused silica. A dielectric filter then reflects the energy in the wavelength range between the absorption edge of our GaSb TPV cells and 4 microns back to the emitter. Given this design, GaSb TPV cells then convert over 30% of the emitted infrared radiation to DC electric power.

A New High Temperature Air-Stable TPV Emitter

Joseph B. Milstein and Ronald G. Roy

New Material Concepts, Inc.
15 Guile Avenue, Tewksbury, MA 01876

Abstract. We have demonstrated a new class of material which appear to provide significant advantages as TPV emitters, including selective emission, high power density by virtue of operation at very high temperature, stability in air, appreciable thermal shock resistance, and the ability to be produced in desired shapes. These materials also afford the possibility of tuning the characteristic emission wavelength to various bands as may be desired, and the possibility of being powered by any one of a variety of power sources, including gas flames and concentrated solar energy.

BACKGROUND

Many emitter and photovoltaic cell combinations for use in TPV applications have been investigated in the past. It seems clear to us that certain fundamental issues must be addressed in order to produce efficient TPV systems. In particular, the issues of power density and efficiency of conversion at reasonable cost are paramount.

We will first review some well known, but occasionally overlooked, physical principles and concepts, and will point out their consequences for an efficient TPV system. We will then describe some of the features of our novel porous rare earth oxide TPV emitter, which was first discussed in part in the paper presented at this conference one year ago.(1) For this discussion we will make one significant assumption, namely that the system will employ silicon solar cells as the receiver.

This assumption is based on the recognition that silicon provides a reasonably efficient cell in either one sun or concentrator designs, these cells are relatively inexpensive compared to many other candidate materials, and these cells are readily available in quantity. Silicon technology provides a lengthy and commercially successful experience base.

RESULTS AND DISCUSSION

Power Density

For a TPV system to operate efficiently, it is necessary to attain high power densities. When one considers possible alternatives to a system or process, it is useful to take note of the successes that have been achieved to date and to understand why they are successful. For example, in providing motive power, internal combustion and jet engines use hydrocarbon fuels which have high power densities on both a unit volume and a unit mass basis. The efficiency of conversion of these fuels to useful work, rather than the total system capacity, has always been the focus of further effort in development. By comparison, systems which are believed to be more efficient in energy utilization often suffer from a serious power density limitation.

In particular, it is well known that thermal emitters have a highly temperature dependent output, or radiance. Specifically, quantum physics requires that, for an oscillator at equilibrium with a radiation field, the amplitude of the radiance at a particular wavelength and temperature is given by the Planck radiation function

$$R_\lambda = (2\pi hc^2/\lambda^5)(e^{hc/\lambda kT} - 1)^{-1}$$

We include here a plot of radiancy versus temperature for a blackbody ($\epsilon = 1$) held at temperatures of 1000 °C, 1200 °C, 1400 °C, 1600 °C, 1800 °C, and 2000 °C, as Fig. 1.

It is apparent that the area under the curve increases dramatically with temperature. This area is the integrated radiance, or the value R_c.

The integrated radiance, or the power emitted per unit area, from a blackbody cavity emitter is given by

$$R_c = \sigma T^4$$

while the radiance of a free surface is given by

$$R = \epsilon \sigma T^4$$

where ϵ is a material-dependent parameter which varies with temperature, surface texture, and surface chemistry, and can vary in the range of zero to 1. The integrated radiance, as a function of temperature is presented in Fig. 2.

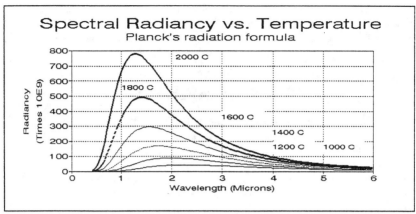

Figure 1. Radiancy of a blackbody as a function of temperature and wavelength.

We can conclude that to attain high power densities, it is necessary to operate at high temperatures. Specifically, the increase in power density is extremely nonlinear, and equal increments in temperature offer ever larger increments in power density. It is to be understood, however, that equal increases in temperature become ever more challenging to attain, both from an operational perspective and with regard to the selection, or availability, of suitable materials.

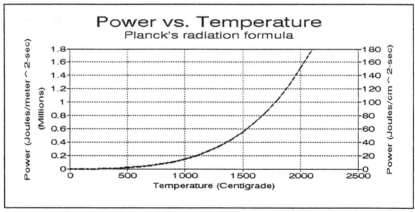

Figure 2. Integrated radiance, or power per unit area, emitted by a blackbody as a function of temperature.

The material that we have prepared is able to provide high levels of optical power. Using concentrated solar illumination, up to approximately 500 watts of emitted power has been observed (2), with fluxes of approximately 8 watts per square centimeter at a distance of a meter from the emitter.

Input Power Sources

When we consider the issues of input power, we have multiple possible sources at our disposal. We have used flames, radio frequency heating, and solar energy as sources of input power. Other possibilities for use as thermal sources can be envisioned, such as hot gases or fluids, or bombardment by particles or other electromagnetic radiation. At the present time, providing a source of sufficient power is not an issue. The thermal stability of the material, both with regard to its resistance to stress and its ability to remain chemically and physically stable at temperature, is often an issue.

Thermal Stress

While thermal stress is often not significant at moderate temperatures, in the range of 1000 to 1200 °C, at higher temperatures it often becomes a major problem. Many materials fail catastrophically when subjected either to high temperatures per se, or when subjected to fluctuations in temperature at elevated temperature. For many applications, it is necessary to have a material which can rapidly be brought to working temperature. Excursions in temperature from room temperature to full working temperature in seconds is likely to be demanded in a practical working system. Further, in instances where the power source is solar energy, fluctuations in input power during operation will be unavoidable. These various fluctuations in temperature should not impose limits on the viability of the emitter material in a practical system. The emitter should have the ability to operate reliably under rather harsh thermal stress conditions.

Less dramatic examples of stressing the material have included the use of gas flames applied to the surface of the material from torches, gas flames which have been sustained by passing fuel through passages in the material with combustion at its surface, and heating the material in a radio frequency furnace containing refractory metal hot zone components. In the latter case, temperatures in the vicinity of 2000 °C can be applied in seconds.

As will be apparent from the video presentation in the following paper by Pitts, et. al.(2), which unfortunately cannot be reproduced in printed form, our material has been subjected to repeated cycling from room temperature to approximately 2000 °C with transition times of less than one second in a solar furnace. Power reduction at similar rates has been performed as well. The material suffered no ill effects from such cycling.

Air Operation

Operation of a power source under ambient conditions is a positive factor in its possible application. Systems which require extensive environmental control for their use, or to limit hazards that they might pose, are at a significant disadvantage. In particular, one of the most difficult and costly environmental control issues is the provision of an ambient other than ordinary room air. One needs only to consider a process which is conducted in another medium to recognize the difficulties and expense which are inherent, both with regard to equipment and with regard to the supply of the medium to be maintained as the ambient. For use in space, rather than in terrestrial applications, operation under vacuum conditions might prove to be the ideal, but this is the only exception which appears to be a consideration in the foreseeable future.

Control of Emitted Wavelength

Choosing a particular photovoltaic detector inherently implies a choice of a specific band gap and absorption characteristic. One may be able to design and fabricate a detector to use a narrow excitation band without having to make compromises in the boundary conditions which would be required if wide band excitation were used. By matching the detector characteristics carefully, an efficient conversion of energy from light to electricity may be possible.

A recent report by Professor M.A. Green and co-workers (3) indicates that 45% conversion efficiency (ratio of electrical power out to power of light supplied) has been measured for illumination of a silicon solar cell with 1.064 micron light from a neodymium-doped yttrium aluminum garnet (Nd:YAG) laser. This report serves to demonstrate that it is possible to achieve quite high conversion efficiencies, especially if one is not required to tune the detector to extract energy from a broad input spectrum, but can build detectors that are optimized for conversion of light at the wavelength of highest absorption of the detector material.

Such an approach places a stringent requirement on the TPV emitter, namely that it emit selectively in a region of the spectrum well matched to the specific detector. In the present case, we have employed as a major constituent of our material ytterbium oxide. This rare earth oxide has a characteristic emission near 1 micron, which is well matched to silicon. In addition, the structure of the material itself tends to control the wavelength of emission by virtue of the characteristic size of the pores in the material. This feature, as well as a discussion of the means of fabricating the material, is discussed in a recent US Patent (4).

We can tune the characteristic emission band of our material in two ways. One means of tuning involves the incorporation of desired chemical elements which have specific emission characteristics, in a manner similar to selection of emitting species in laser crystals. We have determined that it is possible to produce materials which span a large range of chemical compositions and crystallographic phases. For example, in the case of aluminum oxide-rare earth oxide chemical systems, we have been able to make emitters that range from pure aluminum oxide through the garnet composition (62.5% aluminum oxide - 37.5% rare earth oxide) and the rare earth aluminate composition (50% aluminum oxide - 50% rare earth oxide) to essentially pure rare earth oxide. This ability affords us much latitude in selecting an optical characteristic for our emitters.

The second means of selecting the emission characteristic has to do with the dimensions of the pores which are produced in the material. These dimensions, which can be controlled in the course of the manufacturing process, offer the opportunity to build an "optical cavity" which may be tuned to emit in a characteristic manner, and which can suppress emission in regions of the spectrum which are not desired.

Shaping

The mechanical shape of the emitter, such as a hollow cylinder or a flat plate, is an important design consideration for system applications. The emitter may require passages for the conveyance of gaseous fuel or a heating medium to selected locations. In addition, we have determined, as have others, that for emitters powered by gas flames, texturing the surface of the emitter can markedly improve the efficiency of heating and emission of radiation (1). In order to attain these various mechanical parameters, it is useful to have a material that can be fabricated in near net shape form, and that can be modified in shape, for example by machining processes.

Our material is amenable to mechanical shaping in two regards. First, the manufacturing process allows for a near net shaping of the material as it is produced. It is possible to produce the material in the form of rods, plates or cylinders as may be appropriate. Secondly, we have determined that the material itself can be further machined using conventional machining techniques, so as to permit the precise final shape to be attained. For many applications, extreme precision in the shape or in mechanical tolerances generally may not need to be attained, but certain mechanical tolerances will undoubtedly need to be achieved.

Patent Protection

We have received US Patent protection for the class of porous material that we have developed, and which appears to have utility as a selective TPV emitter, and we have been issued a Notice of Allowance for additional US Patent protection relating to the method of manufacture of the material. Other patent applications are presently pending.

CONCLUSIONS

We have demonstrated a new class of material which appear to provide significant advantages as TPV emitters, including selective emission, high power density by virtue of operation at very high temperature, stability in air, appreciable thermal shock resistance, and the ability to be produced in desired shapes. These materials also afford the possibility of tuning the characteristic emission wavelength to various bands as may be desired, and the possibility of being powered by any one of a variety of power sources, including gas flames and concentrated solar energy.

REFERENCES

1. Milstein, J.B., Roy, R.G. and Maurer, D.C., "Some Characteristics of a Novel Direct Thermal to Optical Energy Converter Medium", First NREL Conference on TPV Generation of Electricity, AIP Conf. Proc. 321, pages 276-290, (1994).

2. Pitts, J.R., Bingham, C.E., Milstein, J.B. and Roy, R.G., "Characterization of Selective Emitter Materials in a High Flux Solar Furnace", This Conference.

3. Green, M.A., Zhao, J., Wang, A. and Wenham, S.R., *IEEE Electron Device Letters* **11**, 317-318 (1992).

4. Milstein, J.B., and Roy, R.G., US Patent 5,385,114, January 31, 1995.

Development of Thermophotovoltaic Array Testing Capabilities

James J. Lin, Dale R. Burger, and Robert L. Mueller

Jet Propulsion Laboratory, California Institute of Technology, Pasadena, California 91109

Abstract. The present Jet Propulsion Laboratory (JPL) characterization test method for a single thermophotovoltaic (TPV) cell is to illuminate the cell with black body emission. However, this method is inadequate for the performance testing of a string or an array of cells. This is simply because the black body aperture is too small to supply sufficient illumination for much more than a single, small cell. Alternative light sources for TPV string or array testing would be to use a large area gray body, a diffused high power laser, a high power lamp array, or the Large Area Pulsed Solar Simulator (LAPSS). These methods are analyzed and compared. Conclusions are drawn concerning the needs and methods of TPV string and array testing. The large area gray body source was found to need more development toward larger sizes. The color temperature of this source is limited and a cooling system is required for the test devices. The diffused high power laser source was found to be expensive and power limited. This source would require a special optical system to achieve uniform illumination and also requires a cooling system for the test devices. The high power lamp array source was found to need some development and it would require cooling systems for the lamps and the test devices. The LAPSS was found to be a feasible light source. It required very little development funds and did not heat the test article. In the near term, it was decided to use the LAPSS as the TPV light source. A preliminary technique using the LAPSS with an infrared bandpass filter was developed for high power, 8 watts per square centimeter, TPV testing purposes. Initial results indicate that this system is applicable to single cell, string and array testing of TPV cells because of its high emission power density and well matched emission spectrum. Tests and analyses were performed to establish test plane intensity and uniformity versus test device distance from the lamps. In addition, tests and analyses were performed to determine the spectral transmission characteristics of various infrared bandpass filter combinations. The LAPSS, with the infrared bandpass filter, is now operational at JPL for TPV cell, string and array testing.

INTRODUCTION

The present JPL test method for a single TPV cell is to illuminate the cell under black body emission at various known intensities. This is accomplished by positioning the cell at various distances from the black body aperture, which behaves somewhat like a point source [7, 8]. The intensities at the positions were measured with a broad spectrum laser power meter fitted with either a 0.1 cm^2 or

© 1996 American Institute of Physics

a 1.0 cm^2 water cooled aperture plate. These intensity measurements are used to correct any deviations from point source behavior.

However, this method for single cell testing is insufficient for testing a string or array of cells. A suitable size for a string of cells is about 10 cm x 1 cm and for an array, 10 cm x 10 cm. The string or array of cells cannot be uniformly illuminated at a sufficiently high intensity by the present black body source because the black body aperture is relatively small (1.27 cm diameter) in comparison to the size of the string or array. A source capable of illuminating an array of cells would have similar capabilities for a single cell. Therefore, several different illumination systems will be considered for their capabilities in array testing.

Possible candidate sources would be a large area gray body, diffused high power laser, a high power lamp array, or a configuration using filtered LAPSS emission. Gray body sources provide the advantage of having the same emission spectrum as the black body source, currently used for single cell testing. Lasers offer the advantage of simulating a perfectly selective emitter so peak cell efficiency could be attained if the wavelength of the laser is well matched to the cell bandgap. Lamp arrays are relatively simple and inexpensive compared to the previous two options. Filtered LAPSS emission would utilize existing JPL equipment and the development cost would be minimal. Each of these possibilities are examined in more detail below.

BLACK/GRAY BODY SOURCES

A black body source, with a large enough area to uniformly illuminate a 10 cm x 10 cm array, would need a circular aperture of 14.2 cm diameter. Black body sources [1] are not currently manufactured with such a large aperture because it is difficult to maintain uniform black body radiation at temperatures above 1000°C. An available black body capable of 1200°C typically has a maximum aperture of only 5 cm diameter. TPV cells illuminated with such a black body must be distant from the source for uniform emission, which greatly reduces emission intensity. This is a major drawback for black body sources because TPV cells typically require high power density in a particular wavelength region for efficient energy conversion. In addition, this 1200°C black body source is relatively expensive, costing approximately $15,000. These disadvantages eliminate large aperture black body sources for TPV string and array testing.

Large area planar gray body sources with an emissivity of approximately 0.9 are only produced for low temperature operation, due to material degradation constraints. A maximum temperature of 800°C can be achieved, but this is significantly below the current TPV testing needs. Furthermore, a planar gray body source with this temperature capability is only available with a 2.5 x 2.5 cm surface, which is too small to illuminate an array of cells. These disadvantages, in the near term, eliminate planar gray body sources for TPV string and array testing.

LASERS

High power lasers [2] were also considered for TPV array testing. The desired emission intensity to illuminate TPV cells is a minimum of 2 W/cm^2, therefore, a 200 W laser is necessary to illuminate a 100 cm^2 array. The wavelength of the laser must be about 80%-90% of the cell bandgap wavelength in order to be well matched to the cell bandgap. Most TPV cells will have a bandgap range of from .5 eV to .65 eV (2.48μm to 1.91μm).

Lasers that are bandgap matched include: Nd:YAG @ 1.06 μm and 1.32 μm, Er:Glass @ 1.54 μm, tunable Co:MgF$_2$ from 1.75 μm to 2.5 μm, Tm: YAG @ 2.01 μm, and Ho: YAG @ 2.1 μm.

The Nd:Yag laser is the only high power laser among the group. It is produced as a continuous and pulsed laser. The 1.06μm and 1.32 μm continuous Nd:YAG laser has maximum power outputs of 400 W and 100 W, respectively, and is available from Lee Laser. Although these wavelengths are below the best bandgap match for many TPV cells, the cells will still respond sufficiently to these high power lasers.

The Er:Glass and Co:MgF$_2$ are both pulsed lasers with low average power. They have pulse energies of 650 mJ and 80 mJ for durations of 350 ns (1.86 MW peak) and 80 μs (1000 peak watts), respectively. The short pulse, 350 ns, is currently too fast for the electronics in the JPL LAPSS which requires 1.5 ms to take a full IV sweep and 15 μs for each data point [3]. Although the LAPSS data system is really not suited for pulsed laser-cell operation, it is probable that the LAPSS data system could be modified to be synchronous with the 80 μs laser pulse train.

The major disadvantages of lasers are the problems associated with the optics and the inherent non-uniformity of the beam intensity. The narrow beam must be diverged and recollimated onto the cell test area, which requires optics which could degrade with repeated use, especially with the high power Nd:YAG laser. Since the laser beam is inherently non-uniform, the TPV cells cannot be uniformly illuminated unless some sort of corrective means is used. Obtaining exact comparability of the laser power distribution and optical system characteristics between different laboratories would be difficult to achieve. Although at least one continuous laser is reasonably matched to some TPV cells, the difficulty with optics and non-uniform beam intensity make lasers an unlikely choice for TPV cell or array testing. Additionally, it is not known how well the various TPV cells will respond to the pulsed laser light. The described laser systems cost between $20,000 and $40,000.

LAMP ARRAYS

A simple alternative to using lasers, gray bodies or blackbodies [6] is to use a 30 cm x 30 cm array of high power, low temperature, tubular, tungsten lamps. The result would be a good approximation of gray body emission with color temperatures ranging from 700°C to 2200°C; this is well matched to TPV cell bandgaps. The intensity would be on the order of 2 W/cm^2 for the 900°C color temperature lamp with the cells placed close to the lamps.

These lamp arrays are produced by Research Incorporated of Minneapolis, Minnesota. They are tubular quartz, tungsten, infrared lamps arranged side by side in an array. Intensity uniformity is approximately +/- 10% for an area ranging from 500 to 1000 cm^2. Forced air and water cooling will be necessary to cool the lamps. Water cooling setup for the array of cells is necessary to maintain stable and moderate array temperatures. Although this plan may be feasible, the cooling requirements make it undesirable for a system which may need to handle a number of different configurations.

HIGH POWER INFRARED LAPSS EMISSION

A configuration with the JPL LAPSS system was analyzed for its potential in TPV array testing. It was believed that an infrared filter, placed in front of the LAPSS flash lamps would result in infrared light suitable for testing TPV cells. The analysis for this technique is detailed below and includes tests and discussion concerning emission uniformity, emission intensity, usage of infrared filters, emission characterization, and systems testing.

Emission Uniformity

In order to achieve high intensities in the infrared region, cells must be placed within 60 cm, or less, from the LAPSS flash lamps. At this distance, the emission uniformity is in question since the two lamps, which are 15.25 cm apart, no longer behave as a single point source. The individual contributions from each lamp, and the intensity variance along the lamp length may significantly affect emission uniformity. These characteristics must be verified in order to demonstrate this technique as a feasible method of TPV array testing.

Several tests were performed to show uniformity to be sufficient for string and array testing purposes. More detailed information and test results on LAPSS emission uniformity is provided in the Appendix. However, it is recommended that for greater intensity uniformity, cell strings should be arranged on the test plane in the horizontal direction, and arrays should be constructed in a rectangular configuration and arranged on the test plane horizontally, to take advantage of the non-symmetrical uniformity.

Emission Intensity

The intensity of the LAPSS was re-measured using three concentrator cells in an attempt to generate consistent high intensity measurements as shown in figures 1 and 2 The same method, described in the Appendix, of using the Voc/Intensity relationship was used. A linear curve fit of Isc versus intensity once again demonstrates that the point source equation for intensity is accurate up to 64 suns (8.75 W/cm^2), which is important in assuring accurate intensity values for the Voc/Intensity curve fits. The regression values for the curve fits are between 0.9998 and 0.9999 for Figure 1, and are between 0.9978 (ideal relation) and 0.9997 (non-ideal relation) for Figure 2. Data values for both the Isc and Voc are the average of three measurements taken at the same intensity.

Following the measurements made for Figures 1 and 2, the Voc of the cells was measured 17.8 cm from the flash lamps in an attempt to extrapolate for intensity at this location. Each result is the average of three measurements. Intensity extrapolations were made using both the ideal and the non-ideal

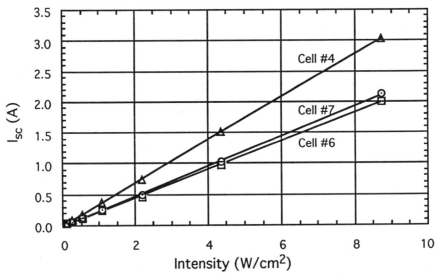

Fig. 1: Short circuit current versus point source intensity for silicon concentrator cells # 4, 6, and 7.

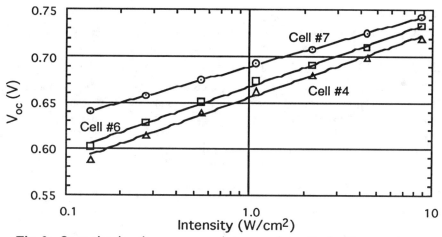

Fig. 2: Open circuit voltage versus point source intensity for Si concentrator cells # 4, 6, and 7

Cell Number	Voc (V)	Intensity Non-ideal (W/cm^2)	Intensity Ideal (W/cm^2)
4	0.7734	99	45
6	0.7990	130	75
7	0.7822	48	43

Table 1: Voc and intensity using the non-ideal and ideal relationships for concentrator cells #4, 6, 7, located 17.8 cm from the flash lamps.

Voc/Intensity curve fit equations derived from Figure 2 data and the results are shown in Table 1.

The values of maximum intensity, 17.8 cm from the flash lamps, vary considerably between the three cells, indicating this method is not usable for measuring such a high intensity. This is due to different series resistance losses in the cells at extremely high intensities and that these concentrator cells were only designed to operate up to 100 suns. These results should, therefore, be disregarded.

Another method of measuring full intensity was performed. The laser power meter used for measuring black body intensity is used to measure LAPSS pulse energy. Power cannot be measured directly using this device because the LAPSS emission is a pulse, but pulse energy can be measured. Pulse energies are directly correlated to intensities because the LAPSS pulse energy profile and pulse duration are nearly constant, making the pulse energy directly proportional to intensity. The relationship of pulse energy to intensity must first be calibrated.

The intensity at the LAPSS 1 sun target plane was set to be 2% higher than AM0 (0.1367 W/cm^2), as measured by the Si reference cell #SS1411. The calibrated intensity at the 1 sun plane is, therefore, 0.1394 W/cm^2. Furthermore, the Isc versus intensity graph (see Figure 1) shows that the point source intensity equation is extremely accurate for extrapolation to 64 suns. Thus, the calibration factor to convert LAPSS pulse energy to LAPSS intensity is determined by measuring the pulse energy and comparing the results to the accurately extrapolated intensities. The results are shown in Table 2.

Distance from Lamps (cm)	Point Source Suns	Point Source Intensity (W/cm^2)	Pulse Energy (mJ)	Calibration Factor (W/cm^2/mJ)
1097.3	1X	0.1394	4	0.03485
775.9	2X	0.2788	8	0.03485
548.6	4X	0.5576	15.5	0.03597
388.0	8X	1.115	33.9	0.03289
274.3	16X	2.23	65	0.03431
194.0	32X	4.46	130.8	0.03410
137.2	64X	8.92	260.7	0.03422
106.7	128X	17.84	592	0.03014
100.6	N/A	N/A	645	N/A
70.1	N/A	N/A	1,260	N/A
39.6	N/A	N/A	3,320	N/A
17.8	N/A	N/A	12,820	N/A

Table 2: Results of AMØ LAPSS (with GG395 Filter) pulse energy compared to intensities calculated from the point source equation.

The laser power meter does not accurately measure pulse energy below 30 mJ due to background light interference and the point source intensity extrapolations are accurate to only 64 suns Therefore, the most accurate calibration factors are from 8x to 64x suns. Averaging the calibration factor from the most accurate pulse energy readings yields a calibration factor of .0339 W/(cm^2 mJ) which gives the following conversion:

Intensity (W/cm^2) = .0339 W/(cm^2 mJ) x Pulse Energy (mJ).

This equation is accurate at all intensities, as long as the laser power meter can accurately measure the pulse energy. Using this equation, full intensity (17.8 cm from the lamps) is 435 W/cm^2 (3182 suns). This value is much larger than the intensities calculated from the Voc/Intensity relationships due to the excessive series resistance losses occurring at extremely high intensities which are generally unaccounted for by the equations derived at lower intensities. It is lower than the point source calculated value of 521 w/cm^2, which is not surprising since the source is not a point. The value of 435 W/cm^2 is, therefore, the most accurate measurement of the LAPSS intensity 17.8 cm from the lamps, with the Schott ultraviolet filter GG 395 in place. This method is a more direct means of measuring full intensity as compared to the Voc/Intensity relationships, and the laser power meter is especially designed to measure high energy pulses.

Spectral Match

The quantity of infrared radiation produced by the LAPSS flash lamps is now analyzed for its suitability for TPV cell testing. The emission spectrum of the lamps is modeled as a gray body with the peak emission occurring at 475 nm; this corresponds to a 6100 K gray body [3]. LAPSS emission is presently only characterized to 1.2 μm, therefore, the gray body model is used to approximate emission beyond 1.2 μm.

The bandgap of TPV cells will probably range from .65 to .5 eV (1.91 - 2.48 μm). Using the 6100 K gray body model, the percentages of illumination falling within a range of emission bands is calculated [4]. The emission bands are 0.5 μm or 1.0 μm wide, starting at a short wave length and stopping at selected cell bandgap wavelengths. The results are shown in Table 3.

At the highest LAPSS operating intensity, 435 W/cm^2, the theoretical band emission intensities range from a high of 85.6 W/cm^2 for a 1.0 μm emission

Cell Bandgap	.5 μm band emission % of total intensity	1.0 μm band emission % of total intensity
.4 eV, 3.1 μm	2.6 - 3.1 μm	2.1 - 3.1 μm
% of total intensity	1.1%	3.4%
.5 eV, 2.48 μm	1.98 - 2.48 μm	1.48 - 2.48 μm
% of total intensity	2.6%	8.9%
.6 eV, 2.07 μm	1.57 - 2.07 μm	1.07 - 2.07 μm
% of total intensity	5.2%	19.7%

Table 3: Estimate of LAPSS intensity for limited infrared emission bands.

band between 1.07 and 2.07 μm, and a low of 4.87 W/cm² for a .5 μm emission band between 2.6 and 3.1 μm. Band emissions at longer wavelengths will require a wider bandpass because the power density will continue to fall off as wavelength is increased from the emission peak at 475 nm. These calculations show that the LAPSS is capable of producing high intensity infrared radiation in the response region of TPV cells if the proper filters are utilized to achieve the desired band emission.

Infrared Filters

In order for the LAPSS to be used as a source for illuminating TPV cells, the emission must be filtered to eliminate excessive quantities of high energy radiation. Ideally, the majority of the resulting emission must be slightly higher in energy than the cell bandgap (see Table 3). In order to achieve this, a pair of filters were placed together to achieve an infrared bandpass filter. The red/infrared filter is a Schott RG 645, and the blue/green filter is a Schott BG 38, which is transparent in the infrared, but red absorbing [5]. The transmission curve for the red/infrared filter is shown in Figure 3 and the transmission curve for the combined blue/green and red/infrared filters, resulting in an infrared bandpass filter is shown in Figure 4. LAPSS emission passes through the Schott BG 38 filter first.

A small spike occurs at 660 nm, as shown in Figure 4, because the transmittance of the filters slightly overlap at that wavelength. This spike can be eliminated by using a black glass, infrared filter, Schott RG 850, instead of the

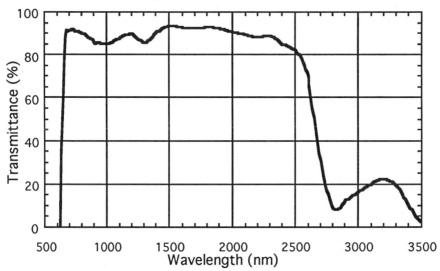

Fig. 3: Transmittance curve for red/infrared filter, Schott RG 645.

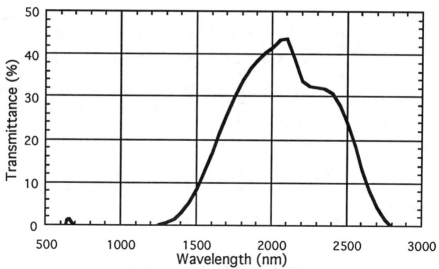

Fig. 4: Transmittance curve for the combined blue/green Schott BG 38 and the red/infrared, Schott RG 645 filters.

red/infrared filter, Schott RG645. The main transmittance peak would remain the same.

The overall transmittance is confirmed by using the laser power meter to measure the intensity with the infrared bandpass filter, at multiple locations from the lamps. Table 4 presents the results. An average of the bandpass transmittance values yields 1.87% of normal LAPSS AMØ illumination (with GG395). The transmittance at the 17.8 cm distance is questionable because the energy meter may have been slightly off position.

An alternative to using this filter combination for the bandpass filter is to

Distance from Bulbs (cm)	LAPSS AMØ Pulse Energy (mJ)	LAPSS AMØ (W/cm^2)	Bandpass Filter Pulse Energy (mJ)	Bandpass Intensity (W/cm^2)	Transmittance of Bandpass Filter (%)
106.7	592	20.07	11.05	0.375	1.867
100.6	645	21.87	12.04	0.408	1.867
70.1	1260	42.71	24.07	0.816	1.910
39.6	3320	112.55	62.1	2.105	1.870
17.8	12,820	434.60	238.0	8.068	1.856

Table 4: Intensities of LAPSS using the infrared bandpass filter compared to the AMØ GG395 filter.

use a Schott infrared filter (RG 1000). The transmittance of this filter is broader in the infrared region, so that TPV cells of greater bandgap can be tested.

Emission Characterization

The emission intensity from the LAPSS with the bandpass filter has been established as 8 W/cm^2, at a distance 17.8 cm from the LAPSS lamps. Multiplying the gray body LAPSS emission model with the bandpass transmittance curve, and normalizing the results for an intensity of 8 W/cm^2 yields the spectral irradiance curve, as shown in Figure 5.

The emission spike at 660 nm is even more dominant than that depicted in the transmittance curve because the gray body LAPSS is more energetic at shorter wavelengths. However, the total energy of the spike is only about 4% of the total emission. The main emission peak occurs at 1800 nm, with the half power points at 1515 nm and 2215 nm.

Infrared LAPSS Testing

TPV cells were tested with the bandpass filtered LAPSS system to ensure that the setup is actually feasible. Two 1 cm^2 InGaAs cells (6-1222-1 and 6-1222-2) from RTI/ASEC were tested with the infrared LAPSS emission at approximately 8 W/cm^2, and the results compared with those from testing with

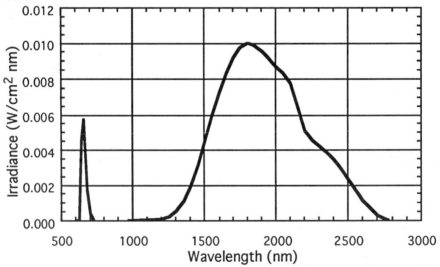

Fig. 5: Spectral irradiance curve for the LAPSS with the Schott RG645 and Schott BG38 combined as a bandpass filter.

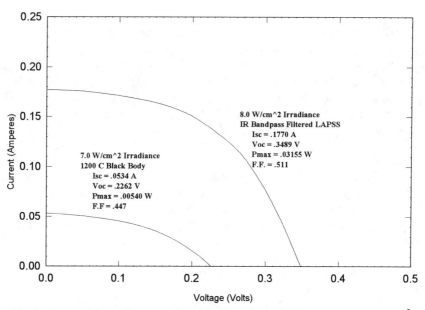

Fig. 6: Current-Voltage Characteristic at 28 C for InGaAs Cell #6-1222-1. Area =1 cm^2

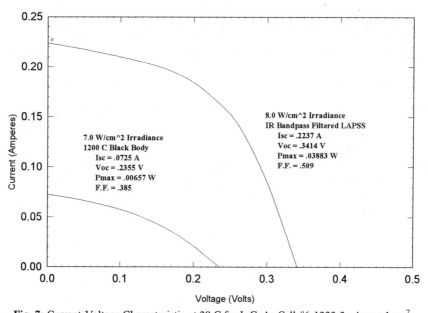

Fig. 7: Current-Voltage Characteristic at 28 C for InGaAs Cell #6-1222-2. Area =1 cm^2

the 1200°C black body emission at approximately 7 W/cm^2. Five separate runs were taken for each test to demonstrate the consistency in results. The upper intensity limit of 7 W/cm^2 for the black body prevented testing the cells at equivalent intensities. The results demonstrate that the narrow band infrared LAPSS emission is far better matched to the bandgap of the cells than is the broad-band black body emission. Performance could be improved if the cell bandgap was lower or if the filtered LAPSS emission peaked at a wavelength slightly higher than 1800 nm. Figures 6 and 7 show the results. The results of these tests indicates that using filtered LAPSS emission is a feasible method for testing TPV cells.

TPV cell 6-1222-2 was next tested at different locations across a 10.16 x 10.16 cm test plane, 17.8 cm from the bandpass filtered infrared LAPSS lamps (1.5 cm from the bandpass filter surface). The results are the average of two tests taken in forward and reverse order to minimize bias from temperature and raw LAPSS intensity variations. The results show that Isc of the cell changes -12.4% from the test plane center to the upper left corner (-5.08 + 5.08 cm location). This is equivalent to a variation of +/- 6.6% from the mean intensity. This Isc change is directly related to the LAPSS intensity uniformity.

This value is somewhat greater than the results from the emission uniformity tests (results shown in the Appendix). However, these results are the most accurate because they were obtained under actual test conditions and at maximum intensity with the infrared bandpass filter in position. Uniformity of intensity can be improved by placing the cell at a greater distance from the flash lamps, however, this will reduce emission intensity. The results are shown in Table 5.

The results also indicate that intensity uniformity is greater in the horizontal direction than in the vertical direction. Intensity changes an average of 4.1% across a 10.16 cm horizontal traverse, and 8.8% across a 10.16 cm vertical traverse. This agrees well with previous uniformity tests. System testing has shown the bandpass filtered infrared LAPSS is a potential method for testing TPV cells and arrays requiring high power density in the infrared spectrum.

The high power infrared LAPSS system was developed with existing JPL equipment so that there was no material or equipment costs; the only cost for the system was research time.

Position (cm.)	(-5.08, +5.08)	(0.0,+5.08)	(+5.08, + 5.08)
Isc (mA)	190.76	198.54	191.46
Voc (V)	0.3155	0.3181	0.3157
Pmax (mW)	25.14	26.67	25.36
Ipmax (mA)	126.75	129.99	124.91
Vpmax (V)	0.1984	0.2052	0.2032
FF	0.418	0.422	0.420
Position (cm.)	(-5.08,0.0)	(0.0,0.0)	(+5.08,0.0)
Isc (mA)	210.21	217.78	208.80
Voc (V)	0.3213	0.3228	0.3200
Pmax (mW)	28.64	30.03	28.46
Ipmax (mA)	136.95	144.49	136.95
Vpmax (V)	0.2077	0.2079	0.2077
FF	0.426	0.427	0.426
Position (cm.)	(-5.08,-5.08)	(0.0,-5.08)	(+5.08, -5.08)
Isc (mA)	195.78	201.92	193.46
Voc (V)	0.3166	0.3182	0.3156
Pmax (mW)	26.06	27.19	25.71
Ipmax (mA)	127.92	133.28	124.96
Vpmax (V)	0.2037	0.2040	0.2058
FF	0.420	0.423	0.421

Table 5: TPV cell #6-1222-2 IV performance with bandpass filtered infrared LAPSS, 8 W/cm^2, 17.8 cm from the lamps.

RECOMMENDATIONS

Much of the emission testing for the infrared LAPSS was completed without a precise positioning system for exact cell and power meter placement from the flash lamps. This limits the accuracy of the results because test cells and the power meter cannot be repositioned with much precision. Great effort was taken to ensure precision positioning, but a long travel, linear positioning slide would be an improvement. In addition, slight changes in the distance to the lamps result in significant changes in intensity. This reinforces the need for a precise positioning system for TPV cell placement.

The present bandpass filter allows a small transmittance spike at 640 nm due to an overlap in the two filters. This can be eliminated by replacing the red/infrared filter with a black glass, infrared filter (RG 850). Also, for testing higher bandgap TPV cells, a single broad band infrared filter (RG 1000) could be substituted for the narrow band infrared filter. These two Schott Glass Technology filters would expand the capabilities of the infrared LAPSS for TPV cell and array testing; at a total filter cost of approximately $300.

The emission characterization is an approximation calculated from a gray body emission model of the LAPSS multiplied by the bandpass transmittance curve and a 8 W/cm^2 correction factor. Actual LAPSS emission does not behave exactly as a gray body. Although it is unlikely that there are major deviations from this model, actual measurement of the spectral irradiance distribution would be desirable.

Although the present filtered infrared LAPSS irradiance characterization is sufficient for present testing purposes, it would be fruitful to completely characterize the LAPSS with and without filter(s) to three or four microns. This can probably be accomplished with loaned equipment and a good UV-VIS-IR grating monochromator.

CONCLUSIONS

A number of methods for high power TPV cell array testing were explored. These methods include using ideal black or gray body sources, high power lasers, lamp arrays, and filtered infrared LAPSS emission. Black or gray bodies provide the same emission as the present JPL single cell testing method. However, in order to illuminate a large area array, a black body with a large aperture area is necessary. Such a system is not presently available in the temperature range of interest; it would be costly to develop (if it is feasible), and a massive cell cooling setup would be necessary. Lasers offer the advantage of providing an extremely narrow band emission, so if the wavelength is well matched to the cell bandgap, peak efficiency could be achieved. However, only one high power laser with a wavelength appropriate for the bandgap of probable TPV cells was located. There is also considerable difficulty in diffusing and recollimating a laser beam so that it uniformly illuminates an array of cells. The difficulties with complicated optics and the high cost of lasers makes it an undesirable choice for array testing at this time. Lamp arrays provide a good approximation for gray body emission which could illuminate the necessary area at sufficient uniformity. However, massive cooling systems for the lamps and cell arrays are necessary, and lamp emission barely meets the 2 W/cm^2 minimum intensity requirement.

A system utilizing the present LAPSS system and an infrared bandpass filter was analyzed for large area TPV cell testing. The uniformity of LAPSS emission was tested and found to vary by a maximum of +/- 6.6% from the mean intensity over a 10.16 x 10.16 cm area directly behind the bandpass filter. The filtered LAPSS intensity at this location (17.8 cm from the flash bulbs, 1.5 cm from the bandpass filter surface) was measured to be about 8 W/cm^2. The emission was characterized by multiplying the transmittance curve of the filter with a gray body irradiance approximation of LAPSS emission. Thus, filtered infrared LAPSS emission is fairly uniform, possesses high intensity, and the emission is well matched for the probable TPV cell bandgap range. The system was tested and found to be a reliable method for TPV cell testing. The high power, infrared LAPSS is now operational at JPL for TPV single cell or array testing.

ACKNOWLEDGMENTS

The authors would like to thank Dr. Carol R. Lewis for her insight and support in the described research, JPL Standards Laboratory for the filter transmittance testing, and Drs. Peter Iles of ASEC and Mike Timmons of RTI for providing the concentrator and TPV cells used in refining the testing system. The work described in this report was conducted by the Jet Propulsion Laboratory, California Institute of Technology, under a contract with the National Aeronautics and Space Administration.

REFERENCES

1. Mikron Instrument Company, catalog M300 Rev. 0 for, "M300 Series Blackbody Radiation Calibration Sources," Wyckoff, New Jersey, 1994.

2. Schwartz Electro-Optics Inc., brochure for, "A Pulsed Multiple Wavelength System," and ,"Cobra 2000 Co:MgF$_2$ Tunable Laser," Orlando, Florida, 1994.

3. R. L. Mueller, "The Large Area Pulsed Solar Simulator (LAPSS)," Jet Propulsion Laboratory Publication 93-22, Pasadena, California. 1993.

4. F. Kreith and M. Bohn, "Principles of Heat Transfer," Harper Collins Publishers, New York, New York. 1986.

5. Schott Glass Technologies Inc., catalog for, "Color Filter Glass Technical Information," Duryea, Pennsylvania, 1983.

6. R. L. Mueller, "Using High Temperature, High Density Infrared Heaters as a 900 C Planar Greybody Source," personal communications, Jet Propulsion Laboratory, Pasadena, California, 1994.

7. J. J. Lin and D. R. Burger, "TPV Cell IV Curve Testing with Varying Black Body Emission Temperatures, Intensities and Cell Temperatures", submitted to Solar Energy Materials and Solar Cells, Aug. 17, 1994.

8. D. R. Burger and R. L. Mueller, "Characterization of Thermophotovoltaic Cells", 1st NREL Conference on Thermophotovoltaic Generation of Electricity, Copper Mountain, CO, July 24-27, 1994 (AIP Conf. Proc. 321, pp. 457-472).

APPENDIX

Emission Uniformity Tests

An indirect method of measuring light pulse intensity using the solar cell Voc/Intensity logarithmic relationship was first attempted. In addition, the solar cell Isc/Intensity relationship was measured. These tests were performed using a typical 2 x 2 cm, BSF, standard Silicon solar cell and the results are shown in Figures 8 and 9.

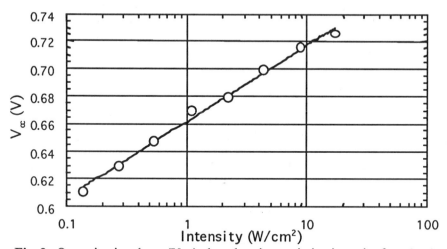

Fig. 8: Open circuit voltage (Voc) plotted against emission intensity for a 2 x 2 cm BSF Silicon solar cell.

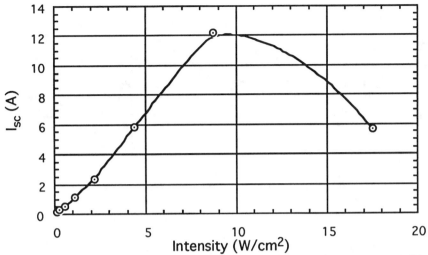

Fig. 9: Short circuit current (Isc) plotted against intensity for a 2 x 2 cm BSF Silicon solar cell.

Voc behaves logarithmically with intensity, where Voc = 0.662 + 0.055 log (Intensity) and the regression value for the curve fit is 0.9968, as shown in Figure 8. Intensity could be approximated with this relationship. The cell temperature was about 25°C, room temperature, and did not change appreciably during the tests because the LAPSS is a pulsed light source.

Figure 9 demonstrates that although Isc is relatively linear with intensity for low intensities, it is nonlinear at higher intensities due to cell series resistance losses, which reduce cell Isc as the LAPSS intensity exceeds about 64 suns. The relationship is no longer linear preventing any intensity predictions. An ASEC concentrator cell was then tested for its Voc and Isc intensity relations. The Isc relationship remained nonlinear with intensity at higher intensities, but the series resistance losses were not as significant as they were with the 2 x 2 cm BSF Silicon solar cell.

Figure 10 shows Voc to be relatively logarithmic with intensity in an ideal relation where Voc = 0.501 + 0.062 log (Intensity), but there are series resistance losses at higher intensities, causing this relationship to become somewhat inaccurate. These losses are taken into account with a precise curve fit including intensity raised to a power factor in a non-ideal relation where Voc = 4.011.9 + 9.367 log (Intensity) - 4011.5 (Intensity)$^{.001}$. The regression value improves (0.9997 vs. .09943) with the non-ideal relationship. The ideal and non-ideal Voc/Intensity relationships are used later to evaluate emission uniformity.

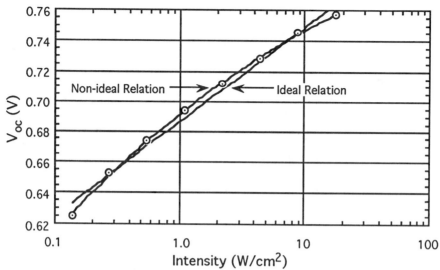

Fig. 10: Open circuit voltage plotted versus intensity for concentrator cell #7 from ASEC

In figures 8-10, the intensities are extrapolations which were calculated using a point source model. The lamp location is assumed to be a point source; and the 1 sun test plane has a calibrated intensity of 136.7 mW/cm² at a location of 10.973 m (36 feet) from the lamps. The approximate Intensity/Location relationship is:

$$\text{Intensity}(mW/cm^2) = (1.646 \times 10^4 \text{ mW-m}^2/cm^2) / \text{Distance}(m)^2$$

An error results from this relationship at distances close to the lamps, because the point source approximation is only accurate to about 64 suns. The extrapolation errors, cell series resistance effects and cell temperature variations could possibly account for the non-ideal Voc/Intensity logarithmic relationship.

Intensity uniformity was now measured using the calibrated Voc/Intensity relationship from Figure 10. The same silicon concentrator cell #7 was now placed about 61 cm from the flash lamps. Voc tabulations are the average of two consecutive measurements taken at the same location. This test was performed for a planar (15.24 x 15.24 cm) area. Measurements were made every 1.27 cm over a 15.24 cm horizontal traverse. This was completed for three rows. One row was centered with the lamps and the other two rows were 7.62 cm above and below the center. The results are shown in Table 6.

Horiz. Position (cm)	-7.62	-6.35	-5.08	-3.81	-2.54	-1.27	0.00
Vertical Position (cm)	Open Circuit Voltage (V)						
7.62	0.7624	0.7627	0.7629	0.7629	0.7629	0.7629	0.7629
0.00	0.7640	0.7640	0.7642	0.7642	0.7643	0.7644	0.7646
-7.62	0.7642	0.7638	0.7635	0.7636	0.7632	0.7628	0.7622

Horiz. Position (cm)	7.62	6.35	5.08	3.81	2.54	1.27	0.00
Vertical Position (cm)	Open Circuit Voltage (V)						
7.62	0.7632	0.7633	0.7632	0.7632	0.7632	0.7634	0.7635
0.00	0.7649	0.7649	0.7651	0.7646	0.7654	0.7659	0.7655
-7.62	0.7636	0.7636	0.7642	0.7641	0.7637	0.764	0.7637

Table 6: Open circuit voltage readings for a 15.24 x 15.24 cm plane located approximately 61 cm from the LAPSS flash lamps.

The test results suggest there is better uniformity horizontally than vertically, as indicated by the smaller variation in Voc values. Voc varied from a high of .7659 V near the center, to a low of .7624 at the top left, a change of -0.46%. This corresponds to intensities of 23.95 and 19.34 W/cm^2 respectively, using the non-ideal Voc/Intensity relationship, for a change of -19.2%. This also corresponds to intensities of 18.89 and 16.59 W/cm^2, using the ideal logarithmic Voc/Intensity relationship, for a change of -12.2%.

An inconsistency was found in the data where the Voc measurements are repeated at the starting locations, (zero cm horizontal versus 7.62, 0, and -7.62 cm vertical) because the data was taken at different times. The sequence in which the data was taken seems to affect the result. This is possibly due to an increase in cell temperature after multiple flashes or variations in raw LAPSS intensity. Voc and intensity variance would, therefore, be less than this test indicates.

A more accurate test was made by performing the same test but measuring Voc at fewer locations to minimize possible cell heating or changes in raw LAPSS intensity, and then normalizing the data. Voc measurements were made for each row at the 7.62, 0 and -7.62 cm horizontal locations; the results are the average of two runs for each location. The Voc at the center location, (0,0), was measured immediately after each set of row readings for comparison with the other results. The results are shown in Table 7.

The effects of varying cell temperature or LAPSS intensity is compensated by assuming a constant Voc value at position (0,0) of .7621 V. The Voc values for each set of row readings are then corrected by the difference between the initial Voc readings and the follow-up measurements at the zero horizontal position. The results are shown in Table 8.

The corrected results show a significant improvement in uniformity in comparison to the uncorrected data and the initial test run. Corrected Voc readings varied from a high of .7623 at position (0.0, +7.62), to a low of .7603 at position (-7.62, 0.0). These correspond to intensities of 19.22 and 17.125 W/cm^2 using the non-ideal Voc/Intensity relationship, and 16.53 and 15.34 W/cm^2 using the ideal logarithmic Voc/Intensity relationship. The -0.26 % change in Voc corresponds to an intensity change of -10.9% by the non-ideal relationship, and -7.2% by the ideal relationship. Again, results show that uniformity is

Horizontal Position (cm)	-7.62	0.0	7.62
Vertical Position (cm)	Open Circuit Voltage (V)		
7.62	0.7604	0.7603	0.7607
0.0 after 7.62		0.7620	
0.0	0.7622	0.7621	0.7623
0.0 after -7.62		0.7606	
-7.62	0.7594	0.7595	0.7595

Table 7: Open circuit voltage in a 15.24 x 15.24 cm plane located approximately 61 cm from the LAPSS flash lamps with additional measurements at position (0,0) for comparison.

Horizontal Position (cm)	-7.62	0.0	7.62
Vertical Position (cm.)	Open Circuit Voltage (V)		
7.62	0.7605	0.7604	0.7608
0.0	0.7622	0.7621	0.7623
-7.62	0.7609	0.7610	0.7610

Table 8: Corrected open circuit voltage in a 15.24 x 15.24 cm plane, 61 cm from the flash lamps.

Voc at Constant (0.0) X-Position			
Vertical Position (cm.)	Forward Order Voc (V)	Reverse Order Voc (V)	Average Voc (V)
7.62	0.7591	0.7586	0.75885
5.08	0.7594	0.7592	0.75930
2.54	0.7597	0.7598	0.75975
0.0	0.7598	0.7600	0.75990
-2.54	0.7596	0.7598	0.75970
-5.08	0.7593	0.7596	0.75945
-7.62	0.7588	0.7582	0.75850

Table 9: Open circuit voltage taken vertically in a 15.24 x 15.24 cm plane, 61 cm from the flash lamps.

significantly better horizontally than vertically.

A more precise test for vertical Voc and intensity variance across the test plane was next conducted. Voc readings were made from position (0.0, +7.62) to (0.0, -7.62) in 2.54 cm intervals, and again, but in a reverse order, from (0.0, -7.62) to (0.0, +7.62). The results are for single measurements to minimize possible cell temperature or LAPSS intensity variations. The Voc readings for identical positions are averaged to cancel out these variances. The results are shown in Table 9.

The average Voc changed -0.16%, in the vertical direction, which corresponds to a change in intensity of -7.2%. A summation of the change in intensity vertically (-7.2%) and horizontally (-2.4%) yields -9.6%, the maximum intensity change occurring between the center and corner positions. This agrees well with the previous data resulting in a maximum intensity change of -10.9% across the 15.24 x 15.24 cm area, 61 cm from the flash lamps.

Since the emission was fairly uniform at this distance, the same test was performed for a 10.16 x 10.16 cm test plane approximately 17.8 cm from the lamps. This is located directly in front of the LAPSS filter, where the highest intensity is achieved. A smaller test plane area was characterized because the LAPSS casing will not allow uniform illumination beyond this size at this location. The results, shown in Table 10, are single measurements to minimize variations due to changing cell temperature or intensity.

The average horizontal Voc ranged from a high of .7800 V to a low of .7795 V which corresponds to intensities of 66.75 and 63.9 W/cm^2, changing -4.2%, by the non-ideal relationship, and 31.87 and 31.29 W/cm^2, changing -1.7%, by the ideal relationship.

The average vertical Voc ranged from a high of .7798 to a low of .77925 V, which corresponds to intensities of 65.6 and 62.5 W/cm^2, changing -4.7% by the

Voc at Constant (0.0) Y-Position			
Horizontal Position (cm)	Forward Order Voc (V)	Reverse Order Voc (V)	Average Voc (V)
5.08	0.7796	0.7795	0.77955
3.81	0.7798	0.7802	0.78000
2.54	0.7796	0.7801	0.77985
1.27	0.7795	0.7800	0.77975
0.00	0.7793	0.7802	0.77975
-1.27	0.7788	0.7802	0.77950
-2.54	0.7789	0.7804	0.77965
-3.81	0.7785	0.7806	0.77955
-5.08	0.7788	0.7805	0.77965

Voc at Constant (0.0) X-Position			
Vertical Position (cm)	Forward Order Voc (V)	Reverse Order Voc (V)	Average Voc (V)
5.08	0.7805	0.7785	0.77950
3.81	0.7804	0.7783	0.77935
2.54	0.7801	0.7784	0.77925
1.27	0.7803	0.7788	0.77955
0.00	0.7801	0.7793	0.77970
-1.27	0.7800	0.7796	0.77980
-2.54	0.7800	0.7794	0.77970
-3.81	0.7797	0.7797	0.77970
-5.08	0.7802	0.7794	0.77980

Table 10: Open circuit voltage in a 10.16 x 10.16 cm plane located 17.8 cm from the LAPSS flash lamps.

non-ideal relationship, and 31.64 and 31.0 W/cm^2, changing -2.1% by the ideal relationship.

Author Index

A

Ahrenkiel, R. K., 434
Ahrenkiel, S. P., 434
Arent, D. J., 434

B

Baldasaro, P. F., 339, 351
Ballantyne, R., 128
Bhat, I., 290, 312, 423
Borrego, J. M., 312, 351, 423
Broman, L., 177
Brown, E. J., 290, 312, 351
Burger, D. R., 502

C

Campbell, B. C., 339
Charache, G. W., 339, 351
Choudhury, N., 290, 312
Chu, C. L., 361, 446
Chubb, D. L., 16, 181, 199, 263
Coutts, T. J., 329

D

Dakshina Murthy, S., 290
DePoy, D. M., 339, 351
Dhere, N. G., 409
Dzeindziel, R., 290, 312

E

Ehsani, H., 312, 423
El-Husseini, A., 3

F

Fatemi, N. S., 109
Ferguson, L., 128, 488
Flood, D. J., 375

Fraas, L. M., 128, 134, 251, 488
Freeman, M. J., 290, 312, 351

G

Garverick, L. M., 109
Good, B. S., 16, 181
Gray, J. L., 3
Guazzoni, G. E., 162
Gutmann, R., 290, 312, 423

H

Hickey, J. P., 278
Hitchcock, C., 423
Hoffman, R. H., 109
Holmquist, G. A., 138, 278
Horne, W. E., 35
Hui, S., 128
Hwang, W., 394

I

Iles, P. A., 361, 446

J

Jain, R., 375
Jain, R. K., 375
Jarefors, K., 177

K

Kochhar, R., 394
Kumar, V., 81

L

Landis, G. A., 375
Langlois, E., 290

Lin, J. J., 502
Linder, E., 446
Lord, S. M., 394
Lowe, R. A., 16, 109, 181, 263
Lund, K. O., 469

M

Marks, J., 177
Mayer, T. S., 98, 394
McCoy, L. G., 488
McLellan, S., 238
Micovic, M., 394
Miller, D. L., 394
Milstein, J. B., 495
Morgan, M. D., 35
Mueller, R. L., 502
Mukunda, M., 55
Mulligan, W. P., 329

N

Nelson, R. E., 221
Nichols, G. J., 351

O

Or, C., 55, 81
Ortabasi, U., 469
Ostrowski, L. J., 251

P

Parrington, J. R., 351
Pernisz, U. C., 251, 488
Postlethwait, M. A., 351

R

Rose, M. F., 162, 213
Roy, R. G., 495

S

Samaras, J., 128
Sarraf, D. B., 98
Scheiman, D., 109
Schock, A., 55, 81
Seal, M., 128
Seshadri, K., 469
Sharps, P. R., 458
Stone, K. W., 199, 238
Sundaram, V. S., 35

T

Timmons, M. L., 458

U

Uppal, P. N., 278

W

Waldman, C. H., 138, 278
Wanlass, M. W., 177, 199
Webb, J. D., 329
West, E., 128
Williams, D. J., 134
Wilt, D. M., 109, 199, 375
Wojtczuk, S., 387
Wong, E. M., 138, 278
Wu, X., 329

X

Xiang, H. H., 128

Z

Zierak, M., 351

AIP Conference Proceedings

Title	L.C. Number	ISBN
No. 215 X-Ray and Inner-Shell Processes (Knoxville, TN 1990)	90-84700	0-88318-790-6
No. 216 Spectral Line Shapes, Vol. 6 (Austin, TX 1990)	90-06278	0-88318-791-4
No. 217 Space Nuclear Power Systems (Albuquerque, NM 1991)	90-56220	0-88318-838-4
No. 218 Positron Beams for Solids and Surfaces (London, Canada 1990)	90-56407	0-88318-842-2
No. 219 Superconductivity and Its Applications (Buffalo, NY 1990)	91-55020	0-88318-835-X
No. 220 High Energy Gamma-Ray Astronomy (Ann Arbor, MI 1990)	91-70876	0-88318-812-0
No. 221 Particle Production Near Threshold (Nashville, IN 1990)	91-55134	0-88318-829-5
No. 222 After the First Three Minutes (College Park, MD 1990)	91-55214	0-88318-828-7
No. 223 Polarized Collider Workshop (University Park, PA 1990)	91-71303	0-88318-826-0
No. 224 LAMPF Workshop on (π, K) Physics (Los Alamos, NM 1990)	91-71304	0-88318-825-2
No. 225 Half Collision Resonance Phenomena in Molecules (Caracas, Venezuela 1990)	91-55210	0-88318-840-6
No. 226 The Living Cell in Four Dimensions (Gif sur Yvette, France 1990)	91-55209	0-88318-794-9
No. 227 Advanced Processing and Characterization Technologies (Clearwater, FL 1991)	91-55194	0-88318-910-0
No. 228 Anomalous Nuclear Effects in Deuterium/Solid Systems (Provo, UT 1990)	91-55245	0-88318-833-3
No. 229 Accelerator Instrumentation (Batavia, IL 1990)	91-55347	0-88318-832-1
No. 230 Nonlinear Dynamics and Particle Acceleration (Tsukuba, Japan 1990)	91-55348	0-88318-824-4
No. 231 Boron-Rich Solids (Albuquerque, NM 1990)	91-53024	0-88318-793-4
No. 232 Gamma-Ray Line Astrophysics (Paris-Saclay, France 1990)	91-55492	0-88318-875-9
No. 233 Atomic Physics 12 (Ann Arbor, MI 1990)	91-55595	088318-811-2
No. 234 Amorphous Silicon Materials and Solar Cells (Denver, CO 1991)	91-55575	088318-831-7

	Title	L.C. Number	ISBN
No. 235	Physics and Chemistry of MCT and Novel IR Detector Materials (San Francisco, CA 1990)	91-55493	0-88318-931-3
No. 236	Vacuum Design of Synchrotron Light Sources (Argonne, IL 1990)	91-55527	0-88318-873-2
No. 237	Kent M. Terwilliger Memorial Symposium (Ann Arbor, MI 1989)	91-55576	0-88318-788-4
No. 238	Capture Gamma-Ray Spectroscopy (Pacific Grove, CA 1990)	91-57923	0-88318-830-9
No. 239	Advances in Biomolecular Simulations (Obernai, France 1991)	91-58106	0-88318-940-2
No. 240	Joint Soviet-American Workshop on the Physics of Semiconductor Lasers (Leningrad, USSR 1991)	91-58537	0-88318-936-4
No. 241	Scanned Probe Microscopy (Santa Barbara, CA 1991)	91-76758	0-88318-816-3
No. 242	Strong, Weak, and Electromagnetic Interactions in Nuclei, Atoms, and Astrophysics: A Workshop in Honor of Stewart D. Bloom's Retirement (Livermore, CA 1991)	91-76876	0-88318-943-7
No. 243	Intersections Between Particle and Nuclear Physics (Tucson, AZ 1991)	91-77580	0-88318-950-X
No. 244	Radio Frequency Power in Plasmas (Charleston, SC 1991)	91-77853	0-88318-937-2
No. 245	Basic Space Science (Bangalore, India 1991)	91-78379	0-88318-951-8
No. 246	Space Nuclear Power Systems (Albuquerque, NM 1992)	91-58793	1-56396-027-3 1-56396-026-5 (pbk.)
No. 247	Global Warming: Physics and Facts (Washington, DC 1991)	91-78423	0-88318-932-1
No. 248	Computer-Aided Statistical Physics (Taipei, Taiwan 1991)	91-78378	0-88318-942-9
No. 249	The Physics of Particle Accelerators (Upton, NY 1989, 1990)	92-52843	0-88318-789-2
No. 250	Towards a Unified Picture of Nuclear Dynamics (Nikko, Japa 1991)	92-70143	0-88318-951-8
No. 251	Superconductivity and its Applications (Buffalo, NY 1991)	92-52726	1-56396-016-8
No. 252	Accelerator Instrumentation (Newport News, VA 1991)	92-70356	0-88318-934-8
No. 253	High-Brightness Beams for Advanced Accelerator Applications (College Park, MD 1991)	92-52705	0-88318-947-X
No. 254	Testing the AGN Paradigm (College Park, MD 1991)	92-52780	1-56396-009-5

Title	L.C. Number	ISBN
No. 255 Advanced Beam Dynamics Workshop on Effects of Errors in Accelerators, Their Diagnosis and Corrections (Corpus Christi, TX 1991)	92-52842	1-56396-006-0
No. 256 Slow Dynamics in Condensed Matter (Fukuoka, Japan 1991)	92-53120	0-88318-938-0
No. 257 Atomic Processes in Plasmas (Portland, ME 1991)	91-08105	0-88318-939-9
No. 258 Synchrotron Radiation and Dynamic Phenomena (Grenoble, France 1991)	92-53790	1-56396-008-7
No. 259 Future Directions in Nuclear Physics with 4π Gamma Detection Systems of the New Generation (Strasbourg, France 1991)	92-53222	0-88318-952-6
No. 260 Computational Quantum Physics (Nashville, TN 1991)	92-71777	0-88318-933-X
No. 261 Rare and Exclusive B&K Decays and Novel Flavor Factories (Santa Monica, CA 1991)	92-71873	1-56396-055-9
No. 262 Molecular Electronics—Science and Technology (St. Thomas, Virgin Islands 1991)	92-72210	1-56396-041-9
No. 263 Stress-Induced Phenomena in Metallization: First International Workshop (Ithaca, NY 1991)	92-72292	1-56396-082-6
No. 264 Particle Acceleration in Cosmic Plasmas (Newark, DE 1991)	92-73316	0-88318-948-8
No. 265 Gamma-Ray Bursts (Huntsville, AL 1991)	92-73456	1-56396-018-4
No. 266 Group Theory in Physics (Cocoyoc, Morelos, Mexico 1991)	92-73457	1-56396-101-6
No. 267 Electromechanical Coupling of the Solar Atmosphere (Capri, Italy 1991)	92-82717	1-56396-110-5
No. 268 Photovoltaic Advanced Research & Development Project (Denver, CO 1992)	92-74159	1-56396-056-7
No. 269 CEBAF 1992 Summer Workshop (Newport News, VA 1992)	92-75403	1-56396-067-2
No. 270 Time Reversal—The Arthur Rich Memorial Symposium (Ann Arbor, MI 1991)	92-83852	1-56396-105-9
No. 271 Tenth Symposium Space Nuclear Power and Propulsion (Vols. I–III) (Albuquerque, NM 1993)	92-75162	1-56396-137-7 (set)
No. 272 Proceedings of the XXVI International Conference on High Energy Physics (Vols. I and II) (Dallas, TX 1992)	93-70412	1-56396-127-X (set)

Title	L.C. Number	ISBN
No. 273 Superconductivity and Its Applications (Buffalo, NY 1992)	93-70502	1-56396-189-X
No. 274 VIth International Conference on the Physics of Highly Charged Ions (Manhattan, KS 1992)	93-70577	1-56396-102-4
No. 275 Atomic Physics 13 (Munich, Germany 1992)	93-70826	1-56396-057-5
No. 276 Very High Energy Cosmic-Ray Interactions: VIIth International Symposium (Ann Arbor, MI 1992)	93-71342	1-56396-038-9
No. 277 The World at Risk: Natural Hazards and Climate Change (Cambridge, MA 1992)	93-71333	1-56396-066-4
No. 278 Back to the Galaxy (College Park, MD 1992)	93-71543	1-56396-227-6
No. 279 Advanced Accelerator Concepts (Port Jefferson, NY 1992)	93-71773	1-56396-191-1
No. 280 Compton Gamma-Ray Observatory (St. Louis, MO 1992)	93-71830	1-56396-104-0
No. 281 Accelerator Instrumentation Fourth Annual Workshop (Berkeley, CA 1992)	93-072110	1-56396-190-3
No. 282 Quantum 1/f Noise & Other Low Frequency Fluctuations in Electronic Devices (St. Louis, MO 1992)	93-072366	1-56396-252-7
No. 283 Earth and Space Science Information Systems (Pasadena, CA 1992)	93-072360	1-56396-094-X
No. 284 US-Japan Workshop on Ion Temperature Gradient-Driven Turbulent Transport (Austin, TX 1993)	93-72460	1-56396-221-7
No. 285 Noise in Physical Systems and 1/f Fluctuations (St. Louis, MO 1993)	93-72575	1-56396-270-5
No. 286 Ordering Disorder: Prospect and Retrospect in Condensed Matter Physics: Proceedings of the Indo-U.S. Workshop (Hyderabad, India 1993)	93-072549	1-56396-255-1
No. 287 Production and Neutralization of Negative Ions and Beams: Sixth International Symposium (Upton, NY 1992)	93-72821	1-56396-103-2
No. 288 Laser Ablation: Mechanismas and Applications-II: Second International Conference (Knoxville, TN 1993)	93-73040	1-56396-226-8
No. 289 Radio Frequency Power in Plasmas: Tenth Topical Conference (Boston, MA 1993)	93-72964	1-56396-264-0

	Title	L.C. Number	ISBN
No. 290	Laser Spectroscopy: XIth International Conference (Hot Springs, VA 1993)	93-73050	1-56396-262-4
No. 291	Prairie View Summer Science Academy (Prairie View, TX 1992)	93-73081	1-56396-133-4
No. 292	Stability of Particle Motion in Storage Rings (Upton, NY 1992)	93-73534	1-56396-225-X
No. 293	Polarized Ion Sources and Polarized Gas Targets (Madison, WI 1993)	93-74102	1-56396-220-9
No. 294	High-Energy Solar Phenomena A New Era of Spacecraft Measurements (Waterville Valley, NH 1993)	93-74147	1-56396-291-8
No. 295	The Physics of Electronic and Atomic Collisions: XVIII International Conference (Aarhus, Denmark, 1993)	93-74103	1-56396-290-X
No. 296	The Chaos Paradigm: Developments an Applications in Engineering and Science (Mystic, CT 1993)	93-74146	1-56396-254-3
No. 297	Computational Accelerator Physics (Los Alamos, NM 1993)	93-74205	1-56396-222-5
No. 298	Ultrafast Reaction Dynamics and Solvent Effects (Royaumont, France 1993)	93-074354	1-56396-280-2
No. 299	Dense Z-Pinches: Third International Conference (London, 1993)	93-074569	1-56396-297-7
No. 300	Discovery of Weak Neutral Currents: The Weak Interaction Before and After (Santa Monica, CA 1993)	94-70515	1-56396-306-X
No. 301	Eleventh Symposium Space Nuclear Power and Propulsion (3 Vols.) (Albuquerque, NM 1994)	92-75162	1-56396-305-1 (Set) 156396-301-9 (pbk. set)
No. 302	Lepton and Photon Interactions/ XVI International Symposium (Ithaca, NY 1993)	94-70079	1-56396-106-7
No. 303	Slow Positron Beam Techniques for Solids and Surfaces Fifth International Workshop (Jackson Hole, WY 1992)	94-71036	1-56396-267-5
No. 304	The Second Compton Symposium (College Park, MD 1993)	94-70742	1-56396-261-6
No. 305	Stress-Induced Phenomena in Metallization Second International Workshop (Austin, TX 1993)	94-70650	1-56396-251-9
No. 306	12th NREL Photovoltaic Program Review (Denver, CO 1993)	94-70748	1-56396-315-9

Title	L.C. Number	ISBN
No. 307 Gamma-Ray Bursts Second Workshop (Huntsville, AL 1993)	94-71317	1-56396-336-1
No. 308 The Evolution of X-Ray Binaries (College Park, MD 1993)	94-76853	1-56396-329-9
No. 309 High-Pressure Science and Technology—1993 (Colorado Springs, CO 1993)	93-72821	1-56396-219-5 (Set)
No. 310 Analysis of Interplanetary Dust (Houston, TX 1993)	94-71292	1-56396-341-8
No. 311 Physics of High Energy Particles in Toroidal Systems (Irvine, CA 1993)	94-72098	1-56396-364-7
No. 312 Molecules and Grains in Space (Mont Sainte-Odile, France 1993)	94-72615	1-56396-355-8
No. 313 The Soft X-Ray Cosmos ROSAT Science Symposium (College Park, MD 1993)	94-72499	1-56396-327-2
No. 314 Advances in Plasma Physics Thomas H. Stix Symposium (Princeton, NJ 1992)	94-72721	1-56396-372-8
No. 315 Orbit Correction and Analysis in Circular Accelerators (Upton, NY 1993)	94-72257	1-56396-373-6
No. 316 Thirteenth International Conference on Thermoelectrics (Kansas City, Missouri 1994)	95-75634	1-56396-444-9
No. 317 Fifth Mexican School of Particles and Fields (Guanajuato, Mexico 1992)	94-72720	1-56396-378-7
No. 318 Laser Interaction and Related Plasma Phenomena 11th International Workshop (Monterey, CA 1993)	94-78097	1-56396-324-8
No. 319 Beam Instrumentation Workshop (Santa Fe, NM 1993)	94-78279	1-56396-389-2
No. 320 Basic Space Science (Lagos, Nigeria 1993)	94-79350	1-56396-328-0
No. 321 The First NREL Conference on Thermophotovoltaic Generation of Electricity (Copper Mountain, CO 1994)	94-72792	1-56396-353-1
No. 322 Atomic Processes in Plasmas Ninth APS Topical Conference (San Antonio, TX)	94-72923	1-56396-411-2
No. 323 Atomic Physics 14 Fourteenth International Conference on Atomic Physics (Boulder, CO 1994)	94-73219	1-56396-348-5
No. 324 Twelfth Symposium on Space Nuclear Power and Propulsion (Albuquerque, NM 1995)	94-73603	1-56396-427-9

Title	L.C. Number	ISBN
No. 325 Conference on NASA Centers for Commercial Development of Space (Albuquerque, NM 1995)	94-73604	1-56396-431-7
No. 326 Accelerator Physics at the Superconducting Super Collider (Dallas, TX 1992-1993)	94-73609	1-56396-354-X
No. 327 Nuclei in the Cosmos III Third International Symposium on Nuclear Astrophysics (Assergi, Italy 1994)	95-75492	1-56396-436-8
No. 328 Spectral Line Shapes, Volume 8 12th ICSLS (Toronto, Canada 1994)	94-74309	1-56396-326-4
No. 329 Resonance Ionization Spectroscopy 1994 Seventh International Symposium (Bernkastel-Kues, Germany 1994)	95-75077	1-56396-437-6
No. 330 E.C.C.C. 1 Computational Chemistry F.E.C.S. Conference (Nancy, France 1994)	95-75843	1-56396-457-0
No. 331 Non-Neutral Plasma Physics II (Berkeley, CA 1994)	95-79630	1-56396-441-4
No. 332 X-Ray Lasers 1994 Fourth International Colloquium (Williamsburg, VA 1994)	95-76067	1-56396-375-2
No. 333 Beam Instrumentation Workshop (Vancouver, B. C., Canada 1994)	95-79635	1-56396-352-3
No. 334 Few-Body Problems in Physics (Williamsburg, VA 1994)	95-76481	1-56396-325-6
No. 335 Advanced Accelerator Concepts (Fontana, WI 1994)	95-78225	1-56396-476-7 (Set) 1-56396-474-0 (Book) 1-56396-475-9 (CD-Rom)
No. 336 Dark Matter (College Park, MD 1994)	95-76538	1-56396-438-4
No. 337 Pulsed RF Sources for Linear Colliders (Montauk, NY 1994)	95-76814	1-56396-408-2
No. 338 Intersections Between Particle and Nuclear Physics 5th Conference (St. Petersburg, FL 1994)	95-77076	1-56396-335-3
No. 339 Polarization Phenomena in Nuclear Physics Eighth International Symposium (Bloomington, IN 1994)	95-77216	1-56396-482-1
No. 340 Strangeness in Hadronic Matter (Tucson, AZ 1995)	95-77477	1-56396-489-9

	Title	L.C. Number	ISBN
No. 341	Volatiles in the Earth and Solar System (Pasadena, CA 1994)	95-77911	1-56396-409-0
No. 342	CAM -94 Physics Meeting (Cacun, Mexico 1994)	95-77851	1-56396-491-0
No. 343	High Energy Spin Physics Eleventh International Symposium (Bloomington, IN 1994)	95-78431	1-56396-374-4
No. 344	Nonlinear Dynamics in Particle Accelerators: Theory and Experiments (Arcidosso, Italy 1994)	95-78135	1-56396-446-5
No. 345	International Conference on Plasma Physics ICPP 1994 (Foz do Iguaçu, Brazil 1994)	95-78438	1-56396-496-1
No. 346	International Conference on Accelerator-Driven Transmutation Technologies and Applications (Las Vegas, NV 1994)	95-78691	1-56396-505-4
No. 347	Atomic Collisions: A Symposium in Honor of Christopher Bottcher (1945-1993) (Oak Ridge, TN 1994)	95-78689	1-56396-322-1
No. 348	Unveiling the Cosmic Infrared Background (College Park, MD, 1995)	95-83477	1-56396-508-9
No. 349	Workshop on the Tau/Charm Factory (Argonne, IL 1995)	95-81467	1-56396-523-2
No. 350	International Symposium on Vector Boson Self-Interactions (Los Angeles, CA 1995)	95-79865	1-56396-520-8
No. 351	The Physics of Beams Andrew Sessler Symposium (Los Angeles, CA 1993)	95-80479	1-56396-376-0
No. 352	Physics Potential and Development of $\mu^+ \mu^-$ Colliders: Second Workshop (Sausalito, CA 1994)	95-81413	1-56396-506-2
No. 353	13th NREL Photovoltaic Program Review (Lakewood, CO 1995)	95-80662	1-56396-510-0
No. 355	Eleventh Topical Conference on Radio Frequency Power in Plasmas (Palm Springs, CA 1995)	95-80867	1-56396-536-4
No. 357	10th Topical Workshop on Proton-Antiproton Collider Physics (Batavia, IL 1995)	95-83078	1-56396-543-7
No. 358	The Second NREL Conference on Thermophotovoltaic Generation of Electricity	95-83335	1-56396-509-7